A MEMBER OF THE INTERNATIONAL CODE FAMILY®

INTERNATIONAL EXISTING BUILDING CODE®

2009

2009 International Existing Building Code®

First Printing: January 2009
Second Printing: April 2009
Third Printing: April 2010
Fourth Printing: October 2010
Fifth Printing: February 2011
Sixth Printing: October 2011
Seventh Printing: April 2012

ISBN: 978-1-58001-737-4 (soft-cover edition)
ISBN: 978-1-58001-736-7 (loose-leaf edition)

37876-T015448

PRINTED IN THE U.S.A.

PREFACE

Introduction

Internationally, code officials recognize the need for a modern, up-to-date code addressing repair, alteration, addition or change of occupancy in existing buildings. The *International Existing Building Code*®, in this 2009 edition, is designed to meet this need through model code regulations that safeguard the public health and safety in all communities, large and small.

This comprehensive existing building code establishes minimum regulations for existing buildings using prescriptive and performance-related provisions. It is founded on broad-based principles intended to encourage the use and reuse of existing buildings while requiring reasonable upgrades and improvements. This 2009 edition is fully compatible with all the *International Codes*® (I-Codes®) published by the International Code Council (ICC)®, including the *International Building Code*®, *International Energy Conservation Code*®, *International Fire Code*®, *International Fuel Gas Code*®, *International Mechanical Code*®, ICC *Performance Code*® *for Buildings and Facilities*®, *International Plumbing Code*®, *International Private Sewage Disposal Code*®, *International Property Maintenance Code*®, *International Residential Code*®, *International Wildland-Urban Interface Code*™ and *International Zoning Code*®.

The *International Existing Building Code* provisions provide many benefits, including the model code development process, which offers an international forum for building professionals to discuss performance and prescriptive code requirements. This forum provides an excellent arena to debate proposed revisions. This model code also encourages international consistency in the application of provisions.

Development

The first edition of the *International Existing Building Code* (2003) was the culmination of an effort initiated in 2000 by a development committee appointed by the ICC and consisting of representatives of the three statutory members of the International Code Council at that time, including: Building Officials and Code Administrators International, Inc. (BOCA), International Conference of Building Officials (ICBO) and Southern Building Code Congress International (SBCCI). The intent was to draft a comprehensive set of regulations for existing buildings consistent with and inclusive of the scope of the existing model codes. Technical content of the latest model codes promulgated by BOCA, ICBO and SBCCI as well as other rehabilitation codes was utilized as the basis for the development, followed by a public forum in 2001 and the publication of the 2001 Final Draft. This 2009 edition presents the code as originally issued in 2003 with the changes reflected in the 2006 edition, and with further changes approved through the ICC code development process through 2008. A new edition such as this is promulgated every three years.

This code is founded on principles intended to encourage the use and reuse of existing buildings that adequately protect public health, safety and welfare; provisions that do not unnecessarily increase construction costs; provisions that do not restrict the use of new materials, products or methods of construction; and provisions that do not give preferential treatment to particular types or classes of materials, products or methods of construction.

Adoption

The *International Existing Building Code* is available for adoption and use by jurisdictions internationally. Its use within a governmental jurisdiction is intended to be accomplished through adoption by reference in accordance with proceedings establishing the jurisdiction's laws. At the time of adoption, jurisdictions should insert the appropriate information in provisions requiring specific local information, such as the name of the adopting jurisdiction. These locations are shown in bracketed words in small capital letters in the code and in the sample ordinance. The sample adoption ordinance on page ix addresses several key elements of a code adoption ordinance, including the information required for insertion into the code text.

Maintenance

The *International Existing Building Code* is kept up to date through the review of proposed changes submitted by code enforcement officials, industry representatives, design professionals, and other interested parties. Proposed changes are carefully considered through an open code development process in which all interested and affected parties may participate.

The contents of this work are subject to change both through the code development cycles and the governmental body that enacts the code into law. For more information regarding the code development process, contact the Code and Standard Development Department of the International Code Council.

While the development procedure of the *International Existing Building Code* assures the highest degree of care, ICC, its members, and those participating in the development of this code do not accept any liability resulting from compliance or noncompliance with these provisions, because ICC does not have the power or authority to police or enforce compliance with the contents of this code. Only the governmental body that enacts the code into law has such authority.

Letter Designations in Front of Section Numbers

In each code development cycle, proposed changes to this code are considered at the Code Development Hearings by the International Existing Building Code Development Committee, whose action constitutes a recommendation to the voting membership for final action on the proposed changes. Proposed changes to a code section having a number beginning with a letter in brackets are considered by a different code development committee. For example, proposed changes to code sections that are preceded by the letter [F] (e.g., [F] 1404.2), are considered by the International Fire Code Development Committee at the Code Development Hearings.

The content of sections in this code that begin with a letter designation is maintained by another code development committee in accordance with the following:

[B] = International Building Code Development Committee;

[F] = International Fire Code Development Committee;

[P] = International Plumbing Code Development Committee;

[FG] = International Fuel Gas Code Development Committee;

[EC] = International Energy Conservation Code Development Committee; and

[M] = International Mechanical Code Development Committee.

Marginal Markings

Solid vertical lines in the margins within the body of the code indicate a technical change from the requirements of the 2006 edition. Deletion indicators in the form of an arrow (➡) are provided in the margin where an entire section, paragraph, exception or table has been deleted or an item in a list of items or a table has been deleted.

Italicized Terms

Selected terms set forth in Chapter 2, Definitions, are italicized where they appear in code text. Such terms are not italicized where the definition set forth in Chapter 2 does not impart the intended meaning in the use of the term. The terms selected have definitions which the user should read carefully to facilitate better understanding of the code.

Effective Use of the International Existing Building Code

The *International Existing Building Code* is a model code in the *International Code* family of codes intended to provide alternative approaches to remodeling, repair or alteration of existing buildings. A large number of existing buildings and structures do not comply with the current building code requirements for new construction. Although many of these buildings are potentially salvageable, rehabilitation is often cost-prohibitive because compliance with all the requirements for new construction could require extensive changes that go well beyond the value of the building or the original scope of the rehabilitation. At the same time, it is necessary to regulate construction in existing buildings that undergo additions, alterations, renovations, extensive repairs or change of occupancy. Such activity represents an opportunity to ensure that new construction complies with the current building codes and that existing conditions are maintained, at a minimum, to their current level of compliance or are improved as required to meet basic safety levels. To accomplish this objective, and to make the rehabilitation process easier, this code allows for options for controlled departure from full compliance with the *International Codes* dealing with new construction, while maintaining basic levels for fire prevention, structural and life safety features of the rehabilitated building.

This code provides three main options for a designer in dealing with rehabilitation of existing buildings. These are laid out in Section 101.5 of this code:

OPTION 1: Work for alteration, repair, change of occupancy, addition or relocation of all existing buildings shall be done in accordance with the Prescriptive Compliance Method given in Chapter 3. It should be noted that this same method is provided in Chapter 34 of the *International Building Code*.

OPTION 2: Work for alteration, repair, change of occupancy, addition or relocation of all existing buildings shall be done in accordance with the Work Area Compliance Method given in Chapters 4 through 12.

OPTION 3: Work for alteration, repair, change of occupancy, addition or relocation of all existing buildings shall be done in accordance with the Performance Compliance Method given in Chapter 13. It should be noted that this option is also provided in Chapter 34 of the *International Building Code*.

Under limited circumstances, a building alteration can be made to comply with the laws under which the building was originally built, as long as there has been no substantial structural damage and there will be limited structural alteration.

Arrangement and Format of the 2009 IEBC

Before applying the requirements of the IEBC it is beneficial to understand its arrangement and format. The IEBC, like other codes published by ICC, is arranged and organized to follow logical steps that generally occur during a plan review or inspection. The IEBC is divided as follows:

Chapters	Subjects
1–2	Administrative Requirements and Definitions
3	Prescriptive Compliance Method for Existing Buildings
4–12	Work Area Compliance Method for Existing Buildings
13	Performance Compliance Method for Existing Buildings
14	Construction Safeguard
15	Referenced Standards
Appendix A	Guidelines for Seismic Retrofit of Existing Buildings
Appendix B	Supplementary Accessibility Requirements for Existing Buildings
Resource A	Information on Fire Resistance of Archaic Materials and Assemblies

The following is a chapter-by-chapter synopsis of the scope and intent of the provisions of the *International Existing Building Code*:

Chapter 1 Scope and Administration. This chapter contains provisions for the application, enforcement and administration of subsequent requirements of the code. In addition to establishing the scope of the code, Chapter 1 identifies which buildings and structures come under its purview. Chapter 1 is largely concerned with maintaining "due process of law" in enforcing the regulations contained in the body of the code. Only through careful observation of the administrative provisions can the code official reasonably expect to demonstrate that "equal protection under the law" has been provided.

Chapter 2 Definitions. All terms that are defined in the code are listed alphabetically in Chapter 2. While a defined term may be used in one chapter or another, the meaning provided in Chapter 2 is applicable throughout the code.

Where understanding of a term's definition is especially key to or necessary for understanding of a particular code provision, the term is show in italics wherever it appears in the code. This is true only for those terms that have a meaning that is unique to the code. In other words, the generally understood meaning of a term or phrase might not be sufficient or consistent with the meaning prescribed by the code; therefore, it is essential that the code-defined meaning be known.

Guidance regarding tense, gender and plurality of defined terms as well as guidance regarding terms not defined in this code is also provided.

Chapter 3 Prescriptive Compliance Method. This chapter provides one of the three main options of compliance available in the IEBC for buildings and structures undergoing repair, alteration, addition or change in occupancy. This chapter duplicates the provisions that are predominantly in Chapter 34 of the IBC—Sections 3403 through 3411. There are also provisions from the other I-Codes dealing with system installations (Electrical, Energy, Fuel Gas, Mechanical and Plumbing) which have been duplicated in the IEBC as well.

Chapter 4 Classification of Work. This chapter provides an overview of the Work Area Method available as an option for rehabilitation of a building. The chapter defines the different classifications of alterations and provides general requirements for repairs, change of occupancy, additions, historic buildings and relocated buildings. Detailed requirements for all of these are given in subsequent Chapters 5 through 12.

Chapter 5 Repairs. Chapter 5 governs the repair of existing buildings. The provisions define conditions under which repairs may be made using materials and methods like those of the original construction or the extent to which repairs must comply with requirements for new buildings.

Chapter 6 Alterations—Level 1. This chapter provides the technical requirements for those existing buildings that undergo Level 1 alterations as described in Section 403, which includes replacement or covering of existing materials, elements, equipment or fixtures using new materials for the same purpose. This chapter, similar to other chapters of this code, covers all building-related subjects, such as structural, mechanical, plumbing, electrical and accessibility as well as the fire and life safety issues when the alterations are classified as Level 1. The purpose of this chapter is to provide detailed requirements and provisions to identify the required improvements in the existing building elements, building spaces and building structural system. This chapter is distinguished from Chapters 7 and 8 by only involving replacement of building components with new components. In contrast, Level 2 alterations involve more space reconfiguration and Level 3 alterations involve more extensive space reconfiguration, exceeding 50 percent of the building area.

Chapter 7 Alterations—Level 2. Like Chapter 6, the purpose of this chapter is to provide detailed requirements and provisions to identify the required improvements in the existing building elements, building spaces and building structural system. This chapter is distinguished from Chapters 6 and 8 by involving space reconfiguration that could be up to and including 50 percent of the area of the building. In contrast, Level 1 alterations (Chapter 6) do not involve space reconfiguration and Level 3 alterations (Chapter 8) involve extensive space reconfiguration that exceeds 50 percent of the building area. Depending on the nature of alteration work, its location within the building and whether it encompasses one or more tenants, improvements and upgrades could be required for the open floor penetrations, sprinkler system or the installation of additional means of egress such as stairs or fire escapes.

Chapter 8 Alterations—Level 3. This chapter provides the technical requirements for those existing buildings that undergo Level 3 alterations. The purpose of this chapter is to provide detailed requirements and provisions to identify the required improvements in the existing building elements, building spaces and building structural system. This chapter is distinguished from Chapters 6 and 7 by involving alterations that cover 50 percent of the aggregate area of the building. In contrast, Level 1 alterations do not involve space reconfiguration and Level 2 alterations involve extensive space reconfiguration that does not exceed 50 percent of the building area. Depending on the nature of alteration work, its location within the building and whether it encompasses one or more tenants, improvements and upgrades could be required for the open floor penetrations, sprinkler system or the installation of additional means of egress such as stairs or fire escapes. At times and under certain situations, this chapter also intends to improve the safety of certain building features beyond the work area and in other parts of the building where no alteration work might be taking place.

Chapter 9 Change of Occupancy. The purpose of this chapter is to provide regulations for the circumstances when an existing building is subject to a change in occupancy or a change in occupancy classification. A change of occupancy is not to be confused with a change of occupancy classification. The *International Building Code*® (IBC®) defines different occupancy classifications in Chapter 3, and special occupancy requirements in Chapter 4. Within specific occupancy classifications there can be many different types of actual activities that can take place. For instance, a Group A-3 occupancy classification deals with a wide variation of different types of activities, including bowling alleys and courtrooms, indoor tennis courts and dance halls. When a facility changes use from, for example, a bowling alley to a dance hall, the occupancy classification remains A-3, but the different uses could lead to drastically different code requirements. Therefore, this chapter deals with the special circumstances that are associated with a change in the use of a building within the same occupancy classification as well as a change of occupancy classification.

Chapter 10 Additions. Chapter 10 provides the requirements for additions, which correlate to the code requirements for new construction. There are, however, some exceptions that are specifically stated within this chapter. An "Addition" is defined in Chapter 2 as "an extension or increase in the floor area, number of stories or height of a building or structure." Chapter 10 contains the minimum requirements for an addition that is not separated from the existing building by a fire wall.

Chapter 11 Historic Buildings. This chapter provides some exceptions from code requirements when the building in question has historic value. The most important criterion for application of this chapter is that the building must be essentially accredited as being of historic significance by a state or local authority after careful review of the historical value of the building. Most, if not all, states have such authorities, as do many local jurisdictions. The agencies with such authority can be located at the state or local government level or through the local chapter of the American Institute of Architects (AIA). Other considerations include the structural condition of the building (i.e., is the building structurally sound), its proposed use, its impact on life safety and how the intent of the code, if not the letter, will be achieved.

Chapter 12 Relocated or Moved Buildings. Chapter 12 is applicable to any building that is moved or relocated.

Chapter 13 Performance Compliance Methods. This chapter, a duplicate of IBC Section 3412, Compliance Alternatives, allows for existing buildings to be evaluated so as to show that alterations, while not meeting new construction requirements, will improve the current existing situation. Provisions are based on a numerical scoring system involving 18 various safety parameters and the degree of code compliance for each issue.

Chapter 14 Construction Safeguards. The building construction process involves a number of known and unanticipated hazards. Chapter 14 establishes specific regulations in order to minimize the risk to the public and adjacent property. Some construction failures have resulted during the initial stages of grading, excavation and demolition. During these early stages, poorly designed and installed sheeting and shoring have resulted in ditch and embankment cave-ins. Also, inadequate underpinning of adjoining existing structures or careless removal of existing structures has produced construction failures.

Chapter 15 Referenced Standards. The code contains numerous references to standards that are used to regulate materials and methods of construction. Chapter 15 contains a comprehensive list of all standards that are referenced in the code, including the appendices. The standards are part of the code to the extent of the reference to the standard. Compliance with the referenced standard is necessary for compliance with this code. By providing specifically adopted standards, the construction and installation requirements necessary for compliance with the code can be readily determined. The basis for code compliance is, therefore, established and available on an equal basis to the building code official, contractor, designer and owner.

Chapter 15 is organized in a manner that makes it easy to locate specific standards. It lists all of the referenced standards, alphabetically, by acronym of the promulgating agency of the standard. Each agency's standards are then listed in either alphabetical or numeric order based upon the standard identification. The list also contains the title of the standard; the edition (date) of the standard referenced; any addenda included as part of the ICC adoption; and the section or sections of this code that reference the standard.

Appendix A Guidelines for the Seismic Retrofit of Existing Buildings. Appendix A provides guidelines for upgrading the seismic resistance capacity of different types of existing buildings. It is organized into separate chapters which deal with buildings of different types, including unreinforced masonry buildings, reinforced concrete and reinforced masonry wall buildings, and light-frame wood buildings.

Appendix B Supplementary Accessibility Requirements for Existing Buildings and Facilities. Chapter 11 of the *International Building Code*® (IBC®) contains provisions that set forth requirements for accessibility to buildings and their associated sites and facilities for people with physical disabilities. Sections 406, 506, 606, 706, 806, 905, 1004, 1005 and 1308 in the code address accessibility provisions and alternatives permitted in existing buildings. Appendix B was added to address accessibility in construction for items that are not typically enforceable through the traditional building code enforcement process.

Resource A Guidelines on Fire Ratings of Archaic Materials and Assemblies. In the process of repair and alteration of existing buildings, based on the nature and the extent of the work, the IEBC might require certain upgrades in the fire-resistance rating of building elements, at which time it becomes critical for the designers and the code officials to be able to determine the fire-resistance rating of the existing building elements as part of the overall evaluation for the assessment of the need for improvements. This resource document provides a guideline for such an evaluation for fire-resistance rating of archaic materials that is not typically found in the modern model building codes.

ORDINANCE

The *International Codes* are designed and promulgated to be adopted by reference by ordinance. Jurisdictions wishing to adopt the 2009 *International Existing Building Code* as an enforceable performance-based regulation governing structures and premises should ensure that certain factual information is included in the adopting ordinance at the time adoption is being considered by the appropriate governmental body. The following sample adoption ordinance addresses several key elements of a code adoption ordinance, including the information required for insertion into the code text.

SAMPLE ORDINANCE FOR ADOPTION OF THE *INTERNATIONAL EXISTING BUILDING CODE* *ORDINANCE NO.*_______

An ordinance of the **[JURISDICTION]** adopting the 2009 edition of the *International Existing Building Code,* regulating and governing the repair, alteration, change of occupancy, addition and relocation of existing buildings, including historic buildings, in the **[JURISDICTION]**; providing for the issuance of permits and collection of fees therefor; repealing Ordinance No. ______ of the **[JURISDICTION]** and all other ordinances and parts of the ordinances in conflict therewith.

The **[GOVERNING BODY]** of the **[JURISDICTION]** does ordain as follows:

Section 1. That a certain document, three (3) copies of which are on file in the office of the **[TITLE OF JURISDICTION'S KEEPER OF RECORDS]** of **[NAME OF JURISDICTION]**, being marked and designated as the *International Existing Building Code,* 2009 edition, including Appendix Chapters **[FILL IN THE APPENDIX CHAPTERS BEING ADOPTED]** (see *International Existing Building Code* Section 101.7, 2009 edition), as published by the International Code Council, be and is hereby adopted as the Existing Building Code of the **[JURISDICTION]**, in the State of **[STATE NAME]** for regulating and governing the repair, alteration, change of occupancy, addition and relocation of existing buildings, including historic buildings, as herein provided; providing for the issuance of permits and collection of fees therefor; and each and all of the regulations, provisions, penalties, conditions and terms of said Existing Building Code on file in the office of the **[JURISDICTION]** are hereby referred to, adopted, and made a part hereof, as if fully set out in this ordinance, with the additions, insertions, deletions and changes, if any, prescribed in Section 2 of this ordinance.

Section 2. The following sections are hereby revised:

Section 101.1 Insert: **[NAME OF JURISDICTION]**

Section 1301.2 Insert: **[DATE IN ONE LOCATION]**

Section 3. That Ordinance No. ______ of **[JURISDICTION]** entitled **[FILL IN HERE THE COMPLETE TITLE OF THE ORDINANCE OR ORDINANCES IN EFFECT AT THE PRESENT TIME SO THAT THEY WILL BE REPEALED BY DEFINITE MENTION]** and all other ordinances or parts of ordinances in conflict herewith are hereby repealed.

Section 4. That if any section, subsection, sentence, clause or phrase of this ordinance is, for any reason, held to be unconstitutional, such decision shall not affect the validity of the remaining portions of this ordinance. The **[GOVERNING BODY]** hereby declares that it would have passed this ordinance, and each section, subsection, clause or phrase thereof, irrespective of the fact that any one or more sections, subsections, sentences, clauses and phrases be declared unconstitutional.

Section 5. That nothing in this ordinance or in the Existing Building Code hereby adopted shall be construed to affect any suit or proceeding impending in any court, or any rights acquired, or liability incurred, or any cause or causes of action acquired or existing, under any act or ordinance hereby repealed as cited in Section 3 of this ordinance; nor shall any just or legal right or remedy of any character be lost, impaired or affected by this ordinance.

Section 6. That the **[JURISDICTION'S KEEPER OF RECORDS]** is hereby ordered and directed to cause this ordinance to be published. (An additional provision may be required to direct the number of times the ordinance is to be published and to specify that it is to be in a newspaper in general circulation. Posting may also be required.)

Section 7. That this ordinance and the rules, regulations, provisions, requirements, orders and matters established and adopted hereby shall take effect and be in full force and effect **[TIME PERIOD]** from and after the date of its final passage and adoption.

TABLE OF CONTENTS

CHAPTER 1

SCOPE AND ADMINISTRATION

PART 1—SCOPE AND APPLICATION

SECTION 101 GENERAL

101.1 Title. These regulations shall be known as the *Existing Building Code* of [NAME OF JURISDICTION], hereinafter referred to as "this code."

101.2 Scope. The provisions of the *International Existing Building Code* shall apply to the *repair, alteration, change of occupancy, addition* and relocation of *existing buildings*.

101.3 Intent. The intent of this code is to provide flexibility to permit the use of alternative approaches to achieve compliance with minimum requirements to safeguard the public health, safety and welfare insofar as they are affected by the *repair, alteration, change of occupancy, addition* and relocation of *existing buildings*.

101.4 Applicability. This code shall apply to the *repair, alteration, change of occupancy, addition* and relocation of all *existing buildings*, regardless of occupancy, subject to the criteria of Sections 101.4.1 and 101.4.2.

101.4.1 Buildings not previously occupied. A building or portion of a building that has not been previously occupied or used for its intended purpose in accordance with the laws in existence at the time of its completion shall comply with the provisions of the *International Building Code* or *International Residential Code*, as applicable, for new construction or with any current permit for such occupancy.

101.4.2 Buildings previously occupied. The legal occupancy of any building existing on the date of adoption of this code shall be permitted to continue without change, except as is specifically covered in this code, the *International Fire Code*, or the *International Property Maintenance Code*, or as is deemed necessary by the *code official* for the general safety and welfare of the occupants and the public.

101.5 Compliance methods. The *repair, alteration, change of occupancy, addition* or relocation of all *existing buildings* shall comply with one of the methods listed in Sections 101.5.1 through 101.5.3 as selected by the applicant. Application of a method shall be the sole basis for assessing the compliance of work performed under a single permit unless otherwise approved by the *code official*. Sections 101.5.1 through 101.5.3 shall not be applied in combination with each other. Where this code requires consideration of the seismic-force-resisting system of an *existing building* subject to *repair, alteration, change of occupancy, addition* or relocation of *existing buildings*, the seismic evaluation and design shall be based on Section 101.5.4 regardless of which compliance method is used.

Exception: Subject to the approval of the *code official*, *alterations* complying with the laws in existence at the time the building or the affected portion of the building was built shall be considered in compliance with the provisions of this code unless the building is undergoing more than a limited structural *alteration* as defined in Section 807.4.3. New structural members added as part of the *alteration* shall comply with the *International Building Code. Alterations* of *existing buildings* in *flood hazard areas* shall comply with Section 601.3.

101.5.1 Prescriptive compliance method. *Repairs, alterations, additions* and changes of occupancy complying with Chapter 3 of this code in buildings complying with the *International Fire Code* shall be considered in compliance with the provisions of this code.

101.5.2 Work area compliance method. *Repairs, alterations, additions*, changes in occupancy and relocated buildings complying with the applicable requirements of Chapters 4 through 12 of this code shall be considered in compliance with the provisions of this code.

101.5.3 Performance compliance method. *Repairs, alterations, additions*, changes in occupancy and relocated buildings complying with Chapter 13 of this code shall be considered in compliance with the provisions of this code.

101.5.4 Evaluation and design procedures. The seismic evaluation and design shall be based on the procedures specified in the *International Building Code*, ASCE 31 or ASCE 41. The procedures contained in Appendix A of this code shall be permitted to be used as specified in Section 101.5.4.2.

101.5.4.1 Compliance with IBC level seismic forces. Where compliance with the seismic design provisions of the *International Building Code* is required, the procedures shall be in accordance with one of the following:

1. One-hundred percent of the values in the *International Building Code*. Where the existing seismic force-resisting system is a type that can be designated as "Ordinary," values of R, Ω_0, and C_d used for analysis in accordance with Chapter 16 of the *International Building Code* shall be those specified for structural systems classified as "Ordinary" in accordance with Table 12.2-1 of ASCE 7, unless it is demonstrated that the structural system will

provide performance equivalent to that of a "Detailed," "Intermediate" or "Special" system.

2. Compliance with ASCE 41 using both the BSE-1 and BSE-2 earthquake hazard levels and the corresponding performance levels shown in Table 101.5.4.1.

**TABLE 101.5.4.1
PERFORMANCE CRITERIA FOR IBC LEVEL SEISMIC FORCES**

OCCUPANCY CATEGORY (Based on IBC Table 1604.5)	PERFORMANCE LEVEL FOR USE WITH ASCE 41 BSE-1 EARTHQUAKE HAZARD LEVEL	PERFORMANCE LEVEL FOR USE WITH ASCE 41 BSE-2 EARTHQUAKE HAZARD LEVEL
I	Life safety (LS)	Collapse prevention (CP)
II	Life safety (LS)	Collapse prevention (CP)
III	Note a	Note a
IV	Immediate occupancy (IO)	Life safety (LS)

a. Acceptable criteria for Occupancy Category III shall be taken as 80 percent of the acceptance criteria specified for Occupancy Category II performance levels, but need not be less than the acceptance criteria specified for Occupancy Category IV performance levels.

101.5.4.2 Compliance with reduced IBC level seismic forces. Where seismic evaluation and design is permitted to meet reduced *International Building Code* seismic force levels, the procedures used shall be in accordance with one of the following:

1. The *International Building Code* using 75 percent of the prescribed forces. Values of R, Ω_0 and C_d used for analysis shall be as specified in Section 101.5.4.1 of this code.

2. Structures or portions of structures that comply with the requirements of the applicable chapter in Appendix A as specified in Items 2.1 through 2.5 shall be deemed to comply with this section.

   2.1. The seismic evaluation and design of unreinforced masonry bearing wall buildings in Occupancy Category I or II are permitted to be based on the procedures specified in Appendix Chapter A1.

   2.2. Seismic evaluation and design of the wall anchorage system in reinforced concrete and reinforced masonry wall buildings with flexible diaphragms in Occupancy Category I or II are permitted to be based on the procedures specified in Chapter A2.

   2.3. Seismic evaluation and design of cripple walls and sill plate anchorage in residential buildings of light-frame wood construction in Occupancy Category I or II are permitted to be based on the procedures specified in Chapter A3.

   2.4. Seismic evaluation and design of soft, weak, or open-front wall conditions in multiunit residential buildings of wood construction in Occupancy Category I or II are permitted to be based on the procedures specified in Chapter A4.

   2.5. Seismic evaluation and design of concrete buildings in all occupancy categories are permitted to be based on the procedures specified in Chapter A5.

3. Compliance with ASCE 31 based on the applicable performance level as shown in Table 101.5.4.2. It shall be permitted to use the BSE-1 earthquake hazard level as defined in ASCE 41 and subject to the limitations in Item 4 below.

4. Compliance with ASCE 41 using the BSE-1 Earthquake Hazard Level and the performance level shown in Table 101.5.4.2. The design spectral response acceleration parameters S_{XS} and S_{XI} specified in ASCE 41 shall not be taken less than 75 percent of the respective design spectral response acceleration parameters S_{DS} and S_{DI} defined by the *International Building Code*.

**TABLE 101.5.4.2
PERFORMANCE CRITERIA FOR REDUCED IBC LEVEL SEISMIC FORCES**

OCCUPANCY CATEGORY (Based on IBC Table 1604.5)	PERFORMANCE LEVEL FOR USE WITH ASCE 31	PERFORMANCE LEVEL FOR USE WITH ASCE 41 BSE-1 EARTHQUAKE HAZARD LEVEL
I	Life safety (LS)	Life safety (LS)
II	Life safety (LS)	Life safety (LS)
III	Notes a, b	Note a
IV	Immediate occupancy (IO)	Immediate occupancy (IO)

a. Acceptable criteria for Occupancy Category III shall be taken as 80 percent of the acceptance criteria specified for Occupancy Category II performance levels, but need not be less than the acceptance criteria specified for Occupancy Category IV performance levels.

b. For Occupancy Category III, the ASCE 31 screening phase checklists shall be based on the life safety performance level.

101.6 Safeguards during construction. All construction work covered in this code, including any related demolition, shall comply with the requirements of Chapter 14.

101.7 Appendices. The *code official* is authorized to require rehabilitation and retrofit of buildings, structures or individual structural members in accordance with the appendices of this code if such appendices have been individually adopted.

101.8 Correction of violations of other codes. *Repairs* or *alterations* mandated by any property, housing, or fire safety maintenance code or mandated by any licensing rule or ordinance adopted pursuant to law shall conform only to the requirements of that code, rule, or ordinance and shall not be required to conform to this code unless the code requiring such *repair* or *alteration* so provides.

SECTION 102
APPLICABILITY

102.1 General. Where there is a conflict between a general requirement and a specific requirement, the specific requirement shall be applicable. Where in any specific case different sections of this code specify different materials, methods of construction or other requirements, the most restrictive shall govern.

102.2 Other laws. The provisions of this code shall not be deemed to nullify any provisions of local, state, or federal law.

102.3 Application of references. References to chapter or section numbers or to provisions not specifically identified by number shall be construed to refer to such chapter, section, or provision of this code.

102.4 Referenced codes and standards. The codes and standards referenced in this code shall be considered part of the requirements of this code to the prescribed extent of each such reference. Where differences occur between provisions of this code and referenced codes and standards, the provisions of this code shall govern.

Exception: Where enforcement of a code provision would violate the conditions of the listing of the equipment or appliance, the conditions of the listing shall govern.

102.5 Partial invalidity. In the event that any part or provision of this code is held to be illegal or void, this shall not have the effect of making void or illegal any of the other parts or provisions.

PART 2—ADMINISTRATION AND ENFORCEMENT

SECTION 103
DEPARTMENT OF BUILDING SAFETY

103.1 Creation of enforcement agency. The Department of Building Safety is hereby created, and the official in charge thereof shall be known as the *code official.*

103.2 Appointment. The *code official* shall be appointed by the chief appointing authority of the jurisdiction.

103.3 Deputies. In accordance with the prescribed procedures of this jurisdiction and with the concurrence of the appointing authority, the *code official* shall have the authority to appoint a deputy *code official*, the related technical officers, inspectors, plan examiners, and other employees. Such employees shall have powers as delegated by the *code official.*

SECTION 104
DUTIES AND POWERS OF CODE OFFICIAL

104.1 General. The *code official* is hereby authorized and directed to enforce the provisions of this code. The *code official* shall have the authority to render interpretations of this code and to adopt policies and procedures in order to clarify the application of its provisions. Such interpretations, policies, and procedures shall be in compliance with the intent and purpose of this code. Such policies and procedures shall not have the effect of waiving requirements specifically provided for in this code.

104.2 Applications and permits. The *code official* shall receive applications, review construction documents, and issue permits for the *repair*, *alteration*, *addition*, demolition, *change of occupancy*, and relocation of buildings; inspect the premises for which such permits have been issued; and enforce compliance with the provisions of this code.

104.2.1 Preliminary meeting. When requested by the permit applicant or the *code official*, the *code official* shall meet with the permit applicant prior to the application for a construction permit to discuss plans for the proposed work or *change of occupancy* in order to establish the specific applicability of the provisions of this code.

Exception: *Repairs* and Level 1 *alterations.*

104.2.1.1 Building evaluation. The *code official* is authorized to require an *existing building* to be investigated and evaluated by a registered design professional based on the circumstances agreed upon at the preliminary meeting. The design professional shall notify the *code official* if any potential nonconformance with the provisions of this code is identified.

104.3 Notices and orders. The *code official* shall issue all necessary notices or orders to ensure compliance with this code.

104.4 Inspections. The *code official* shall make all of the required inspections, or the *code official* shall have the authority to accept reports of inspection by approved agencies or individuals. Reports of such inspections shall be in writing and be certified by a responsible officer of such approved agency or by the responsible individual. The *code official* is authorized to engage such expert opinion as deemed necessary to report upon unusual technical issues that arise, subject to the approval of the appointing authority.

104.5 Identification. The *code official* shall carry proper identification when inspecting structures or premises in the performance of duties under this code.

104.6 Right of entry. Where it is necessary to make an inspection to enforce the provisions of this code, or where the *code official* has reasonable cause to believe that there exists in a structure or upon a premises a condition which is contrary to or in violation of this code which makes the structure or premises unsafe, *dangerous*, or hazardous, the *code official* is authorized to enter the structure or premises at reasonable times to inspect or to perform the duties imposed by this code, provided that if such structure or premises be occupied that credentials be presented to the occupant and entry requested. If such structure or premises be unoccupied, the *code official* shall first make a reasonable effort to locate the owner or other person having charge or control of the structure or premises and request entry. If entry is refused, the *code official* shall have recourse to the remedies provided by law to secure entry.

104.7 Department records. The *code official* shall keep official records of applications received, permits and certificates issued, fees collected, reports of inspections, and notices and orders issued. Such records shall be retained in the official records for the period required for retention of public records.

104.8 Liability. The *code official*, member of the Board of Appeals, or employee charged with the enforcement of this code, while acting for the jurisdiction in good faith and without malice in the discharge of the duties required by this code or other pertinent law or ordinance, shall not thereby be rendered liable personally and is hereby relieved from personal liability for any damage accruing to persons or property as a result of any act or by reason of an act or omission in the discharge of official duties. Any suit instituted against an officer or employee because of an act performed by that officer or employee in the lawful discharge of duties and under the provisions of this code shall be defended by legal representative of the jurisdiction until the final termination of the proceedings. The *code official* or any subordinate shall not be liable for cost in any action, suit, or proceeding that is instituted in pursuance of the provisions of this code.

104.9 Approved materials and equipment. Materials, equipment, and devices approved by the *code official* shall be constructed and installed in accordance with such approval.

104.9.1 Used materials and equipment. The use of used materials that meet the requirements of this code for new materials is permitted. Used equipment and devices shall be permitted to be reused subject to the approval of the *code official.*

104.10 Modifications. Wherever there are practical difficulties involved in carrying out the provisions of this code, the code official shall have the authority to grant modifications for individual cases upon application of the owner or owner's representative, provided the code official shall first find that special individual reason makes the strict letter of this code impractical and the modification is in compliance with the intent and purpose of this code, and that such modification does not lessen health, accessibility, life and fire safety, or structural requirements. The details of action granting modifications shall be recorded and entered in the files of the Department of Building Safety.

104.10.1 Flood hazard areas. For *existing buildings* located in *flood hazard areas* for which *repairs*, *alterations* and *additions* constitute *substantial improvement*, the code official shall not grant modifications to provisions related to flood resistance unless a determination is made that:

1. The applicant has presented good and sufficient cause that the unique characteristics of the size, configuration or topography of the site render compliance with the flood-resistant construction provisions inappropriate.
2. Failure to grant the modification would result in exceptional hardship.
3. The granting of the modification will not result in increased flood heights, additional threats to public safety, extraordinary public expense nor create nuisances, cause fraud on or victimization of the public or conflict with existing laws or ordinances.
4. The modification is the minimum necessary to afford relief, considering the flood hazard.
5. A written notice will be provided to the applicant specifying, if applicable, the difference between the design flood elevation and the elevation to which the building is to be built, stating that the cost of flood insurance will be commensurate with the increased risk resulting from the reduced floor elevation and that construction below the design flood elevation increases risks to life and property.

104.11 Alternative materials, design and methods of construction, and equipment. The provisions of this code are not intended to prevent the installation of any material or to prohibit any design or method of construction not specifically prescribed by this code, provided that any such alternative has been approved. An alternative material, design, or method of construction shall be approved where the code official finds that the proposed design is satisfactory and complies with the intent of the provisions of this code, and that the material, method, or work offered is, for the purpose intended, at least the equivalent of that prescribed in this code in quality, strength, effectiveness, fire resistance, durability, and safety.

104.11.1 Research reports. Supporting data, where necessary to assist in the approval of materials or assemblies not specifically provided for in this code, shall consist of valid research reports from approved sources.

104.11.2 Tests. Whenever there is insufficient evidence of compliance with the provisions of this code or evidence that a material or method does not conform to the requirements of this code, or in order to substantiate claims for alternative materials or methods, the code official shall have the authority to require tests as evidence of compliance to be made at no expense to the jurisdiction. Test methods shall be as specified in this code or by other recognized test standards. In the absence of recognized and accepted test methods, the code official shall approve the testing procedures. Tests shall be performed by an approved agency. Reports of such tests shall be retained by the code official for the period required for retention.

SECTION 105
PERMITS

105.1 Required. Any owner or authorized agent who intends to *repair*, add to, alter, relocate, demolish, or change the occupancy of a building or to *repair*, install, add, alter, remove, convert, or replace any electrical, gas, mechanical, or plumbing system, the installation of which is regulated by this code, or to cause any such work to be done, shall first make application to the *code official* and obtain the required permit.

105.1.1 Annual permit. In lieu of an individual permit for each *alteration* to an already approved electrical, gas, mechanical, or plumbing installation, the *code official* is authorized to issue an annual permit upon application therefor to any person, firm, or corporation regularly employing one or more qualified trade persons in the building, structure, or on the premises owned or operated by the applicant for the permit.

105.1.2 Annual permit records. The person to whom an annual permit is issued shall keep a detailed record of *alterations* made under such annual permit. The *code official* shall have access to such records at all times, or such records shall be filed with the *code official* as designated.

105.2 Work exempt from permit. Exemptions from permit requirements of this code shall not be deemed to grant authorization for any work to be done in any manner in violation of the provisions of this code or any other laws or ordinances of this jurisdiction. Permits shall not be required for the following:

Building:

1. Sidewalks and driveways not more than 30 inches (762 mm) above grade and not over any basement or story below and that are not part of an accessible route.
2. Painting, papering, tiling, carpeting, cabinets, counter tops, and similar finish work.
3. Temporary motion picture, television, and theater stage sets and scenery.
4. Shade cloth structures constructed for nursery or agricultural purposes, and not including service systems.
5. Window awnings supported by an exterior wall of Group R-3 or Group U occupancies.
6. Movable cases, counters, and partitions not over 69 inches (1753 mm) in height.

Electrical:

Repairs and maintenance: Minor *repair* work, including the replacement of lamps or the connection of approved portable electrical equipment to approved permanently installed receptacles.

Radio and television transmitting stations: The provisions of this code shall not apply to electrical equipment used for radio and television transmissions, but do apply to equipment and wiring for power supply, the installations of towers, and antennas.

Temporary testing systems: A permit shall not be required for the installation of any temporary system required for the testing or servicing of electrical equipment or apparatus.

Gas:

1. Portable heating appliance.
2. Replacement of any minor part that does not alter approval of equipment or make such equipment unsafe.

Mechanical:

1. Portable heating appliance.
2. Portable ventilation equipment.
3. Portable cooling unit.
4. Steam, hot, or chilled water piping within any heating or cooling equipment regulated by this code.
5. Replacement of any part that does not alter its approval or make it unsafe.
6. Portable evaporative cooler.
7. Self-contained refrigeration system containing 10 pounds (4.54 kg) or less of refrigerant and actuated by motors of 1 horsepower (746 W) or less.

Plumbing:

1. The stopping of leaks in drains, water, soil, waste, or vent pipe; provided, however, that if any concealed trap, drainpipe, water, soil, waste, or vent pipe becomes defective and it becomes necessary to remove and replace the same with new material, such work shall be considered as new work, and a permit shall be obtained and inspection made as provided in this code.
2. The clearing of stoppages or the repairing of leaks in pipes, valves, or fixtures, and the removal and reinstallation of water closets, provided such *repairs* do not involve or require the replacement or rearrangement of valves, pipes, or fixtures.

105.2.1 Emergency repairs. Where equipment replacements and *repairs* must be performed in an emergency situation, the permit application shall be submitted within the next working business day to the *code official.*

105.2.2 Repairs. Application or notice to the *code official* is not required for ordinary *repairs* to structures and items listed in Section 105.2. Such *repairs* shall not include the cutting away of any wall, partition, or portion thereof, the removal or cutting of any structural beam or load-bearing support, or the removal or change of any required means of egress or rearrangement of parts of a structure affecting the egress requirements; nor shall ordinary *repairs* include *addition* to, *alteration* of, replacement, or relocation of any standpipe, water supply, sewer, drainage, drain leader, gas, soil, waste, vent, or similar piping, electric wiring, or mechanical or other work affecting public health or general safety.

105.2.3 Public service agencies. A permit shall not be required for the installation, *alteration*, or *repair* of generation, transmission, distribution, or metering or other related equipment that is under the ownership and control of public service agencies by established right.

105.3 Application for permit. To obtain a permit, the applicant shall first file an application therefor in writing on a form furnished by the Department of Building Safety for that purpose. Such application shall:

1. Identify and describe the work in accordance with Chapter 3 to be covered by the permit for which application is made.
2. Describe the land on which the proposed work is to be done by legal description, street address, or similar description that will readily identify and definitely locate the proposed building or work.

3. Indicate the use and occupancy for which the proposed work is intended.
4. Be accompanied by construction documents and other information as required in Section 106.3.
5. State the valuation of the proposed work.
6. Be signed by the applicant or the applicant's authorized agent.
7. Give such other data and information as required by the *code official.*

105.3.1 Action on application. The *code official* shall examine or cause to be examined applications for permits and amendments thereto within a reasonable time after filing. If the application or the construction documents do not conform to the requirements of pertinent laws, the *code official* shall reject such application in writing, stating the reasons therefor. If the *code official* is satisfied that the proposed work conforms to the requirements of this code and laws and ordinances applicable thereto, the *code official* shall issue a permit therefor as soon as practicable.

105.3.2 Time limitation of application. An application for a permit for any proposed work shall be deemed to have been abandoned 180 days after the date of filing, unless such application has been pursued in good faith or a permit has been issued; except that the *code official* is authorized to grant one or more extensions of time for additional periods not exceeding 90 days each. The extension shall be requested in writing and justifiable cause demonstrated.

105.4 Validity of permit. The issuance or granting of a permit shall not be construed to be a permit for, or an approval of, any violation of any of the provisions of this code or of any other ordinance of the jurisdiction. Permits presuming to give authority to violate or cancel the provisions of this code or other ordinances of the jurisdiction shall not be valid. The issuance of a permit based on construction documents and other data shall not prevent the *code official* from requiring the correction of errors in the construction documents and other data. The *code official* is also authorized to prevent occupancy or use of a structure where in violation of this code or of any other ordinances of this jurisdiction.

105.5 Expiration. Every permit issued shall become invalid unless the work on the site authorized by such permit is commenced within 180 days after its issuance, or if the work authorized on the site by such permit is suspended or abandoned for a period of 180 days after the time the work is commenced. The *code official* is authorized to grant, in writing, one or more extensions of time for periods not more than 180 days each. The extension shall be requested in writing and justifiable cause demonstrated.

105.6 Suspension or revocation. The *code official* is authorized to suspend or revoke a permit issued under the provisions of this code wherever the permit is issued in error or on the basis of incorrect, inaccurate, or incomplete information or in violation of any ordinance or regulation or any of the provisions of this code.

105.7 Placement of permit. The building permit or copy shall be kept on the site of the work until the completion of the project.

SECTION 106
CONSTRUCTION DOCUMENTS

106.1 General. Submittal documents consisting of construction documents, special inspection and structural observation programs, investigation and evaluation reports, and other data shall be submitted in two or more sets with each application for a permit. The construction documents shall be prepared by a registered design professional where required by the statutes of the jurisdiction in which the project is to be constructed. Where special conditions exist, the *code official* is authorized to require additional construction documents to be prepared by a registered design professional.

Exception: The *code official* is authorized to waive the submission of construction documents and other data not required to be prepared by a registered design professional if it is found that the nature of the work applied for is such that reviewing of construction documents is not necessary to obtain compliance with this code.

106.2 Construction documents. Construction documents shall be in accordance with Sections 106.2.1 through 106.2.5.

106.2.1 Construction documents. Construction documents shall be dimensioned and drawn upon suitable material. Electronic media documents are permitted to be submitted when approved by the *code official.* Construction documents shall be of sufficient clarity to indicate the location, nature and extent of the work proposed and show in detail that it will conform to the provisions of this code and relevant laws, ordinances, rules and regulations, as determined by the *code official.* The work areas shall be shown.

106.2.2 Fire protection system(s) shop drawings. Shop drawings for the fire protection system(s) shall be submitted to indicate conformance with this code and the construction documents and shall be approved prior to the start of system installation. Shop drawings shall contain all information as required by the referenced installation standards in Chapter 9 of the *International Building Code.*

106.2.3 Means of egress. The construction documents for *Alterations*–Level 2, *Alterations*–Level 3, *additions* and changes of occupancy shall show in sufficient detail the location, construction, size and character of all portions of the means of egress in compliance with the provisions of this code. The construction documents shall designate the number of occupants to be accommodated in every *work area* of every floor and in all affected rooms and spaces.

106.2.4 Exterior wall envelope. Construction documents for all work affecting the exterior wall envelope shall describe the exterior wall envelope in sufficient detail to determine compliance with this code. The construction documents shall provide details of the exterior wall envelope as required, including windows, doors, flashing, intersections with dissimilar materials, corners, end details, control joints, intersections at roof, eaves, or parapets, means of drainage, water-resistive membrane, and details around openings.

The construction documents shall include manufacturer's installation instructions that provide supporting documentation that the proposed penetration and opening details described in the construction documents maintain the wind and weather resistance of the exterior wall envelope. The supporting documentation shall fully describe the exterior wall system which was tested, where applicable, as well as the test procedure used.

106.2.5 Site plan. The construction documents submitted with the application for permit shall be accompanied by a site plan showing to scale the size and location of new construction and existing structures on the site, distances from lot lines, the established street grades, and the proposed finished grades; and it shall be drawn in accordance with an accurate boundary line survey. In the case of demolition, the site plan shall show construction to be demolished and the location and size of existing structures and construction that are to remain on the site or plot. The *code official* is authorized to waive or modify the requirement for a site plan when the application for permit is for *alteration*, *repair* or *change of occupancy*.

106.3 Examination of documents. The *code official* shall examine or cause to be examined the submittal documents and shall ascertain by such examinations whether the construction or occupancy indicated and described is in accordance with the requirements of this code and other pertinent laws or ordinances.

106.3.1 Approval of construction documents. When the *code official* issues a permit, the construction documents shall be approved in writing or by stamp as "Reviewed for Code Compliance." One set of construction documents so reviewed shall be retained by the *code official*. The other set shall be returned to the applicant, shall be kept at the site of work, and shall be open to inspection by the *code official* or a duly authorized representative.

106.3.2 Previous approval. This code shall not require changes in the construction documents, construction or designated occupancy of a structure for which a lawful permit has been issued and the construction of which has been pursued in good faith within 180 days after the effective date of this code and has not been abandoned.

106.3.3 Phased approval. The *code official* is authorized to issue a permit for the construction of foundations or any other part of a building before the construction documents for the whole building or structure have been submitted, provided that adequate information and detailed statements have been filed complying with pertinent requirements of this code. The holder of such permit for the foundation or other parts of a building shall proceed at the holder's own risk with the building operation and without assurance that a permit for the entire structure will be granted.

106.3.4 Deferred submittals. For the purposes of this section, deferred submittals are defined as those portions of the design that are not submitted at the time of the application and that are to be submitted to the *code official* within a specified period.

Deferral of any submittal items shall have the prior approval of the *code official*. The registered design professional in responsible charge shall list the deferred submittals on the construction documents for review by the *code official*.

Submittal documents for deferred submittal items shall be submitted to the registered design professional in responsible charge who shall review them and forward them to the *code official* with a notation indicating that the deferred submittal documents have been reviewed and that they have been found to be in general conformance to the design of the building. The deferred submittal items shall not be installed until their deferred submittal documents have been approved by the *code official*.

106.4 Amended construction documents. Work shall be installed in accordance with the reviewed construction documents, and any changes made during construction that are not in compliance with the approved construction documents shall be resubmitted for approval as an amended set of construction documents.

106.5 Retention of construction documents. One set of approved construction documents shall be retained by the *code official* for a period of not less than the period required for retention of public records.

106.6 Design professional in responsible charge. When it is required that documents be prepared by a registered design professional, the *code official* shall be authorized to require the owner to engage and designate on the building permit application a registered design professional who shall act as the registered design professional in responsible charge. If the circumstances require, the owner shall designate a substitute registered design professional in responsible charge who shall perform the duties required of the original registered design professional in responsible charge. The *code official* shall be notified in writing by the owner if the registered design professional in responsible charge is changed or is unable to continue to perform the duties. The registered design professional in responsible charge shall be responsible for reviewing and coordinating submittal documents prepared by others, including phased and deferred submittal items, for compatibility with the design of the building. Where structural observation is required, the inspection program shall name the individual or firms who are to perform structural observation and describe the stages of construction at which structural observation is to occur.

SECTION 107
TEMPORARY STRUCTURES AND USES

107.1 General. The *code official* is authorized to issue a permit for temporary uses. Such permits shall be limited as to time of service but shall not be permitted for more than 180 days. The *code official* is authorized to grant extensions for demonstrated cause.

107.2 Conformance. Temporary uses shall conform to the structural strength, fire safety, means of egress, accessibility, light, ventilation and sanitary requirements of this code as necessary to ensure the public health, safety and general welfare.

107.3 Temporary power. The *code official* is authorized to give permission to temporarily supply and use power in part of an electric installation before such installation has been fully completed and the final certificate of completion has been issued. The part covered by the temporary certificate shall comply with the requirements specified for temporary lighting, heat or power in NFPA 70.

107.4 Termination of approval. The *code official* is authorized to terminate such permit for a temporary use and to order the temporary use to be discontinued.

SECTION 108 FEES

108.1 Payment of fees. A permit shall not be valid until the fees prescribed by law have been paid. Nor shall an amendment to a permit be released until the additional fee, if any, has been paid.

108.2 Schedule of permit fees. On buildings, electrical, gas, mechanical, and plumbing systems or *alterations* requiring a permit, a fee for each permit shall be paid as required in accordance with the schedule as established by the applicable governing authority.

108.3 Building permit valuations. The applicant for a permit shall provide an estimated permit value at time of application. Permit valuations shall include total value of work including materials and labor for which the permit is being issued, such as electrical, gas, mechanical, plumbing equipment, and permanent systems. If, in the opinion of the *code official*, the valuation is underestimated on the application, the permit shall be denied unless the applicant can show detailed estimates to meet the approval of the *code official*. Final building permit valuation shall be set by the *code official*.

108.4 Work commencing before permit issuance. Any person who commences any work before obtaining the necessary permits shall be subject to an additional fee established by the *code official* that shall be in addition to the required permit fees.

108.5 Related fees. The payment of the fee for the construction, *alteration*, removal, or demolition of work done in connection to or concurrently with the work authorized by a building permit shall not relieve the applicant or holder of the permit from the payment of other fees that are prescribed by law.

108.6 Refunds. The *code official* is authorized to establish a refund policy.

SECTION 109 INSPECTIONS

109.1 General. Construction or work for which a permit is required shall be subject to inspection by the *code official*, and such construction or work shall remain accessible and exposed for inspection purposes until approved. Approval as a result of an inspection shall not be construed to be an approval of a violation of the provisions of this code or of other ordinances of the jurisdiction. Inspections presuming to give authority to violate or cancel the provisions of this code or of other ordinances of the jurisdiction shall not be valid. It shall be the duty of the permit applicant to cause the work to remain accessible and exposed for inspection purposes. Neither the *code official* nor the jurisdiction shall be liable for expense entailed in the removal or replacement of any material required to allow inspection.

109.2 Preliminary inspection. Before issuing a permit, the *code official* is authorized to examine or cause to be examined buildings and sites for which an application has been filed.

109.3 Required inspections. The *code official*, upon notification, shall make the inspections set forth in Sections 109.3.1 through 109.3.9.

109.3.1 Footing or foundation inspection. Footing and foundation inspections shall be made after excavations for footings are complete and any required reinforcing steel is in place. For concrete foundations, any required forms shall be in place prior to inspection. Materials for the foundation shall be on the job, except where concrete is ready-mixed in accordance with ASTM C 94, the concrete need not be on the job.

109.3.2 Concrete slab or under-floor inspection. Concrete slab and under-floor inspections shall be made after in-slab or under-floor reinforcing steel and building service equipment, conduit, piping accessories, and other ancillary equipment items are in place but before any concrete is placed or floor sheathing installed, including the sub floor.

109.3.3 Lowest floor elevation. For *additions* and *substantial improvements* to *existing buildings* in *flood hazard areas*, upon placement of the lowest floor, including basement, and prior to further vertical construction, the elevation documentation required in the *International Building Code* shall be submitted to the *code official.*

109.3.4 Frame inspection. Framing inspections shall be made after the roof deck or sheathing, all framing, fire blocking, and bracing are in place and pipes, chimneys, and vents to be concealed are complete and the rough electrical, plumbing, heating wires, pipes, and ducts are approved.

109.3.5 Lath or gypsum board inspection. Lath and gypsum board inspections shall be made after lathing and gypsum board, interior and exterior, is in place but before any plastering is applied or before gypsum board joints and fasteners are taped and finished.

Exception: Gypsum board that is not part of a fire-resistance-rated assembly or a shear assembly.

109.3.6 Fire and smoke-resistant penetrations. Protection of joints and penetrations in fire-resistance-rated assemblies, smoke barriers and smoke partitions shall not be concealed from view until inspected and approved.

109.3.7 Other inspections. In addition to the inspections specified above, the *code official* is authorized to make or require other inspections of any construction work to ascertain compliance with the provisions of this code and other laws that are enforced by the Department of Building Safety.

109.3.8 Special inspections. Special inspections shall be required in accordance with the *International Building Code*.

109.3.9 Final inspection. The final inspection shall be made after all work required by the building permit is completed.

109.4 Inspection agencies. The *code official* is authorized to accept reports of approved inspection agencies, provided such agencies satisfy the requirements as to qualifications and reliability.

109.5 Inspection requests. It shall be the duty of the holder of the building permit or their duly authorized agent to notify the *code official* when work is ready for inspection. It shall be the duty of the permit holder to provide access to and means for any inspections of such work that are required by this code.

109.6 Approval required. Work shall not be done beyond the point indicated in each successive inspection without first obtaining the approval of the *code official*. The *code official*, upon notification, shall make the requested inspections and shall either indicate the portion of the construction that is satisfactory as completed or shall notify the permit holder or an agent of the permit holder wherein the same fails to comply with this code. Any portions that do not comply shall be corrected and such portion shall not be covered or concealed until authorized by the *code official*.

SECTION 110 CERTIFICATE OF OCCUPANCY

110.1 Altered area use and occupancy classification change. No altered area of a building and no relocated building shall be used or occupied, and no change in the existing occupancy classification of a building or portion thereof shall be made until the code official has issued a certificate of occupancy therefor as provided herein. Issuance of a certificate of occupancy shall not be construed as an approval of a violation of the provisions of this code or of other ordinances of the jurisdiction.

110.2 Certificate issued. After the *code official* inspects the building and finds no violations of the provisions of this code or other laws that are enforced by the Department of Building Safety, the *code official* shall issue a certificate of occupancy that shall contain the following:

1. The building permit number.
2. The address of the structure.
3. The name and address of the owner.
4. A description of that portion of the structure for which the certificate is issued.
5. A statement that the described portion of the structure has been inspected for compliance with the requirements of this code for the occupancy and division of occupancy and the use for which the proposed occupancy is classified.
6. The name of the *code official*.
7. The edition of the code under which the permit was issued.
8. The use and occupancy in accordance with the provisions of the *International Building Code*.
9. The type of construction as defined in the *International Building Code*.
10. The design occupant load and any impact the *alteration* has on the design occupant load of the area not within the scope of the work.
11. If fire protection systems are provided, whether the fire protection systems are required.
12. Any special stipulations and conditions of the building permit.

110.3 Temporary occupancy. The *code official* is authorized to issue a temporary certificate of occupancy before the completion of the entire work covered by the permit, provided that such portion or portions shall be occupied safely. The *code official* shall set a time period during which the temporary certificate of occupancy is valid.

110.4 Revocation. The *code official* is authorized to, in writing, suspend or revoke a certificate of occupancy or completion issued under the provisions of this code wherever the certificate is issued in error or on the basis of incorrect information supplied, or where it is determined that the building or structure or portion thereof is in violation of any ordinance or regulation or any of the provisions of this code.

SECTION 111 SERVICE UTILITIES

111.1 Connection of service utilities. No person shall make connections from a utility, source of energy, fuel, or power to any building or system that is regulated by this code for which a permit is required, until approved by the *code official*.

111.2 Temporary connection. The *code official* shall have the authority to authorize the temporary connection of the building or system to the utility source of energy, fuel, or power.

111.3 Authority to disconnect service utilities. The *code official* shall have the authority to authorize disconnection of utility service to the building, structure or system regulated by this code and the referenced codes and standards in case of emergency where necessary to eliminate an immediate hazard to life or property or when such utility connection has been made without the approval required by Section 111.1 or 111.2. The code official shall notify the serving utility and, wherever possible, the owner and occupant of the building, structure or service system of the decision to disconnect prior to taking such action. If not notified prior to disconnecting, the owner or occupant of the building, structure or service system shall be notified in writing, as soon as practical thereafter.

SECTION 112
BOARD OF APPEALS

112.1 General. In order to hear and decide appeals of orders, decisions, or determinations made by the code official relative to the application and interpretation of this code, there shall be and is hereby created a board of appeals. The board of appeals shall be appointed by the governing body and shall hold office at its pleasure. The board shall adopt rules of procedure for conducting its business.

112.2 Limitations on authority. An application for appeal shall be based on a claim that the true intent of this code or the rules legally adopted thereunder have been incorrectly interpreted, the provisions of this code do not fully apply, or an equally good or better form of construction is proposed. The board shall have no authority to waive requirements of this code.

112.3 Qualifications. The board of appeals shall consist of members who are qualified by experience and training to pass on matters pertaining to building construction and are not employees of the jurisdiction.

SECTION 113
VIOLATIONS

113.1 Unlawful acts. It shall be unlawful for any person, firm, or corporation to *repair*, alter, extend, add, move, remove, demolish, or change the occupancy of any building or equipment regulated by this code or cause same to be done in conflict with or in violation of any of the provisions of this code.

113.2 Notice of violation. The *code official* is authorized to serve a notice of violation or order on the person responsible for the *repair*, *alteration*, extension, *addition*, moving, removal, demolition, or change in the occupancy of a building in violation of the provisions of this code or in violation of a permit or certificate issued under the provisions of this code. Such order shall direct the discontinuance of the illegal action or condition and the abatement of the violation.

113.3 Prosecution of violation. If the notice of violation is not complied with promptly, the *code official* is authorized to request the legal counsel of the jurisdiction to institute the appropriate proceeding at law or in equity to restrain, correct, or abate such violation or to require the removal or termination of the unlawful occupancy of the building or structure in violation of the provisions of this code or of the order or direction made pursuant thereto.

113.4 Violation penalties. Any person who violates a provision of this code or fails to comply with any of the requirements thereof or who *repairs* or alters or changes the occupancy of a building or structure in violation of the approved construction documents or directive of the *code official* or of a permit or certificate issued under the provisions of this code shall be subject to penalties as prescribed by law.

SECTION 114
STOP WORK ORDER

114.1 Authority. Whenever the *code official* finds any work regulated by this code being performed in a manner contrary to the provisions of this code or in a *dangerous* or unsafe manner, the *code official* is authorized to issue a stop work order.

114.2 Issuance. The stop work order shall be in writing and shall be given to the owner of the property involved or to the owner's agent, or to the person doing the work. Upon issuance of a stop work order, the cited work shall immediately cease. The stop work order shall state the reason for the order and the conditions under which the cited work will be permitted to resume.

114.3 Unlawful continuance. Any person who shall continue any work after having been served with a stop work order, except such work as that person is directed to perform to remove a violation or unsafe condition, shall be subject to penalties as prescribed by law.

SECTION 115
UNSAFE BUILDINGS AND EQUIPMENT

115.1 Conditions. Buildings, structures or equipment that are or hereafter become unsafe, shall be taken down, removed or made safe as the code official deems necessary and as provided for in this code.

115.2 Record. The *code official* shall cause a report to be filed on an unsafe condition. The report shall state the occupancy of the structure and the nature of the unsafe condition.

115.3 Notice. If an unsafe condition is found, the *code official* shall serve on the owner, agent, or person in control of the structure a written notice that describes the condition deemed unsafe and specifies the required *repairs* or improvements to be made to abate the unsafe condition, or that requires the unsafe building to be demolished within a stipulated time. Such notice shall require the person thus notified to declare immediately to the *code official* acceptance or rejection of the terms of the order.

115.4 Method of service. Such notice shall be deemed properly served if a copy thereof is delivered to the owner personally; sent by certified or registered mail addressed to the owner at the last known address with the return receipt requested; or delivered in any other manner as prescribed by local law. If the certified or registered letter is returned showing that the letter was not delivered, a copy thereof shall be posted in a conspicuous place in or about the structure affected by such notice. Service of such notice in the foregoing manner upon the owner's agent or upon the person responsible for the structure shall constitute service of notice upon the owner.

115.5 Restoration. The building or equipment determined to be unsafe by the *code official* is permitted to be restored to a safe condition. To the extent that *repairs*, *alterations*, or *additions* are made or a *change of occupancy* occurs during the restoration of the building, such *repairs*, *alterations*, *additions*, or *change of occupancy* shall comply with the requirements of this code.

SECTION 116
EMERGENCY MEASURES

116.1 Imminent danger. When, in the opinion of the *code official*, there is imminent danger of failure or collapse of a building that endangers life, or when any building or part of a building has fallen and life is endangered by the occupation of the building, or when there is actual or potential danger to the building occupants or those in the proximity of any structure because of explosives, explosive fumes or vapors, or the presence of toxic fumes, gases, or materials, or operation of defective or *dangerous* equipment, the *code official* is hereby authorized and empowered to order and require the occupants to vacate the premises forthwith. The *code official* shall cause to be posted at each entrance to such structure a notice reading as follows: "This Structure Is Unsafe and Its Occupancy Has Been Prohibited by the *Code Official*." It shall be unlawful for any person to enter such structure except for the purpose of securing the structure, making the required *repairs*, removing the hazardous condition, or of demolishing the same.

116.2 Temporary safeguards. Notwithstanding other provisions of this code, whenever, in the opinion of the *code official*, there is imminent danger due to an unsafe condition, the *code official* shall order the necessary work to be done, including the boarding up of openings, to render such structure temporarily safe whether or not the legal procedure herein described has been instituted; and shall cause such other action to be taken as the *code official* deems necessary to meet such emergency.

116.3 Closing streets. When necessary for public safety, the *code official* shall temporarily close structures and close or order the authority having jurisdiction to close sidewalks, streets, public ways, and places adjacent to unsafe structures, and prohibit the same from being utilized.

116.4 Emergency repairs. For the purposes of this section, the *code official* shall employ the necessary labor and materials to perform the required work as expeditiously as possible.

116.5 Costs of emergency repairs. Costs incurred in the performance of emergency work shall be paid by the jurisdiction. The legal counsel of the jurisdiction shall institute appropriate action against the owner of the premises where the unsafe structure is or was located for the recovery of such costs.

116.6 Hearing. Any person ordered to take emergency measures shall comply with such order forthwith. Any affected person shall thereafter, upon petition directed to the appeals board, be afforded a hearing as described in this code.

SECTION 117
DEMOLITION

117.1 General. The *code official* shall order the owner of any premises upon which is located any structure that in the *code official's* judgment is so old, dilapidated, or has become so out of *repair* as to be *dangerous*, unsafe, insanitary, or otherwise unfit for human habitation or occupancy, and such that it is unreasonable to *repair* the structure, to demolish and remove such structure; or if such structure is capable of being made safe by *repairs*, to *repair* and make safe and sanitary or to demolish and remove at the owner's option; or where there has been a cessation of normal construction of any structure for a period of more than two years, to demolish and remove such structure.

117.2 Notices and orders. All notices and orders shall comply with Section 113.

117.3 Failure to comply. If the owner of a premises fails to comply with a demolition order within the time prescribed, the *code official* shall cause the structure to be demolished and removed, either through an available public agency or by contract or arrangement with private persons, and the cost of such demolition and removal shall be charged against the real estate upon which the structure is located and shall be a lien upon such real estate.

117.4 Salvage materials. When any structure has been ordered demolished and removed, the governing body or other designated officer under said contract or arrangement aforesaid shall have the right to sell the salvage and valuable materials at the highest price obtainable. The net proceeds of such sale, after deducting the expenses of such demolition and removal, shall be promptly remitted with a report of such sale or transaction, including the items of expense and the amounts deducted, for the person who is entitled thereto, subject to any order of a court. If such a surplus does not remain to be turned over, the report shall so state.

CHAPTER 2
DEFINITIONS

SECTION 201
GENERAL

201.1 Scope. Unless otherwise expressly stated, the following words and terms shall, for the purposes of this code, have the meanings shown in this chapter.

201.2 Interchangeability. Words used in the present tense include the future; words stated in the masculine gender include the feminine and neuter; the singular number includes the plural and the plural, the singular.

201.3 Terms defined in other codes. Where terms are not defined in this code and are defined in the other *International Codes,* such terms shall have the meanings ascribed to them in those codes.

201.4 Terms not defined. Where terms are not defined through the methods authorized by this chapter, such terms shall have ordinarily accepted meanings such as the context implies.

SECTION 202
GENERAL DEFINITIONS

ADDITION. An extension or increase in floor area, number of stories, or height of a building or structure.

ALTERATION. Any construction or renovation to an existing structure other than a *repair* or *addition*. Alterations are classified as Level 1, Level 2, and Level 3.

CHANGE OF OCCUPANCY. A change in the purpose or level of activity within a building that involves a change in application of the requirements of this code.

CODE OFFICIAL. The officer or other designated authority charged with the administration and enforcement of this code.

DANGEROUS. Any building, structure or portion thereof that meets any of the conditions described below shall be deemed *dangerous*:

1. The building or structure has collapsed, partially collapsed, moved off its foundation or lacks the support of ground necessary to support it.
2. There exists a significant risk of collapse, detachment or dislodgment of any portion, member, appurtenance or ornamentation of the building or structure under service loads.

EQUIPMENT OR FIXTURE. Any plumbing, heating, electrical, ventilating, air conditioning, refrigerating, and fire protection equipment, and elevators, dumb waiters, escalators, boilers, pressure vessels and other mechanical facilities or installations that are related to building services. Equipment or fixture shall not include manufacturing, production, or process equipment, but shall include connections from building service to process equipment.

EXISTING BUILDING. A building erected prior to the date of adoption of the appropriate code, or one for which a legal building permit has been issued.

[B] FLOOD HAZARD AREA. The greater of the following two areas:

1. The area within a flood plain subject to a 1-percent or greater chance of flooding in any year.
2. The area designated as a *flood hazard area* on a community's flood hazard map, or otherwise legally designated.

HISTORIC BUILDING. Any building or structure that is listed in the State or National Register of Historic Places; designated as a historic property under local or state designation law or survey; certified as a contributing resource within a National Register listed or locally designated historic district; or with an opinion or certification that the property is eligible to be listed on the National or State Register of Historic Places either individually or as a contributing building to a historic district by the State Historic Preservation Officer or the Keeper of the National Register of Historic Places.

LOAD BEARING ELEMENT. Any column, girder, beam, joist, truss, rafter, wall, floor or roof sheathing that supports any vertical load in *addition* to its own weight or any lateral load.

NONCOMBUSTIBLE MATERIAL. A material that, under the conditions anticipated, will not ignite or burn when subjected to fire or heat. Materials that pass ASTM E 136 are considered noncombustible materials.

PRIMARY FUNCTION. A *primary function* is a major activity for which the facility is intended. Areas that contain a *primary function* include, but are not limited to, the customer services lobby of a bank, the dining area of a cafeteria, the meeting rooms in a conference center, as well as offices and other work areas in which the activities of the public accommodation or other private entity using the facility are carried out. Mechanical rooms, boiler rooms, supply storage rooms, employee lounges or locker rooms, janitorial closets, entrances, corridors and restrooms are not areas containing a *primary function*.

REGISTERED DESIGN PROFESSIONAL IN RESPONSIBLE CHARGE. A registered design professional engaged by the owner to review and coordinate certain aspects of the project, as determined by the *code official*, for compatibility with the design of the building or structure, including submittal documents prepared by others, deferred submittal documents and phased submittal documents.

REHABILITATION. Any work, as described by the categories of work defined herein, undertaken in an *existing building*.

REHABILITATION, SEISMIC. Work conducted to improve the seismic lateral force resistance of an *existing building*.

REPAIR. The restoration to good or sound condition of any part of an *existing building* for the purpose of its maintenance.

SEISMIC LOADING. The forces prescribed herein, related to the response of the structure to earthquake motions, to be used in the analysis and design of the structure and its components.

[B] SUBSTANTIAL DAMAGE. For the purpose of determining compliance with the flood provisions of this code, damage of any origin sustained by a structure whereby the cost of restoring the structure to its before-damaged condition would equal or exceed 50 percent of the market value of the structure before the damage occurred.

SUBSTANTIAL IMPROVEMENT. For the purpose of determining compliance with the flood provisions of this code, any *repair*, *alteration*, *addition*, or improvement of a building or structure, the cost of which equals or exceeds 50 percent of the market value of the structure, before the improvement or *repair* is started. If the structure has sustained *substantial damage*, any repairs are considered *substantial improvement* regardless of the actual *repair* work performed. The term does not, however, include either:

1. Any project for improvement of a building required to correct existing health, sanitary, or safety code violations identified by the *code official* and that is the minimum necessary to assure safe living conditions, or
2. Any *alteration* of a historic structure, provided that the *alteration* will not preclude the structure's continued designation as a historic structure.

SUBSTANTIAL STRUCTURAL DAMAGE. A condition where:

1. In any story, the vertical elements of the lateral-force-resisting system have suffered damage such that the lateral load-carrying capacity of the structure in any horizontal direction has been reduced by more than 20 percent from its predamaged condition; or
2. The capacity of any vertical gravity load-carrying component, or any group of such components, that supports more than 30 percent of the total area of the structure's floor(s) and roof(s) has been reduced more than 20 percent from its predamaged condition and the remaining capacity of such affected elements, with respect to all dead and live loads, is less than 75 percent of that required by the *International Building Code* for new buildings of similar structure, purpose and location.

TECHNICALLY INFEASIBLE. An *alteration* of a building or a facility that has little likelihood of being accomplished because the existing structural conditions require the removal or *alteration* of a load-bearing member that is an essential part of the structural frame or because other existing physical or site constraints prohibit modification or *addition* of elements, spaces, or features that are in full and strict compliance with the minimum requirements for new construction and that are necessary to provide accessibility.

UNSAFE. Buildings, structures or equipment that are unsanitary, or that are deficient due to inadequate means of egress facilities, inadequate light and ventilation, or that constitute a fire hazard, or in which the structure or individual structural members meet the definition of "*Dangerous*," or that are otherwise *dangerous* to human life or the public welfare, or that involve illegal or improper occupancy or inadequate maintenance shall be deemed unsafe. A vacant structure that is not secured against entry shall be deemed unsafe.

WORK AREA. That portion or portions of a building consisting of all reconfigured spaces as indicated on the construction documents. Work area excludes other portions of the building where incidental work entailed by the intended work must be performed and portions of the building where work not initially intended by the owner is specifically required by this code.

CHAPTER 3

PRESCRIPTIVE COMPLIANCE METHOD

[B] SECTION 301 GENERAL

301.1 Scope. The provisions of this chapter shall control the *alteration*, *repair*, *addition* and *change of occupancy* of existing structures, including historic and moved structures as referenced in Section 101.5.1.

Exception: Existing bleachers, grandstands and folding and telescopic seating shall comply with ICC 300-02.

301.1.1 Compliance with other methods. Alterations, repairs, *additions* and changes of occupancy to existing structures shall comply with the provisions of this chapter or with one of the methods provided in Section 101.5.

301.2 Building materials. Building materials shall comply with the requirements of this section.

301.2.1 Existing materials. Materials already in use in a building in compliance with requirements or approvals in effect at the time of their erection or installation shall be permitted to remain in use unless determined by the *code official* to be *dangerous* to life, health or safety. Where such conditions are determined to be *dangerous* to life, health or safety, they shall be mitigated or made safe.

301.2.2 New and replacement materials. Except as otherwise required or permitted by this code, materials permitted by the applicable code for new construction shall be used. Like materials shall be permitted for repairs and alterations, provided no hazard to life, health or property is created. Hazardous materials shall not be used where the code for new construction would not permit their use in buildings of similar occupancy, purpose and location.

[B] SECTION 302 ADDITIONS

302.1 General. *Additions* to any building or structure shall comply with the requirements of the *International Building Code* for new construction. Alterations to the *existing building* or structure shall be made to ensure that the *existing building* or structure together with the *addition* are no less conforming to the provisions of the *International Building Code* than the *existing building* or structure was prior to the *addition*. An *existing building* together with its *additions* shall comply with the height and area provisions of Chapter 5 of the *International Building Code*.

302.2 Flood hazard areas. For buildings and structures in flood hazard areas established in Section 1612.3 of the *International Building Code*, any *addition* that constitutes *substantial improvement* of the existing structure, as defined in Section 202, shall comply with the flood design requirements for new construction, and all aspects of the existing structure shall be brought into compliance with the requirements for new construction for flood design.

For buildings and structures in flood hazard areas established in Section 1612.3 of the *International Building Code*, any *additions* that do not constitute *substantial improvement* or *substantial damage* of the existing structure, as defined in Section 202, are not required to comply with the flood design requirements for new construction.

302.3 Existing structural elements carrying gravity load. Any existing gravity load-carrying structural element for which an *addition* and its related alterations cause an increase in design gravity load of more than 5 percent shall be strengthened, supplemented, replaced or otherwise altered as needed to carry the increased load required by the *International Building Code* for new structures. Any existing gravity load-carrying structural element whose gravity load-carrying capacity is decreased shall be considered an altered element subject to the requirements of Section 303.3. Any existing element that will form part of the lateral load path for any part of the *addition* shall be considered an existing lateral load-carrying structural element subject to the requirements of Section 302.4.

302.3.1 Design live load. Where the *addition* does not result in increased design live load, existing gravity load-carrying structural elements shall be permitted to be evaluated and designed for live loads approved prior to the *addition*. If the approved live load is less than that required by Section 1607 of the *International Building Code*, the area designed for the nonconforming live load shall be posted with placards of approved design indicating the approved live load. Where the *addition* does result in increased design live load, the live load required by Section 1607 of the *International Building Code* shall be used.

302.4 Existing structural elements carrying lateral load. Where the *addition* is structurally independent of the existing structure, existing lateral load-carrying structural elements shall be permitted to remain unaltered. Where the *addition* is not structurally independent of the existing structure, the existing structure and its *addition* acting together as a single structure shall be shown to meet the requirements of Sections 1609 and 1613 of the *International Building Code*.

Exception: Any existing lateral load-carrying structural element whose demand-capacity ratio with the *addition* considered is no more than 10 percent greater than its demand-capacity ratio with the *addition* ignored shall be permitted to remain unaltered. For purposes of calculating demand-capacity ratios, the demand shall consider applicable load combinations with design lateral loads or forces in accordance with Sections 1609 and 1613 of the *International Building Code*. For purposes of this exception, comparisons of demand-capacity ratios and calculation of design lateral loads, forces and capacities shall account for the cumulative effects of *additions* and alterations since original construction.

302.4.1 Seismic. Seismic requirements for additions shall be in accordance with this section. Where the existing seismic force-resisting system is a type that can be designated ordinary, values of R, Ω_0 and C_d for the existing seismic force-resisting system shall be those specified by the *International Building Code* for an ordinary system unless it is demonstrated that the existing system will provide performance equivalent to that of a detailed, intermediate or special system.

[B] SECTION 303 ALTERATIONS

303.1 General. Except as provided by Section 301.2 or this section, alterations to any building or structure shall comply with the requirements of the *International Building Code* for new construction. Alterations shall be such that the *existing building* or structure is no less conforming to the provisions of the *International Building Code* than the *existing building* or structure was prior to the *alteration*.

Exceptions:

1. An existing stairway shall not be required to comply with the requirements of Section 1009 of the *International Building Code* where the existing space and construction does not allow a reduction in pitch or slope.
2. Handrails otherwise required to comply with Section 1009.12 of the *International Building Code* shall not be required to comply with the requirements of Section 1012.6 of the *International Building Code* regarding full extension of the handrails where such extensions would be hazardous due to plan configuration.

303.2 Flood hazard areas. For buildings and structures in flood hazard areas established in Section 1612.3 of the *International Building Code*, any *alteration* that constitutes *substantial improvement* of the existing structure, as defined in Section 202, shall comply with the flood design requirements for new construction, and all aspects of the existing structure shall be brought into compliance with the requirements for new construction for flood design.

For buildings and structures in flood hazard areas established in Section 1612.3 of the *International Building Code*, any alterations that do not constitute *substantial improvement* or *substantial damage* of the existing structure, as defined in Section 202, are not required to comply with the flood design requirements for new construction.

303.3 Existing structural elements carrying gravity load. Any existing gravity load-carrying structural element for which an *alteration* causes an increase in design gravity load of more than 5 percent shall be strengthened, supplemented, replaced or otherwise altered as needed to carry the increased gravity load required by the *International Building Code* for new structures. Any existing gravity load-carrying structural element whose gravity load-carrying capacity is decreased as part of the *alteration* shall be shown to have the capacity to resist the applicable design gravity loads required by the *International Building Code* for new structures.

303.3.1 Design live load. Where the *alteration* does not result in increased design live load, existing gravity load-carrying structural elements shall be permitted to be evaluated and designed for live loads approved prior to the *alteration*. If the approved live load is less than that required by Section 1607 of the *International Building Code*, the area designed for the nonconforming live load shall be posted with placards of approved design indicating the approved live load. Where the *alteration* does result in increased design live load, the live load required by Section 1607 of the *International Building Code* shall be used.

303.4 Existing structural elements carrying lateral load. Except as permitted by Section 303.5, with the *alteration* increases design lateral loads in accordance with Section 1609 or 1613 of the *International Building Code*, or where the *alteration* results in a structural irregularity as defined in ASCE 7, or where the alteration decreases the capacity of any existing lateral load-carrying structural element, the structure of the altered building or structure shall be shown to meet the requirements of Sections 1609 and 1613 of the *International Building Code*.

Exception: Any existing lateral load-carrying structural element whose demand-capacity ratio with the *alteration* considered is no more than 10 percent greater than its demand-capacity ratio with the *alteration* ignored shall be permitted to remain unaltered. For purposes of calculating demand-capacity ratios, the demand shall consider applicable load combinations with design lateral loads or forces in accordance with Sections 1609 and 1613 of the *International Building Code*. For purposes of this exception, comparisons of demand-capacity ratios and calculation of design lateral loads, forces and capacities shall account for the cumulative effects of *additions* and alterations since original construction.

303.4.1 Seismic. Seismic requirements for alterations shall be in accordance with this section. Where the existing seismic force-resisting system is a type that can be designated ordinary, values of R, Ω_0 and C_d for the existing seismic force-resisting system shall be those specified by this code for an ordinary system unless it is demonstrated that the existing system will provide performance equivalent to that of a detailed, intermediate or special system.

303.5 Voluntary seismic improvements. Alterations to existing structural elements or *additions* of new structural elements that are not otherwise required by this chapter and are initiated for the purpose of improving the performance of the seismic-force-resisting system of an existing structure or the performance of seismic bracing or anchorage of existing nonstructural elements shall be permitted, provided that an engineering analysis is submitted demonstrating all of the following:

1. The altered structure and the altered nonstructural elements are no less conforming to the provisions of the *International Building Code* with respect to earthquake design than they were prior to the *alteration*.
2. New structural elements are detailed and connected to the existing structural elements as required by Chapter 16 of the *International Building Code*.

3. New or relocated nonstructural elements are detailed and connected to existing or new structural elements as required by Chapter 16 of the *International Building Code*.
4. The alterations do not create a structural irregularity as defined in ASCE 7 or make an existing structural irregularity more severe.

303.6 Means of egress capacity factors. Alterations to any *existing building* or structure shall not be subject to the egress width factors in Section 1005.1 of the *International Building Code* for new construction in determining the minimum egress widths or the minimum number of exits in an *existing building* or structure. The minimum egress widths for the components of the means of egress shall be based on the means of egress width factors in the building code under which the building was constructed, and shall be considered as complying means of egress for any *alteration* if, in the opinion of the *code official*, they do not constitute a distinct hazard to life.

[B] SECTION 304 REPAIRS

304.1 General. Buildings and structures, and parts thereof, shall be repaired in conformance to this section and to Section 301.2. Work on nondamaged components that is necessary for the required *repair* of damaged components shall be considered part of the *repair* and shall not be subject to the requirements for alterations in this chapter. Routine maintenance required by Section 301.2, ordinary repairs exempt from permit in accordance with Section 105.2, and abatement of wear due to normal service conditions shall not be subject to the requirements for repairs in this section.

304.1.1 Dangerous conditions. Regardless of the extent of structural or nonstructural damage, the *code official* shall have the authority to require the elimination of conditions deemed *dangerous*.

304.2 Substantial structural damage to vertical elements of the lateral-force-resisting system. A building that has sustained *substantial structural damage* to the vertical elements of its lateral-force-resisting system shall be evaluated and repaired in accordance with the applicable provisions of Sections 304.2.1 through 304.2.3.

304.2.1 Evaluation. The building shall be evaluated by a registered design professional, and the evaluation findings shall be submitted to the *code official*. The evaluation shall establish whether the damaged building, if repaired to its pre-damage state, would comply with the provisions of the *International Building Code* for wind and earthquake loads. Evaluation for earthquake loads shall be required if the *substantial structural damage* was caused by or related to earthquake effects or if the building is in Seismic Design Category C, D, E or F.

Wind loads for this evaluation shall be those prescribed in Section 1609 of the *International Building Code*. Earthquake loads for this evaluation, if required, shall be permitted to be 75 percent of those prescribed in Section 1613 of the *International Building Code*. Values of R, Ω_0 and C_d for the existing seismic force-resisting system shall be those specified by this code for an ordinary system unless it is demonstrated that the existing system will provide performance equivalent to that of an intermediate or special system.

304.2.2 Extent of repair for compliant buildings. If the evaluation establishes compliance of the pre-damage building in accordance with Section 304.2.1, then repairs shall be permitted that restore the building to its pre-damage state using materials and strengths that existed prior to the damage.

304.2.3 Extent of repair for noncompliant buildings. If the evaluation does not establish compliance of the pre-damage building in accordance with Section 304.2.1, then the building shall be rehabilitated to comply with applicable provisions of the *International Building Code* for load combinations, including wind or seismic loads. The wind loads for the *repair* shall be as required by the building code in effect at the time of original construction, unless the damage was caused by wind, in which case the wind loads shall be as required by the building code in effect at the time of original construction or as required by the *International Building Code,* whichever are greater. Earthquake loads for this rehabilitation design shall be those required for the design of the pre-damage building, but not less than 75 percent of those prescribed in Section 1613 of the *International Building Code*. New structural members and connections required by this rehabilitation design shall comply with the detailing provisions of the *International Building Code* for new buildings of similar structure, purpose and location.

304.3 Substantial structural damage to gravity load-carrying components. Gravity load-carrying components that have sustained *substantial structural damage* shall be rehabilitated to comply with the applicable provisions of the *International Building Code* for dead and live loads. Snow loads shall be considered if the *substantial structural damage* was caused by or related to snow load effects. Existing gravity load-carrying structural elements shall be permitted to be designed for live loads approved prior to the damage. Nondamaged gravity load-carrying components that receive dead, live or snow loads from rehabilitated components shall also be rehabilitated or shown to have the capacity to carry the design loads of the rehabilitation design. New structural members and connections required by this rehabilitation design shall comply with the detailing provisions of the *International Building Code* for new buildings of similar structure, purpose and location.

304.3.1 Lateral force-resisting elements. Regardless of the level of damage to vertical elements of the lateral force-resisting system, if *substantial structural damage* to gravity load-carrying components was caused primarily by wind or earthquake effects, then the building shall be evaluated in accordance with Section 304.2.1 and, if noncompliant, rehabilitated in accordance with Section 304.2.3.

304.4 Less than substantial structural damage. For damage less than *substantial structural damage*, repairs shall be allowed that restore the building to its pre-damage state using materials and strengths that existed prior to the damage. New structural members and connections used for this *repair* shall comply with the detailing provisions of the *International Building Code* for new buildings of similar structure, purpose and location.

304.5 Flood hazard areas. For buildings and structures in flood hazard areas established in Section 1612.3 of the *International Building Code*, any *repair* that constitutes *substantial improvement* of the existing structure, as defined in Section 202, shall comply with the flood design requirements for new construction, and all aspects of the existing structure shall be brought into compliance with the requirements for new construction for flood design.

For buildings and structures in flood hazard areas established in Section 1612.3 of the *International Building Code*, any repairs that do not constitute *substantial improvement* or *substantial damage* of the existing structure, as defined in Section 202, are not required to comply with the flood design requirements for new construction.

[B] SECTION 305 FIRE ESCAPES

305.1 Where permitted. Fire escapes shall be permitted only as provided for in Sections 305.1.1 through 305.1.4.

305.1.1 New buildings. Fire escapes shall not constitute any part of the required means of egress in new buildings.

305.1.2 Existing fire escapes. Existing fire escapes shall continue to be accepted as a component in the means of egress in existing buildings only.

305.1.3 New fire escapes. New fire escapes for existing buildings shall be permitted only where exterior stairs cannot be utilized due to lot lines limiting stair size or due to the sidewalks, alleys or roads at grade level. New fire escapes shall not incorporate ladders or access by windows.

305.1.4 Limitations. Fire escapes shall comply with this section and shall not constitute more than 50 percent of the required number of exits nor more than 50 percent of the required exit capacity.

305.2 Location. Where located on the front of the building and where projecting beyond the building line, the lowest landing shall not be less than 7 feet (2134 mm) or more than 12 feet (3658 mm) above grade, and shall be equipped with a counterbalanced stairway to the street. In alleyways and thoroughfares less than 30 feet (9144 mm) wide, the clearance under the lowest landing shall not be less than 12 feet (3658 mm).

305.3 Construction. The fire escape shall be designed to support a live load of 100 pounds per square foot (4788 Pa) and shall be constructed of steel or other approved noncombustible materials. Fire escapes constructed of wood not less than nominal 2 inches (51 mm) thick are permitted on buildings of Type V construction. Walkways and railings located over or supported by combustible roofs in buildings of Type III and IV construction are permitted to be of wood not less than nominal 2 inches (51 mm) thick.

305.4 Dimensions. Stairs shall be at least 22 inches (559 mm) wide with risers not more than, and treads not less than, 8 inches (203 mm) and landings at the foot of stairs not less than 40 inches (1016 mm) wide by 36 inches (914 mm) long, located not more than 8 inches (203 mm) below the door.

305.5 Opening protectives. Doors and windows along the fire escape shall be protected with $^{3}/_{4}$-hour opening protectives.

[B] SECTION 306 GLASS REPLACEMENT

306.1 Conformance. The installation or replacement of glass shall be as required for new installations.

SECTION 307 CHANGE OF OCCUPANCY

[B] 307.1 Conformance. No change shall be made in the use or occupancy of any building that would place the building in a different division of the same group of occupancy or in a different group of occupancies, unless such building is made to comply with the requirements of the *International Building Code* for such division or group of occupancy. Subject to the approval of the building official, the use or occupancy of existing buildings shall be permitted to be changed and the building is allowed to be occupied for purposes in other groups without conforming to all the requirements of this code for those groups, provided the new or proposed use is less hazardous, based on life and fire risk, than the existing use.

[B] 307.2 Certificate of occupancy. A certificate of occupancy shall be issued where it has been determined that the requirements for the new occupancy classification have been met.

[B] 307.3 Stairways. Existing stairways in an existing structure shall not be required to comply with the requirements of a new stairway as outlined in Section 1009 of the *International Building Code* where the existing space and construction will not allow a reduction in pitch or slope.

[B] 307.4 Seismic. When a *change of occupancy* results in a structure being reclassified to a higher occupancy category, the structure shall conform to the seismic requirements for a new structure of the higher occupancy category. Where the existing seismic force-resisting system is a type that can be designated ordinary, values of R, Ω_0 and C_d for the existing seismic force-resisting system shall be those specified by the *International Building Code* for an ordinary system unless it is demonstrated that the existing system will provide performance equivalent to that of a detailed, intermediate or special system.

Exceptions:

1. Specific seismic detailing requirements of Section 1613 of the *International Building Code* for a new structure shall not be required to be met where it can be shown that the level of performance and seismic safety is equivalent to that of a new structure. Such analysis shall consider the regularity, over strength, redundancy and ductility of the structure within the

context of the existing and retrofit (if any) detailing provided.

2. When a change of use results in a structure being reclassified from Occupancy Category I or II to Occupancy Category III and the structure is located where the seismic coefficient, S_{DS}, is less than 0.33, compliance with the seismic requirements of Section 1613 of the *International Building Code* is not required.

[EC] 307.5 Energy. Buildings undergoing a change in occupancy that would result in an increase in demand for either fossil fuel or electrical energy shall comply with the *International Energy Conservation Code*.

307.6 Electrical. It shall be unlawful to make a change in the occupancy of a structure that will subject the structure to the special provisions of the *International Building Code* related to electrical installations applicable to the new occupancy without approval. The *code official* shall certify that the structure meets the intent of the provisions of law governing building construction for the proposed new occupancy and that such *change of occupancy* does not result in any hazard to the public health, safety or welfare.

[FG] 307.7 Fuel gas. It shall be unlawful to make a change in the occupancy of a structure that will subject the structure to the special provisions of the *International Fuel Gas Code* applicable to the new occupancy without approval. The *code official* shall certify that the structure meets the intent of the provisions of law governing building construction for the proposed new occupancy and that such *change of occupancy* does not result in any hazard to the public health, safety or welfare.

[M] 307.8 Mechanical. It shall be unlawful to make a change in the occupancy of a structure that will subject the structure to the special provisions of the *International Mechanical Code* applicable to the new occupancy without approval. The code official shall certify that the structure meets the intent of the provisions of law governing building construction for the proposed new occupancy and that such *change of occupancy* does not result in any hazard to the public health, safety or welfare.

[P] 307.9 Plumbing. It shall be unlawful to make a change in the occupancy of a structure that will subject the structure to the special provisions of the *International Plumbing Code* applicable to the new occupancy without approval. The *code official* shall certify that the structure meets the intent of the provisions of law governing building construction for the proposed new occupancy and that such *change of occupancy* does not result in any hazard to the public health, safety or welfare.

[B] SECTION 308 HISTORIC BUILDINGS

308.1 Historic buildings. The provisions of this code relating to the construction, *repair*, *alteration*, *addition*, restoration and movement of structures, and *change of occupancy* shall not be mandatory for historic buildings where such buildings are judged by the building official to not constitute a distinct life safety hazard.

308.2 Flood hazard areas. Within flood hazard areas established in accordance with Section 1612.3 of the *International Building Code*, where the work proposed constitutes *substantial improvement* as defined in Section 1612.2 of the *International Building Code*, the building shall be brought into compliance with Section 1612 of the *International Building Code*.

Exception: Historic buildings that are:

1. Listed or preliminarily determined to be eligible for listing in the National Register of Historic Places;
2. Determined by the Secretary of the U.S. Department of Interior as contributing to the historical significance of a registered historic district or a district preliminarily determined to qualify as an historic district; or
3. Designated as historic under a state or local historic preservation program that is approved by the Department of Interior.

[B] SECTION 309 MOVED STRUCTURES

309.1 Conformance. Structures moved into or within the jurisdiction shall comply with the provisions of this code for new structures.

[B] SECTION 310 ACCESSIBILITY FOR EXISTING BUILDINGS

310.1 Scope. The provisions of Sections 310.1 through 310.9 apply to maintenance, *change of occupancy*, *additions* and alterations to existing buildings, including those identified as historic buildings.

Exception: Type B dwelling or sleeping units required by Section 1107 of the *International Building Code* are not required to be provided in existing buildings and facilities being altered or undergoing a *change of occupancy*.

310.2 Maintenance of facilities. A building, facility or element that is constructed or altered to be accessible shall be maintained accessible during occupancy.

310.3 Extent of application. An *alteration* of an existing element, space or area of a building or facility shall not impose a requirement for greater accessibility than that which would be required for new construction.

Alterations shall not reduce or have the effect of reducing accessibility of a building, portion of a building or facility.

310.4 Change of occupancy. Existing buildings that undergo a change of group or occupancy shall comply with this section.

310.4.1 Partial change in occupancy. Where a portion of the building is changed to a new occupancy classification, any alterations shall comply with Sections 310.6, 310.7 and 310.8.

310.4.2 Complete change of occupancy. Where an entire building undergoes a *change of occupancy*, it shall comply with Section 310.4.1 and shall have all of the following accessible features:

1. At least one accessible building entrance.

2. At least one accessible route from an accessible building entrance to *primary function* areas.
3. Signage complying with Section 1110 of the *International Building Code.*
4. Accessible parking, where parking is being provided.
5. At least one accessible passenger loading zone, when loading zones are provided.
6. At least one accessible route connecting accessible parking and accessible passenger loading zones to an accessible entrance.

Where it is *technically infeasible* to comply with the new construction standards for any of these requirements for a change of group or occupancy, the above items shall conform to the requirements to the maximum extent technically feasible.

310.5 Additions. Provisions for new construction shall apply to *additions*. An *addition* that affects the accessibility to, or contains an area of, a *primary function* shall comply with the requirements in Section 310.7.

310.6 Alterations. A building, facility or element that is altered shall comply with the applicable provisions in Chapter 11 of the *International Building Code* and ICC A117.1, unless *technically infeasible*. Where compliance with this section is *technically infeasible*, the *alteration* shall provide access to the maximum extent technically feasible.

Exceptions:

1. The altered element or space is not required to be on an accessible route, unless required by Section 310.7.
2. Accessible means of egress required by Chapter 10 of the *International Building Code* are not required to be provided in existing buildings and facilities.
3. The *alteration* to Type A individually owned dwelling units within a Group R-2 occupancy shall meet the provision for a Type B dwelling unit and shall comply with the applicable provisions in Chapter 11 of the *International Building Code* and ICC A117.1.

310.7 Alterations affecting an area containing a primary function. Where an *alteration* affects the accessibility to, or contains an area of, a *primary function*, the route to the *primary function* area shall be accessible. The accessible route to the *primary function* area shall include toilet facilities or drinking fountains serving the area of *primary function*.

Exceptions:

1. The costs of providing the accessible route are not required to exceed 20 percent of the costs of the alterations affecting the area of *primary function*.
2. This provision does not apply to alterations limited solely to windows, hardware, operating controls, electrical outlets and signs.
3. This provision does not apply to alterations limited solely to mechanical systems, electrical systems, installation or *alteration* of fire protection systems and abatement of hazardous materials.
4. This provision does not apply to alterations undertaken for the primary purpose of increasing the accessibility of an *existing building*, facility or element.

310.8 Scoping for alterations. The provisions of Sections 310.8.1 through 310.8.14 shall apply to alterations to existing buildings and facilities.

310.8.1 Entrances. Accessible entrances shall be provided in accordance with Section 1105 of the *International Building Code*.

Exception: Where an *alteration* includes alterations to an entrance, and the building or facility has an accessible entrance, the altered entrance is not required to be accessible, unless required by Section 310.7. Signs complying with Section 1110 of the *International Building Code* shall be provided.

310.8.2 Elevators. Altered elements of existing elevators shall comply with ASME A17.1 and ICC A117.1. Such elements shall also be altered in elevators programmed to respond to the same hall call control as the altered elevator.

310.8.3 Platform lifts. Platform (wheelchair) lifts complying with ICC A117.1 and installed in accordance with ASME A18.1 shall be permitted as a component of an accessible route.

310.8.4 Stairs and escalators in existing buildings. In *alterations, change of occupancy* or *additions* where an escalator or stair is added where none existed previously and major structural modifications are necessary for installation, an accessible route shall be provided between the levels served by the escalator or stairs in accordance with Sections 1104.4 and 1104.5 of the *International Building Code*.

310.8.5 Ramps. Where steeper slopes than allowed by Section 1010.2 of the *International Building Code* are necessitated by space limitations, the slope of ramps in or providing access to existing buildings or facilities shall comply with Table 310.8.5.

TABLE 310.8.5
RAMPS

SLOPE	MAXIMUM RISE
Steeper than 1:10 but not steeper than 1:8	3 inches
Steeper than 1:12 but not steeper than 1:10	6 inches

For SI: 1 inch = 25.4 mm.

310.8.6 Performance areas. Where it is *technically infeasible* to alter performance areas to be on an accessible route, at least one of each type of performance area shall be made accessible.

310.8.7 Accessible dwelling or sleeping units. Where Group I-1, I-2, I-3, R-1, R-2 or R-4 dwelling or sleeping units are being altered or added, the requirements of Section 1107 of the *International Building Code* for Accessible units apply only to the quantity of spaces being altered or added.

310.8.8 Type A dwelling or sleeping units. Where more than 20 Group R-2 dwelling or sleeping units are being added, the requirements of Section 1107 of the *Interna-*

tional Building Code for Type A units apply only to the quantity of the spaces being added.

310.8.9 Type B dwelling or sleeping units. Where four or more Group I-1, I-2, R-1, R-2, R-3 or R-4 dwelling or sleeping units are being added, the requirements of Section 1107 of the *International Building Code* for Type B units apply only to the quantity of the spaces being added.

310.8.10 Jury boxes and witness stands. In alterations, accessible wheelchair spaces are not required to be located within the defined area of raised jury boxes or witness stands and shall be permitted to be located outside these spaces where the ramp or lift access restricts or projects into the means of egress.

310.8.11 Toilet rooms. Where it is *technically infeasible* to alter existing toilet and bathing facilities to be accessible, an accessible family or assisted-use toilet or bathing facility constructed in accordance with Section 1109.2.1 of the *International Building Code* is permitted. The family or assisted-use facility shall be located on the same floor and in the same area as the existing facilities.

310.8.12 Dressing, fitting and locker rooms. Where it is *technically infeasible* to provide accessible dressing, fitting or locker rooms at the same location as similar types of rooms, one accessible room on the same level shall be provided. Where separate-sex facilities are provided, accessible rooms for each sex shall be provided. Separate-sex facilities are not required where only unisex rooms are provided.

310.8.13 Fuel dispensers. Operable parts of replacement fuel dispensers shall be permitted to be 54 inches (1370 mm) maximum, measuring from the surface of the vehicular way where fuel dispensers are installed on existing curbs.

310.8.14 Thresholds. The maximum height of thresholds at doorways shall be $^3/_4$ inch (19.1 mm). Such thresholds shall have beveled edges on each side.

310.9 Historic buildings. These provisions shall apply to buildings and facilities designated as historic structures that undergo alterations or a *change of occupancy*, unless *technically infeasible*. Where compliance with the requirements for accessible routes, entrances or toilet facilities would threaten or destroy the historic significance of the building or facility, as determined by the applicable governing authority, the alternative requirements of Sections 310.9.1 through 310.9.4 for that element shall be permitted.

310.9.1 Site arrival points. At least one accessible route from a site arrival point to an accessible entrance shall be provided.

310.9.2 Multilevel buildings and facilities. An accessible route from an accessible entrance to public spaces on the level of the accessible entrance shall be provided.

310.9.3 Entrances. At least one main entrance shall be accessible.

Exceptions:

1. If a main entrance cannot be made accessible, an accessible nonpublic entrance that is unlocked while the building is occupied shall be provided; or
2. If a main entrance cannot be made accessible, a locked accessible entrance with a notification system or remote monitoring shall be provided.

Signs complying with Section 1110 of the *International Building Code* shall be provided at the primary entrance and the accessible entrance.

310.9.4 Toilet and bathing facilities. Where toilet rooms are provided, at least one accessible family or assisted-use toilet room complying with Section 1109.2.1 of the *International Building Code* shall be provided.

CHAPTER 4
CLASSIFICATION OF WORK

SECTION 401
GENERAL

401.1 Scope. The provisions of this chapter shall be used in conjunction with Chapters 5 through 12 and shall apply to the *alteration*, *repair*, *addition* and *change of occupancy* of existing structures, including historic and moved structures, as referenced in Section 101.5.2. The work performed on an *existing building* shall be classified in accordance with this chapter.

401.1.1 Compliance with other alternatives. *Alterations*, *repairs*, *additions* and changes of occupancy to existing structures shall comply with the provisions of Chapters 4 through 12 or with one of the alternatives provided in Section 101.5.

401.2 Work area. The *work area*, as defined in Chapter 2, shall be identified on the construction documents.

401.3 Occupancy and use. When determining the appropriate application of the referenced sections of this code, the occupancy and use of a building shall be determined in accordance with Chapter 3 of the *International Building Code*.

SECTION 402
REPAIRS

402.1 Scope. *Repairs,* as defined in Chapter 2, include the patching or restoration or replacement of damaged materials, elements, equipment or fixtures for the purpose of maintaining such components in good or sound condition with respect to existing loads or performance requirements.

402.2 Application. *Repairs* shall comply with the provisions of Chapter 5.

402.3 Related work. Work on nondamaged components that is necessary for the required *repair* of damaged components shall be considered part of the *repair* and shall not be subject to the provisions of Chapter 6, 7, 8, 9 or 10.

SECTION 403
ALTERATION—LEVEL 1

403.1 Scope. Level 1 alterations include the removal and replacement or the covering of existing materials, elements, equipment, or fixtures using new materials, elements, equipment, or fixtures that serve the same purpose.

403.2 Application. Level 1 *alterations* shall comply with the provisions of Chapter 6.

SECTION 404
ALTERATION—LEVEL 2

404.1 Scope. Level 2 *alterations* include the reconfiguration of space, the *addition* or elimination of any door or window, the reconfiguration or extension of any system, or the installation of any additional equipment.

404.2 Application. Level 2 alterations shall comply with the provisions of Chapter 6 for Level 1 *alterations* as well as the provisions of Chapter 7.

SECTION 405
ALTERATION—LEVEL 3

405.1 Scope. Level 3 *alterations* apply where the *work area* exceeds 50 percent of the aggregate area of the building.

405.2 Application. Level 3 alterations shall comply with the provisions of Chapters 6 and 7 for Level 1 and 2 *alterations*, respectively, as well as the provisions of Chapter 8.

SECTION 406
CHANGE OF OCCUPANCY

406.1 Scope. *Change of occupancy* provisions apply where the activity is classified as a *change of occupancy* as defined in Chapter 2.

406.2 Application. Changes of occupancy shall comply with the provisions of Chapter 9.

SECTION 407
ADDITIONS

407.1 Scope. Provisions for *additions* shall apply where work is classified as an *addition* as defined in Chapter 2.

407.2 Application. *Additions* to existing buildings shall comply with the provisions of Chapter 10.

SECTION 408
HISTORIC BUILDINGS

408.1 Scope. Historic buildings provisions shall apply to buildings classified as historic as defined in Chapter 2.

408.2 Application. Except as specifically provided for in Chapter 11, historic buildings shall comply with applicable provisions of this code for the type of work being performed.

SECTION 409
RELOCATED BUILDINGS

409.1 Scope. Relocated buildings provisions shall apply to relocated or moved buildings.

409.2 Application. Relocated buildings shall comply with the provisions of Chapter 12.

CHAPTER 5
REPAIRS

SECTION 501
GENERAL

501.1 Scope. Repairs as described in Section 402 shall comply with the requirements of this chapter. Repairs to historic buildings shall comply with this chapter, except as modified in Chapter 11.

501.2 Conformance. The work shall not make the building less conforming than it was before the *repair* was undertaken.

501.3 Flood hazard areas. In flood hazard areas, repairs that constitute *substantial improvement* shall require that the building comply with Section 1612 of the *International Building Code.*

SECTION 502
BUILDING ELEMENTS AND MATERIALS

502.1 Existing building materials. Materials already in use in a building in conformance with requirements or approvals in effect at the time of their erection or installation shall be permitted to remain in use unless determined by the *code official* to render the building or structure unsafe or *dangerous* as defined in Chapter 2.

502.2 New and replacement materials. Except as otherwise required or permitted by this code, materials permitted by the applicable code for new construction shall be used. Like materials shall be permitted for repairs and alterations, provided no *dangerous* or *unsafe* condition, as defined in Chapter 2, is created. Hazardous materials, such as asbestos and lead-based paint, shall not be used where the code for new construction would not permit their use in buildings of similar occupancy, purpose and location.

502.3 Glazing in hazardous locations. Replacement glazing in hazardous locations shall comply with the safety glazing requirements of the *International Building Code* or *International Residential Code* as applicable.

Exception: Glass block walls, louvered windows, and jalousies repaired with like materials.

SECTION 503
FIRE PROTECTION

503.1 General. Repairs shall be done in a manner that maintains the level of fire protection provided.

SECTION 504
MEANS OF EGRESS

504.1 General. Repairs shall be done in a manner that maintains the level of protection provided for the means of egress.

SECTION 505
ACCESSIBILITY

505.1 General. Repairs shall be done in a manner that maintains the level of accessibility provided.

SECTION 506
STRUCTURAL

506.1 General. Structural repairs shall be in compliance with this section and Section 501.2. Regardless of the extent of structural or nonstructural damage, *dangerous* conditions shall be eliminated. Regardless of the scope of *repair*, new structural members and connections used for *repair* or rehabilitation shall comply with the detailing provisions of the *International Building Code* for new buildings of similar structure, purpose and location.

506.2 Repairs to damaged buildings. Repairs to damaged buildings shall comply with this section.

506.2.1 Repairs for less than substantial structural damage. For damage less than *substantial structural damage*, the damaged elements shall be permitted to be restored to their predamage condition.

506.2.2 Repairs for substantial structural damage to vertical elements of the lateral-force-resisting system. A building that has sustained *substantial structural damage* to the vertical elements of its lateral-force-resisting system shall be evaluated in accordance with Section 506.2.2.1, and either repaired in accordance with Section 506.2.2.2 or repaired and rehabilitated in accordance with Section 506.2.2.3 depending on the results of the evaluation.

506.2.2.1 Evaluation. The building shall be evaluated by a registered design professional, and the evaluation findings shall be submitted to the *code official*. The evaluation shall establish whether the damaged building, if repaired to its predamaged state, would comply with the provisions of the *International Building Code*, except that the seismic design criteria shall be the reduced IBC level seismic forces specified in Section 101.5.4.2.

506.2.2.2 Extent of repair for compliant buildings. If the evaluation establishes that the building in its predamage condition complies with the provisions of Section 506.2.2.1, then the damaged elements shall be permitted to be restored to their predamage condition.

506.2.2.3 Extent of repair for noncompliant buildings. If the evaluation does not establish that the building in its predamage condition complies with the provisions of Section 506.2.2.1, then the building shall be rehabilitated to comply with the provisions of this section. The wind load for the *repair* and rehabilitation shall be those required by the building code in effect at the time of original construction, unless the damage was caused by wind, in which case the wind loads shall be in accordance with the *International Building Code*. The seismic loads for this rehabilitation design shall be those required by the building code in effect at the time of original construction, but not less than the reduced-level seismic forces specified in Section 101.5.4.2.

506.2.3 Substantial structural damage to gravity load-carrying components. Gravity load-carrying components that have sustained *substantial structural damage* shall be rehabilitated to comply with the applicable provisions for dead and live loads in the *International Building Code*. Snow loads shall be considered if the *substantial structural damage* was caused by or related to snow load effects. Undamaged gravity load-carrying components that receive dead, live or snow loads from rehabilitated components shall also be rehabilitated if required to comply with the design loads of the rehabilitation design.

506.2.3.1 Lateral-force-resisting elements. Regardless of the level of damage to gravity elements of the lateral-force-resisting system, if substantial structural damage gravity load-carrying components was caused primarily by wind or seismic effects, then the building shall be evaluated in accordance with Section 506.2.2.1 and, if noncompliant, rehabilitated in accordance with Section 506.2.2.3.

506.2.4 Flood hazard areas. In flood hazard areas, buildings that have sustained *substantial damage* shall be brought into compliance with Section 1612 of the *International Building Code*.

SECTION 507
ELECTRICAL

507.1 Material. Existing electrical wiring and equipment undergoing *repair* shall be allowed to be repaired or replaced with like material.

507.1.1 Receptacles. Replacement of electrical receptacles shall comply with the applicable requirements of Section 406.3(D) of NFPA 70.

507.1.2 Plug fuses. Plug fuses of the Edison-base type shall be used for replacements only where there is no evidence of over fusing or tampering per applicable requirements of Section 240.51(B) of NFPA 70.

507.1.3 Nongrounding-type receptacles. For replacement of nongrounding-type receptacles with grounding-type receptacles and for branch circuits that do not have an equipment grounding conductor in the branch circuitry, the grounding conductor of a grounding-type receptacle outlet shall be permitted to be grounded to any accessible point on the grounding electrode system or to any accessible point on the grounding electrode conductor in accordance with Section 250.130(C) of NFPA 70.

507.1.4 Group I-2 receptacles. Non-"hospital grade" receptacles in patient bed locations of Group I-2 shall be replaced with "hospital grade" receptacles, as required by NFPA 99 and Article 517 of NFPA 70.

507.1.5 Grounding of appliances. Frames of electric ranges, wall-mounted ovens, counter-mounted cooking units, clothes dryers and outlet or junction boxes that are part of the existing branch circuit for these appliances shall be permitted to be grounded to the grounded circuit conductor in accordance with Section 250.140 of NFPA 70.

SECTION 508
MECHANICAL

508.1 General. Existing mechanical systems undergoing *repair* shall not make the building less conforming than it was before the *repair* was undertaken.

508.2 Mechanical draft systems for manually fired appliances and fireplaces. A mechanical draft system shall be permitted to be used with manually fired appliances and fireplaces where such a system complies with all of the following requirements:

1. The mechanical draft device shall be listed and installed in accordance with the manufacturer's installation instructions.
2. A device shall be installed that produces visible and audible warning upon failure of the mechanical draft device or loss of electrical power at any time that the mechanical draft device is turned on. This device shall be equipped with a battery backup if it receives power from the building wiring.
3. A smoke detector shall be installed in the room with the appliance or fireplace. This device shall be equipped with a battery backup if it receives power from the building wiring.

SECTION 509
PLUMBING

509.1 Materials. Plumbing materials and supplies shall not be used for repairs that are prohibited in the *International Plumbing Code*.

509.2 Water closet replacement. The maximum water consumption flow rates and quantities for all replaced water closets shall be 1.6 gallons (6 L) per flushing cycle.

Exception: Blowout-design water closets [3.5 gallons (13 L) per flushing cycle].

CHAPTER 6

ALTERATIONS—LEVEL 1

SECTION 601 GENERAL

601.1 Scope. Level 1 alterations as described in Section 403 shall comply with the requirements of this chapter. Level 1 alterations to historic buildings shall comply with this chapter, except as modified in Chapter 11.

601.2 Conformance. An *existing building* or portion thereof shall not be altered such that the building becomes less safe than its existing condition.

Exception: Where the current level of safety or sanitation is proposed to be reduced, the portion altered shall conform to the requirements of the *International Building Code*.

601.3 Flood hazard areas. In flood hazard areas, alterations that constitute *substantial improvement* shall require that the building comply with Section 1612 of the *International Building Code*.

SECTION 602 BUILDING ELEMENTS AND MATERIALS

602.1 Interior finishes. All newly installed interior wall and ceiling finishes shall comply with Chapter 8 of the *International Building Code*.

602.2 Interior floor finish. New interior floor finish, including new carpeting used as an interior floor finish material, shall comply with Section 804 of the *International Building Code*.

602.3 Interior trim. All newly installed interior trim materials shall comply with Section 806 of the *International Building Code*.

602.4 Materials and methods. All new work shall comply with materials and methods requirements in the *International Building Code*, *International Energy Conservation Code*, *International Mechanical Code*, and *International Plumbing Code*, as applicable, that specify material standards, detail of installation and connection, joints, penetrations, and continuity of any element, component, or system in the building.

[FG] 602.4.1 International Fuel Gas Code. The following sections of the *International Fuel Gas Code* shall constitute the fuel gas materials and methods requirements for Level 1 alterations.

1. All of Chapter 3, entitled "General Regulations," except Sections 303.7 and 306.
2. All of Chapter 4, entitled "Gas Piping Installations," except Sections 401.8 and 402.3.
   2.1. Sections 401.8 and 402.3 shall apply when the work being performed increases the load on the system such that the existing pipe does not meet the size required by code. Existing systems that are modified shall not require resizing as long as the load on the system is not increased and the system length is not increased even if the altered system does not meet code minimums.
3. All of Chapter 5, entitled "Chimneys and Vents."
4. All of Chapter 6, entitled "Specific Appliances."

SECTION 603 FIRE PROTECTION

603.1 General. Alterations shall be done in a manner that maintains the level of fire protection provided.

SECTION 604 MEANS OF EGRESS

604.1 General. Alterations shall be done in a manner that maintains the level of protection provided for the means of egress.

SECTION 605 ACCESSIBILITY

605.1 General. A building, facility or element that is altered shall comply with the applicable provisions in Sections 605.1.1 through 605.1.14, Chapter 11 of the *International Building Code* and ICC A117.1 unless it is *technically infeasible*. Where compliance with this section is *technically infeasible*, the *alteration* shall provide access to the maximum extent that is technically feasible.

A building, facility or element that is constructed or altered to be accessible shall be maintained accessible during occupancy.

Exceptions:

1. The altered element or space is not required to be on an accessible route unless required by Section 605.2.
2. Accessible means of egress required by Chapter 10 of the *International Building Code* are not required to be provided in existing buildings and facilities.
3. Type B dwelling or sleeping units required by Section 1107 of the *International Building Code* are not required to be provided in existing buildings and facilities.
4. The *alteration* to Type A individually owned dwelling units within a Group R-2 occupancy shall meet the provisions for Type B dwelling units and shall comply with the applicable provisions in Chapter 11 of the *International Building Code* and ICC A117.1.

605.1.1 Entrances. Where an *alteration* includes alterations to an entrance, and the building or facility has an accessible entrance on an accessible route, the altered entrance is not required to be accessible unless required by

Section 605.2. Signs complying with Section 1110 of the *International Building Code* shall be provided.

605.1.2 Elevators. Altered elements of existing elevators shall comply with ASME A17.1 and ICC A117.1. Such elements shall also be altered in elevators programmed to respond to the same hall call control as the altered elevator.

605.1.3 Platform lifts. Platform (wheelchair) lifts complying with ICC A117.1 and installed in accordance with ASME A18.1 shall be permitted as a component of an accessible route.

605.1.4 Ramps. Where steeper slopes than allowed by Section 1010.2 of the *International Building Code* are necessitated by space limitations, the slope of ramps in or providing access to existing buildings or facilities shall comply with Table 605.1.4.

TABLE 605.1.4
RAMPS

SLOPE	MAXIMUM RISE
Steeper than 1:10 but not steeper than 1:8	3 inches
Steeper than 1:12 but not steeper than 1:10	6 inches

For SI: 1 inch = 25.4 mm.

605.1.5 Dining areas. An accessible route to raised or sunken dining areas or to outdoor seating areas is not required provided that the same services and decor are provided in an accessible space usable by any occupant and not restricted to use by people with a disability.

605.1.6 Performance areas. Where it is *technically infeasible* to alter performance areas to be on an accessible route, at least one of each type of performance area shall be made accessible.

605.1.7 Jury boxes and witness stands. In alterations, accessible wheelchair spaces are not required to be located within the defined area of raised jury boxes or witness stands and shall be permitted to be located outside these spaces where ramp or lift access poses a hazard by restricting or projecting into a required means of egress.

605.1.8 Accessible dwelling or sleeping units. Where Group I-1, I-2, I-3, R-1, R-2 or R-4 dwelling or sleeping units are being altered, the requirements of Section 1107 of the *International Building Code* for accessible units and Chapter 9 of the *International Building Code* for visible alarms apply only to the quantity of the spaces being altered.

605.1.9 Type A dwelling or sleeping units. Where more than 20 Group R-2 dwelling or sleeping units are being altered, the requirements of Section 1107 of the *International Building Code* for Type A units and Chapter 9 of the *International Building Code* for visible alarms apply only to the quantity of the spaces being altered.

605.1.10 Toilet rooms. Where it is *technically infeasible* to alter existing toilet and bathing facilities to be accessible, an accessible family or assisted-use toilet or bathing facility constructed in accordance with Section 1109.2.1 of the *International Building Code* is permitted. The family or assisted-use facility shall be located on the same floor and in the same area as the existing facilities.

605.1.11 Dressing, fitting and locker rooms. Where it is *technically infeasible* to provide accessible dressing, fitting, or locker rooms at the same location as similar types of rooms, one accessible room on the same level shall be provided. Where separate sex facilities are provided, accessible rooms for each sex shall be provided. Separate sex facilities are not required where only unisex rooms are provided.

605.1.12 Fuel dispensers. Operable parts of replacement fuel dispensers shall be permitted to be 54 inches (1370 mm) maximum measured from the surface of the vehicular way where fuel dispensers are installed on existing curbs.

605.1.13 Thresholds. The maximum height of thresholds at doorways shall be $^3/_4$ inch (19.1 mm). Such thresholds shall have beveled edges on each side.

605.1.14 Extent of application. An *alteration* of an existing element, space, or area of a building or facility shall not impose a requirement for greater accessibility than that which would be required for new construction. Alterations shall not reduce or have the effect of reducing accessibility of a building, portion of a building, or facility.

605.2 Alterations affecting an area containing a primary function. Where an *alteration* affects the accessibility to a, or contains an area of, *primary function*, the route to the *primary function* area shall be accessible. The accessible route to the *primary function* area shall include toilet facilities or drinking fountains serving the area of *primary function*.

Exceptions:

1. The costs of providing the accessible route are not required to exceed 20 percent of the costs of the alterations affecting the area of *primary function*.
2. This provision does not apply to alterations limited solely to windows, hardware, operating controls, electrical outlets and signs.
3. This provision does not apply to alterations limited solely to mechanical systems, electrical systems, installation or *alteration* of fire protection systems and abatement of hazardous materials.
4. This provision does not apply to alterations undertaken for the primary purpose of increasing the accessibility of an *existing building*, facility or element.

SECTION 606
STRUCTURAL

606.1 General. Where alteration work includes replacement of equipment that is supported by the building or where a reroofing permit is required, the provisions of this section shall apply.

606.2 Addition or replacement of roofing or replacement of equipment. Where addition or replacement of roofing or replacement of equipment results in additional dead loads, structural components supporting such reroofing or equipment shall comply with the gravity load requirements of the *International Building Code*.

Exceptions:

1. Structural elements where the additional dead load from the roofing or equipment does not increase the force in the element by more than 5 percent.
2. Buildings constructed in accordance with the *International Residential Code* or the conventional light-frame construction methods of the *International Building Code* and where the dead load from the roofing or equipment is not increased by more than 5 percent.
3. Addition of a second layer of roof covering weighing 3 pounds per square foot (0.1437 kN/m^2) or less over an existing, single layer of roof covering.

606.2.1 Wall anchors for concrete and masonry buildings. Where a permit is issued for reroofing more than 25 percent of the roof area of a building assigned to Seismic Design Category D, E or F with a structural system consisting of concrete or reinforced masonry walls with a flexible roof diaphragm or unreinforced masonry walls with any type of roof diaphragms, the work shall include installation of wall anchors at the roof line to resist the reduced *International Building Code* level seismic forces as specified in Section 101.5.4.2 of this code and design procedures of Section 101.5.4, unless an evaluation demonstrates compliance of existing wall anchorage.

606.3 Additional requirements for reroof permits. The requirements of this section shall apply to *alteration* work requiring reroof permits.

606.3.1 Bracing for unreinforced masonry bearing wall parapets. Where a permit is issued for reroofing for more than 25 percent of the roof area of a building assigned to Seismic Design Category D, E or F that has parapets constructed of unreinforced masonry, the work shall include installation of parapet bracing to resist the reduced *International Building Code* seismic forces level as specified in Section 101.5.4.2 of this code, unless an evaluation demonstrates compliance of such items.

606.3.2 Roof diaphragms resisting wind loads in high-wind regions. Where roofing materials are removed from more than 50 percent of the roof diaphragm of a building or section of a building located where the basic wind speed is greater than 90 mph or in a special wind region, as defined in Section 1609 of the *International Building Code*, roof diaphragms and connections that are part of the main wind-force resisting system shall be evaluated for the wind loads specified in the *International Building Code*, including wind uplift. If the diaphragms and connections in their current condition do not comply with those wind provisions, they shall be replaced or strengthened in accordance with the loads specified in the *International Building Code*.

SECTION 607 ENERGY CONSERVATION

607.1 Minimum requirements. Level 1 alterations to existing buildings or structures are permitted without requiring the entire building or structure to comply with the energy requirements of the *International Energy Conservation Code* or *International Residential Code*. The alterations shall conform to the energy requirements of the *International Energy Conservation Code* or *International Residential Code* as they relate to new construction only.

CHAPTER 7
ALTERATIONS—LEVEL 2

SECTION 701 GENERAL

701.1 Scope. Level 2 *alterations* as described in Section 404 shall comply with the requirements of this chapter.

Exception: Buildings in which the reconfiguration is exclusively the result of compliance with the accessibility requirements of Section 605.2 shall be permitted to comply with Chapter 6.

701.2 Alteration Level 1 compliance. In addition to the requirements of this chapter, all work shall comply with the requirements of Chapter 6.

701.3 Compliance. All new construction elements, components, systems, and spaces shall comply with the requirements of the *International Building Code*.

Exceptions:

1. Windows may be added without requiring compliance with the light and ventilation requirements of the *International Building Code*.
2. Newly installed electrical equipment shall comply with the requirements of Section 708.
3. The length of dead-end corridors in newly constructed spaces shall only be required to comply with the provisions of Section 705.6.
4. The minimum ceiling height of the newly created habitable and occupiable spaces and corridors shall be 7 feet (2134 mm).

SECTION 702 SPECIAL USE AND OCCUPANCY

702.1 General. *Alteration* of buildings classified as special use and occupancy as described in the *International Building Code* shall comply with the requirements of Section 701.1 and the scoping provisions of Chapter 1 where applicable.

SECTION 703 BUILDING ELEMENTS AND MATERIALS

703.1 Scope. The requirements of this section are limited to work areas in which Level 2 *alterations* are being performed, and shall apply beyond the *work area* where specified.

703.2 Vertical openings. Existing vertical openings shall comply with the provisions of Sections 703.2.1, 703.2.2, and 703.2.3.

703.2.1 Existing vertical openings. All existing interior vertical openings connecting two or more floors shall be enclosed with approved assemblies having a fire-resistance rating of not less than 1 hour with approved opening protectives.

Exceptions:

1. Where vertical opening enclosure is not required by the *International Building Code* or the *International Fire Code*.
2. Interior vertical openings other than stairways may be blocked at the floor and ceiling of the *work area* by installation of not less than 2 inches (51 mm) of solid wood or equivalent construction.
3. The enclosure shall not be required where:
   - 3.1. Connecting the main floor and mezzanines; or
   - 3.2. All of the following conditions are met:
     - 3.2.1. The communicating area has a low hazard occupancy or has a moderate hazard occupancy that is protected throughout by an automatic sprinkler system.
     - 3.2.2. The lowest or next to the lowest level is a street floor.
     - 3.2.3. The entire area is open and unobstructed in a manner such that it may be assumed that a fire in any part of the interconnected spaces will be readily obvious to all of the occupants.
     - 3.2.4. Exit capacity is sufficient to provide egress simultaneously for all the occupants of all levels by considering all areas to be a single floor area for the determination of required exit capacity.
     - 3.2.5. Each floor level, considered separately, has at least one half of its individual required exit capacity pro- vided by an exit or exits leading directly out of that level without having to traverse another communicating floor level or be exposed to the smoke or fire spreading from another communicating floor level.
4. In Group A occupancies, a minimum 30-minute enclosure shall be provided to protect all vertical openings not exceeding three stories.
5. In Group B occupancies, a minimum 30-minute enclosure shall be provided to protect all vertical openings not exceeding three stories. This enclo-

sure, or the enclosure specified in Section 703.2.1, shall not be required in the following locations:

5.1. Buildings not exceeding 3,000 square feet (279 m^2) per floor.

5.2. Buildings protected throughout by an approved automatic fire sprinkler system.

6. In Group E occupancies, the enclosure shall not be required for vertical openings not exceeding three stories when the building is protected throughout by an approved automatic fire sprinkler system.

7. In Group F occupancies, the enclosure shall not be required in the following locations:

7.1. Vertical openings not exceeding three stories.

7.2. Special purpose occupancies where necessary for manufacturing operations and direct access is provided to at least one protected stairway.

7.3. Buildings protected throughout by an approved automatic sprinkler system.

8. In Group H occupancies, the enclosure shall not be required for vertical openings not exceeding three stories where necessary for manufacturing operations and every floor level has direct access to at least two remote enclosed stairways or other approved exits.

9. In Group M occupancies, a minimum 30-minute enclosure shall be provided to protect all vertical openings not exceeding three stories. This enclosure, or the enclosure specified in Section 703.2.1, shall not be required in the following locations:

9.1. Openings connecting only two floor levels.

9.2. Occupancies protected throughout by an approved automatic sprinkler system.

10. In Group R-1 occupancies, the enclosure shall not be required for vertical openings not exceeding three stories in the following locations:

10.1. Buildings protected throughout by an approved automatic sprinkler system.

10.2. Buildings with less than 25 dwelling units or sleeping units where every sleeping room above the second floor is provided with direct access to a fire escape or other approved second exit by means of an approved exterior door or window having a sill height of not greater than 44 inches (1118 mm) and where:

10.2.1. Any exit access corridor exceeding 8 feet (2438 mm) in length that serves two means of egress, one of which is an unprotected vertical opening, shall have at least one of the means of egress separated from the vertical opening by a 1-hour fire barrier; and

10.2.2. The building is protected through- out by an automatic fire alarm system, installed and supervised in accordance with the *International Building Code*.

11. In Group R-2 occupancies, a minimum 30-minute enclosure shall be provided to protect all vertical openings not exceeding three stories. This enclosure, or the enclosure specified in Section 703.2.1, shall not be required in the following locations:

11.1. Vertical openings not exceeding two stories with not more than four dwelling units per floor.

11.2. Buildings protected throughout by an approved automatic sprinkler system.

11.3. Buildings with not more than four dwelling units per floor where every sleeping room above the second floor is provided with direct access to a fire escape or other approved second exit by means of an approved exterior door or window having a sill height of not greater than 44 inches (1118 mm) and the building is protected throughout by an automatic fire alarm system complying with Section 704.4.

12. One- and two-family dwellings.

13. Group S occupancies where connecting not more than two floor levels or where connecting not more than three floor levels and the structure is equipped throughout with an approved automatic sprinkler system.

14. Group S occupancies where vertical opening protection is not required for open parking garages and ramps.

703.2.2 Supplemental shaft and floor opening enclosure requirements. Where the *work area* on any floor exceeds 50 percent of that floor area, the enclosure requirements of Section 703.2 shall apply to vertical openings other than stairways throughout the floor.

Exception: Vertical openings located in tenant spaces that are entirely outside the *work area*.

703.2.3 Supplemental stairway enclosure requirements. Where the *work area* on any floor exceeds 50 percent of that floor area, stairways that are part of the means of egress serving the *work area* shall, at a minimum, be enclosed with smoke-tight construction on the highest *work area* floor and all floors below.

Exception: Where stairway enclosure is not required by the *International Building Code* or the *International Fire Code*.

703.3 Smoke barriers. Smoke barriers in Group I-2 occupancies shall be installed where required by Sections 703.3.1 and 703.3.2.

703.3.1 Compartmentation. Where the *work area* is on a story used for sleeping rooms for more than 30 patients, the story shall be divided into not less than two compartments by smoke barrier walls complying with Section 703.3.2 such that each compartment does not exceed 22,500 square feet (2093 m^2), and the travel distance from any point to reach a door in the required smoke barrier shall not exceed 200 feet (60 960 mm).

Exception: Where neither the length nor the width of the smoke compartment exceeds 150 feet (45 720 mm), the travel distance to reach the smoke barrier door shall not be limited.

703.3.2 Fire-resistance rating. The smoke barriers shall be fire-resistance rated for 30 minutes and constructed in accordance with the *International Building Code*.

703.4 Interior finish. The interior finish of walls and ceilings in exits and corridors in any *work area* shall comply with the requirements of the *International Building Code*.

Exception: Existing interior finish materials that do not comply with the interior finish requirements of the *International Building Code* shall be permitted to be treated with an approved fire-retardant coating in accordance with the manufacturer's instructions to achieve the required rating.

703.4.1 Supplemental interior finish requirements. Where the *work area* on any floor exceeds 50 percent of the floor area, Section 703.4 shall also apply to the interior finish in exits and corridors serving the *work area* throughout the floor.

Exception: Interior finish within tenant spaces that are entirely outside the *work area*.

703.5 Guards. The requirements of Sections 703.5.1 and 703.5.2 shall apply in all *work areas*.

703.5.1 Minimum requirement. Every portion of a floor, such as a balcony or a loading dock, that is more than 30 inches (762 mm) above the floor or grade below and is not provided with guards, or those in which the existing guards are judged to be in danger of collapsing, shall be provided with guards.

703.5.2 Design. Where there are no guards or where existing guards must be replaced, the guards shall be designed and installed in accordance with the *International Building Code*.

SECTION 704
FIRE PROTECTION

704.1 Scope. The requirements of this section shall be limited to work areas in which Level 2 alterations are being performed, and where specified they shall apply throughout the floor on which the work areas are located or otherwise beyond the *work area*.

704.1.1 Corridor ratings. Where an approved automatic sprinkler system is installed throughout the story, the required fire-resistance rating for any corridor located on the story shall be permitted to be reduced in accordance with the *International Building Code*. In order to be considered for a corridor rating reduction, such system shall provide coverage for the stairwell landings serving the floor and the intermediate landings immediately below.

704.2 Automatic sprinkler systems. Automatic sprinkler systems shall be provided in accordance with the requirements of Sections 704.2.1 through 704.2.5. Installation requirements shall be in accordance with the *International Building Code*.

704.2.1 High-rise buildings. In high-rise buildings, work areas that have exits or corridors shared by more than one tenant or that have exits or corridors serving an occupant load greater than 30 shall be provided with automatic sprinkler protection in the entire *work area* where the *work area* is located on a floor that has a sufficient sprinkler water supply system from an existing standpipe or a sprinkler riser serving that floor.

704.2.1.1 Supplemental automatic sprinkler system requirements. Where the *work area* on any floor exceeds 50 percent of that floor area, Section 704.2.1 shall apply to the entire floor on which the *work area* is located.

Exception: Tenant spaces that are entirely outside the *work area*.

704.2.2 Groups A, B, E, F-1, H, I, M, R-1, R-2, R-4, S-1 and S-2. In buildings with occupancies in Groups A, B, E, F-1, H, I, M, R-1, R-2, R-4, S-1 and S-2, work areas that have exits or corridors shared by more than one tenant or that have exits or corridors serving an occupant load greater than 30 shall be provided with automatic sprinkler protection where all of the following conditions occur:

1. The *work area* is required to be provided with automatic sprinkler protection in accordance with the *International Building Code* as applicable to new construction;
2. The *work area* exceeds 50 percent of the floor area; and
3. The building has sufficient municipal water supply for design of a fire sprinkler system available to the floor without installation of a new fire pump.

704.2.2.1 Mixed uses. In work areas containing mixed uses, one or more of which requires automatic sprinkler protection in accordance with Section 704.2.2, such protection shall not be required throughout the *work area* provided that the uses requiring such protection are separated from those not requiring protection by fire-resistance-rated construction having a minimum 2-hour rating for Group H and a minimum 1-hour rating for all other occupancy groups.

704.2.3 Windowless stories. Work located in a windowless story, as determined in accordance with the *International Building Code*, shall be sprinklered where the *work area* is required to be sprinklered under the provisions of the *International Building Code* for newly constructed buildings and the building has a sufficient municipal water supply without installation of a new fire pump.

704.2.4 Other required suppression systems. In buildings and areas listed in Table 903.2.11.6 of the *International Building Code*, *work areas* that have exits or corridors shared by more than one tenant or that have exits or corridors serving an occupant load greater than 30 shall be provided with sprinkler protection under the following conditions:

1. The *work area* is required to be provided with automatic sprinkler protection in accordance with the *International Building Code* applicable to new construction; and
2. The building has sufficient municipal water supply for design of a fire sprinkler system available to the floor without installation of a new fire pump.

704.2.5 Supervision. Fire sprinkler systems required by this section shall be supervised by one of the following methods:

1. Approved central station system in accordance with NFPA 72;
2. Approved proprietary system in accordance with NFPA 72;
3. Approved remote station system of the jurisdiction in accordance with NFPA 72; or
4. When approved by the *code official*, approved local alarm service that will cause the sounding of an alarm in accordance with NFPA 72.

Exception: Supervision is not required for the following:

1. Underground gate valve with roadway boxes.
2. Halogenated extinguishing systems.
3. Carbon dioxide extinguishing systems.
4. Dry and wet chemical extinguishing systems.
5. Automatic sprinkler systems installed in accordance with NFPA 13R where a common supply main is used to supply both domestic and automatic sprinkler systems and a separate shutoff valve for the automatic sprinkler system is not provided.

704.3 Standpipes. Where the *work area* includes exits or corridors shared by more than one tenant and is located more than 50 feet (15 240 mm) above or below the lowest level of fire department access, a standpipe system shall be provided. Standpipes shall have an approved fire department connection with hose connections at each floor level above or below the lowest level of fire department access. Standpipe systems shall be installed in accordance with the *International Building Code*.

Exceptions:

1. No pump shall be required provided that the standpipes are capable of accepting delivery by fire department apparatus of a minimum of 250 gallons per minute (gpm) at 65 pounds per square inch (psi) (946 L/m at 448KPa) to the topmost floor in buildings equipped throughout with an automatic sprinkler system or a minimum of 500 gpm at 65 psi (1892 L/m at 448KPa) to the topmost floor in all other buildings. Where the standpipe terminates below the topmost floor, the standpipe shall be designed to meet (gpm/psi) (L/m/KPa) requirements of this exception for possible future extension of the standpipe.
2. The interconnection of multiple standpipe risers shall not be required.

704.4 Fire alarm and detection. An approved fire alarm system shall be installed in accordance with Sections 704.4.1 through 704.4.3. Where automatic sprinkler protection is provided in accordance with Section 704.2 and is connected to the building fire alarm system, automatic heat detection shall not be required.

An approved automatic fire detection system shall be installed in accordance with the provisions of this code and NFPA 72. Devices, combinations of devices, appliances, and equipment shall be approved. The automatic fire detectors shall be smoke detectors, except that an approved alternative type of detector shall be installed in spaces such as boiler rooms, where products of combustion are present during normal operation in sufficient quantity to actuate a smoke detector.

704.4.1 Occupancy requirements. A fire alarm system shall be installed in accordance with Sections 704.4.1.1 through 704.4.1.7. Existing alarm-notification appliances shall be automatically activated throughout the building. Where the building is not equipped with a fire alarm system, alarm-notification appliances within the *work area* shall be provided and automatically activated.

Exceptions:

1. Occupancies with an existing, previously approved fire alarm system.
2. Where selective notification is permitted, alarm-notification appliances shall be automatically activated in the areas selected.

704.4.1.1 Group E. A fire alarm system shall be installed in *work areas* of Group E occupancies as required by the *International Fire Code* for existing Group E occupancies.

704.4.1.2 Group I-1. A fire alarm system shall be installed in *work areas* of Group I-1 residential care/assisted living facilities as required by the *International Fire Code* for existing Group I-1 occupancies.

704.4.1.3 Group I-2. A fire alarm system shall be installed in *work areas* of Group I-2 occupancies as required by the *International Fire Code* for existing Group I-2 occupancies.

704.4.1.4 Group I-3. A fire alarm system shall be installed in *work areas* of Group I-3 occupancies as required by the *International Fire Code* for existing Group I-3 occupancies.

704.4.1.5 Group R-1. A fire alarm system shall be installed in Group R-1 occupancies as required by the *International Fire Code* for existing Group R-1 occupancies.

704.4.1.6 Group R-2. A fire alarm system shall be installed in *work areas* of Group R-2 apartment buildings as required by the *International Fire Code* for existing Group R-2 occupancies.

704.4.1.7 Group R-4. A fire alarm system shall be installed in *work areas* of Group R-4 residential care/assisted living facilities as required by the *International Fire Code* for existing Group R-4 occupancies.

704.4.2 Supplemental fire alarm system requirements. Where the *work area* on any floor exceeds 50 percent of that floor area, Section 704.4.1 shall apply throughout the floor.

Exception: Alarm-initiating and notification appliances shall not be required to be installed in tenant spaces outside of the *work area.*

704.4.3 Smoke alarms. Individual sleeping units and individual dwelling units in any *work area* in Group R-1, R-2, R-3, R-4, and I-1 occupancies shall be provided with smoke alarms in accordance with the *International Fire Code.*

Exception: Interconnection of smoke alarms outside of the rehabilitation *work area* shall not be required.

SECTION 705
MEANS OF EGRESS

705.1 Scope. The requirements of this section shall be limited to work areas that include exits or corridors shared by more than one tenant within the *work area* in which Level 2 alterations are being performed, and where specified they shall apply throughout the floor on which the work areas are located or otherwise beyond the *work area.*

705.2 General. The means of egress shall comply with the requirements of this section.

Exceptions:

1. Where the *work area* and the means of egress serving it complies with NFPA 101.
2. Means of egress conforming to the requirements of the building code under which the building was constructed shall be considered compliant means of egress if, in the opinion of the *code official,* they do not constitute a distinct hazard to life.

705.3 Number of exits. The number of exits shall be in accordance with Sections 705.3.1 through 705.3.3.

705.3.1 Minimum number. Every story utilized for human occupancy on which there is a *work area* that includes exits or corridors shared by more than one tenant within the *work area* shall be provided with the minimum number of exits based on the occupancy and the occupant load in accordance with the *International Building Code.* In addition, the exits shall comply with Sections 705.3.1.1 and 705.3.1.2.

705.3.1.1 Single-exit buildings. Only one exit is required from buildings and spaces of the following occupancies:

1. In Group A, B, E, F, M, U and S occupancies, a single exit is permitted in the story at the level of exit discharge when the occupant load of the story does not exceed 50 and the exit access travel distance does not exceed 75 feet (22 860 mm).
2. Group B, F-2, and S-2 occupancies not more than two stories in height that are not greater than 3,500 square feet per floor (326 m^2), when the exit access travel distance does not exceed 75 feet (22 860 mm). The minimum fire-resistance rating of the exit enclosure and of the opening protection shall be 1 hour.
3. Open parking structures where vehicles are mechanically parked.
4. In community residences for the developmentally disabled, the maximum occupant load excluding staff is 12.
5. Groups R-1 and R-2 not more than two stories in height, when there are not more than four dwelling units per floor and the exit access travel distance does not exceed 50 feet (15 240 mm). The minimum fire-resistance rating of the exit enclosure and of the opening protection shall be 1 hour.
6. In multilevel dwelling units in buildings of occupancy Group R-1 or R-2, an exit shall not be required from every level of the dwelling unit provided that one of the following conditions is met:
   6.1. The travel distance within the dwelling unit does not exceed 75 feet (22 860 mm); or
   6.2. The building is not more than three stories in height and all third-floor space is part of one or more dwelling units located in part on the second floor; and no habitable room within any such dwelling unit shall have a travel distance that exceeds 50 feet (15 240 mm) from the outside of the habitable room entrance door to the inside of the entrance door to the dwelling unit.
7. In Group R-2, H-4, H-5 and I occupancies and in rooming houses and child care centers, a single exit is permitted in a one-story building with a maximum occupant load of 10 and the exit access

travel distance does not exceed 75 feet (22 860 mm).

8. In buildings of Group R-2 occupancy that are equipped throughout with an automatic fire sprinkler system, a single exit shall be permitted from a basement or story below grade if every dwelling unit on that floor is equipped with an approved window providing a clear opening of at least 5 square feet (0.47 m^2) in area, a minimum net clear opening of 24 inches (610 mm) in height and 20 inches (508 mm) in width, and a sill height of not more than 44 inches (1118 mm) above the finished floor.
9. In buildings of Group R-2 occupancy of any height with not more than four dwelling units per floor; with a smokeproof enclosure or outside stair as an exit; and with such exit located within 20 feet (6096 mm) of travel to the entrance doors to all dwelling units served thereby.
10. In buildings of Group R-3 occupancy equipped throughout with an automatic fire sprinkler system, only one exit shall be required from basements or stories below grade.

705.3.1.2 Fire escapes required. When more than one exit is required, an existing or newly constructed fire escape complying with Section 705.3.1.2.1 shall be accepted as providing one of the required means of egress.

705.3.1.2.1 Fire escape access and details. Fire escapes shall comply with all of the following requirements:

1. Occupants shall have unobstructed access to the fire escape without having to pass through a room subject to locking.
2. Access to a new fire escape shall be through a door, except that windows shall be permitted to provide access from single dwelling units or sleeping units in Group R-1, R-2 and I-1 occupancies or to provide access from spaces having a maximum occupant load of 10 in other occupancy classifications.
   - 2.1. The window shall have a minimum net clear opening of 5.7 square feet (0.53 m^2) or 5 square feet (0.46 m^2) where located at grade.
   - 2.2. The minimum net clear opening height shall be 24 inches (610 mm) and net clear opening width shall be 20 inches (508 mm).
   - 2.3. The bottom of the clear opening shall not be greater than 44 inches (1118 mm) above the floor.
   - 2.4. The operation of the window shall comply with the operational constraints of the *International Building Code*.
3. Newly constructed fire escapes shall be permitted only where exterior stairs cannot be utilized because of lot lines limiting the stair size or because of the sidewalks, alleys, or roads at grade level.
4. Openings within 10 feet (3048 mm) of fire escape stairs shall be protected by fire assemblies having minimum $^3/_4$-hour fire-resistance ratings.

   Exception: Opening protection shall not be required in buildings equipped throughout with an approved automatic sprinkler system.
5. In all buildings of Group E occupancy, up to and including the 12th grade, buildings of Group I occupancy, rooming houses and childcare centers, ladders of any type are prohibited on fire escapes used as a required means of egress.

705.3.1.2.2 Construction. The fire escape shall be designed to support a live load of 100 pounds per square foot (4788 Pa) and shall be constructed of steel or other approved noncombustible materials. Fire escapes constructed of wood not less than nominal 2 inches (51 mm) thick are permitted on buildings of Type V construction. Walkways and railings located over or supported by combustible roofs in buildings of Types III and IV construction are permitted to be of wood not less than nominal 2 inches (51 mm) thick.

705.3.1.2.3 Dimensions. Stairs shall be at least 22 inches (559 mm) wide with risers not more than, and treads not less than, 8 inches (203 mm). Landings at the foot of stairs shall not be less than 40 inches (1016 mm) wide by 36 inches (914 mm) long and located not more than 8 inches (203 mm) below the door.

705.3.2 Mezzanines. Mezzanines in the *work area* and with an occupant load of more than 50 or in which the travel distance to an exit exceeds 75 feet (22 860 mm) shall have access to at least two independent means of egress.

Exception: Two independent means of egress are not required where the travel distance to an exit does not exceed 100 feet (30 480 mm) and the building is protected throughout with an automatic sprinkler system.

705.3.3 Main entrance—Group A. All buildings of Group A with an occupant load of 300 or more shall be provided with a main entrance capable of serving as the main exit with an egress capacity of at least one half of the total occupant load. The remaining exits shall be capable of providing one half of the total required exit capacity.

Exception: Where there is no well-defined main exit or where multiple main exits are provided, exits shall be permitted to be distributed around the perimeter of the building provided that the total width of egress is not less than 100 percent of the required width.

705.4 Egress doorways. Egress doorways in any *work area* shall comply with Sections 705.4.1 through 705.4.5.

705.4.1 Two egress doorways required. Work areas shall be provided with two egress doorways in accordance with the requirements of Sections 705.4.1.1 and 705.4.1.2.

705.4.1.1 Occupant load and travel distance. In any *work area*, all rooms and spaces having an occupant load greater than 50 or in which the travel distance to an exit exceeds 75 feet (22 860 mm) shall have a minimum of two egress doorways.

Exceptions:

1. Storage rooms having a maximum occupant load of 10.
2. Where the *work area* is served by a single exit in accordance with Section 705.3.1.1.

705.4.1.2 Group I-2. In buildings of Group I-2 occupancy, any patient sleeping room or suite of patient rooms greater than 1,000 square feet (93 m^2) within the *work area* shall have a minimum of two egress doorways.

705.4.2 Door swing. In the *work area* and in the egress path from any *work area* to the exit discharge, all egress doors serving an occupant load greater than 50 shall swing in the direction of exit travel.

705.4.2.1 Supplemental requirements for door swing. Where the *work area* exceeds 50 percent of the floor area, door swing shall comply with Section 705.4.2 throughout the floor.

Exception: Means of egress within or serving only a tenant space that is entirely outside the *work area.*

705.4.3 Door closing. In any *work area*, all doors opening onto an exit passageway at grade or an exit stair shall be self-closing or automatically closing by listed closing devices.

Exceptions:

1. Where exit enclosure is not required by the *International Building Code*.
2. Means of egress within or serving only a tenant space that is entirely outside the *work area*.

705.4.3.1 Supplemental requirements for door closing. Where the *work area* exceeds 50 percent of the floor area, doors shall comply with Section 705.4.3 throughout the exit stair from the *work area* to, and including, the level of exit discharge.

705.4.4 Panic hardware. In any *work area*, and in the egress path from any *work area* to the exit discharge, in buildings or portions thereof of Group A assembly occupancies with an occupant load greater than 100, all required exit doors equipped with latching devices shall be equipped with approved panic hardware.

705.4.4.1 Supplemental requirements for panic hardware. Where the *work area* exceeds 50 percent of the floor area, panic hardware shall comply with Section 705.4.4 throughout the floor.

Exception: Means of egress within a tenant space that is entirely outside the *work area*.

705.4.5 Emergency power source in Group I-3. *Work areas* in buildings of Group I-3 occupancy having remote power unlocking capability for more than 10 locks shall be provided with an emergency power source for such locks. Power shall be arranged to operate automatically upon failure of normal power within 10 seconds and for a duration of not less than 1 hour.

705.5 Openings in corridor walls. Openings in corridor walls in any *work area* shall comply with Sections 705.5.1 through 705.5.4.

Exception: Openings in corridors where such corridors are not required to be rated in accordance with the *International Building Code*.

705.5.1 Corridor doors. Corridor doors in the *work area* shall not be constructed of hollow core wood and shall not contain louvers. All dwelling unit or sleeping unit corridor doors in work areas in buildings of Groups R-1, R-2, and I-1 shall be at least $1^3/_8$-inch (35 mm) solid core wood or approved equivalent and shall not have any glass panels, other than approved wired glass or other approved glazing material in metal frames. All dwelling unit or sleeping unit corridor doors in *work areas* in buildings of Groups R-1, R-2, and I-1 shall be equipped with approved door closers. All replacement doors shall be $1^3/_4$-inch (45 mm) solid bonded wood core or approved equivalent, unless the existing frame will accommodate only a $1^3/_8$-inch (35 mm) door.

Exceptions:

1. Corridor doors within a dwelling unit or sleeping unit.
2. Existing doors meeting the requirements of *HUD Guideline on Fire Ratings of Archaic Materials and Assemblies* (IEBC Resource A) for a rating of 15 minutes or more shall be accepted as meeting the provisions of this requirement.
3. Existing doors in buildings protected throughout with an approved automatic sprinkler system shall be required only to resist smoke, be reasonably tight fitting, and shall not contain louvers.
4. In group homes with a maximum of 15 occupants and that are protected with an approved automatic detection system, closing devices may be omitted.
5. Door assemblies having a fire-protection rating of at least 20 minutes.

705.5.2 Transoms. In all buildings of Group I-1, R-1, and R-2 occupancy, all transoms in corridor walls in work areas shall either be glazed with $^1/_4$-inch (6.4 mm) wired glass set in metal frames or other glazing assemblies having a fire-protection rating as required for the door and permanently secured in the closed position or sealed with materials consistent with the corridor construction.

705.5.3 Other corridor openings. In any *work area*, any other sash, grille, or opening in a corridor and any window in a corridor not opening to the outside air shall be sealed with materials consistent with the corridor construction.

705.5.3.1 Supplemental requirements for other corridor opening. Where the *work area* exceeds 50 percent of the floor area, Section 705.5.3 shall be applicable to all corridor windows, grills, sashes, and other openings on the floor.

Exception: Means of egress within or serving only a tenant space that is entirely outside the *work area.*

705.5.4 Supplemental requirements for corridor openings. Where the *work area* on any floor exceeds 50 percent of the floor area, the requirements of Sections 705.5.1 through 705.5.3 shall apply throughout the floor.

705.6 Dead-end corridors. Dead-end corridors in any *work area* shall not exceed 35 feet (10 670 mm).

Exceptions:

1. Where dead-end corridors of greater length are permitted by the *International Building Code*.
2. In other than Group A and H occupancies, the maximum length of an existing dead-end corridor shall be 50 feet (15 240 mm) in buildings equipped throughout with an automatic fire alarm system installed in accordance with the *International Building Code*.
3. In other than Group A and H occupancies, the maximum length of an existing dead-end corridor shall be 70 feet (21 356 mm) in buildings equipped throughout with an automatic sprinkler system installed in accordance with the *International Building Code*.
4. In other than Group A and H occupancies, the maximum length of an existing, newly constructed, or extended dead-end corridor shall not exceed 50 feet (15 240 mm) on floors equipped with an automatic sprinkler system installed in accordance with the *International Building Code*.

705.7 Means-of-egress lighting. Means-of-egress lighting shall be in accordance with this section, as applicable.

705.7.1 Artificial lighting required. Means of egress in all work areas shall be provided with artificial lighting in accordance with the requirements of the *International Building Code.*

705.7.2 Supplemental requirements for means-of-egress lighting. Where the *work area* on any floor exceeds 50 percent of that floor area, means of egress throughout the floor shall comply with Section 705.7.1.

Exception: Means of egress within or serving only a tenant space that is entirely outside the *work area.*

705.8 Exit signs. Exit signs shall be in accordance with this section, as applicable.

705.8.1 Work areas. Means of egress in all work areas shall be provided with exit signs in accordance with the requirements of the *International Building Code*.

705.8.2 Supplemental requirements for exit signs. Where the *work area* on any floor exceeds 50 percent of that floor area, means of egress throughout the floor shall comply with Section 705.8.1.

Exception: Means of egress within a tenant space that is entirely outside the *work area.*

705.9 Handrails. The requirements of Sections 705.9.1 and 705.9.2 shall apply to handrails from the *work area* floor to, and including, the level of exit discharge.

705.9.1 Minimum requirement. Every required exit stairway that is part of the means of egress for any *work area* and that has three or more risers and is not provided with at least one handrail, or in which the existing handrails are judged to be in danger of collapsing, shall be provided with handrails for the full length of the run of steps on at least one side. All exit stairways with a required egress width of more than 66 inches (1676 mm) shall have handrails on both sides.

705.9.2 Design. Handrails required in accordance with Section 705.9.1 shall be designed and installed in accordance with the provisions of the *International Building Code*.

705.10 Guards. The requirements of Sections 705.10.1 and 705.10.2 shall apply to guards from the *work area* floor to, and including, the level of exit discharge but shall be confined to the egress path of any *work area*.

705.10.1 Minimum requirement. Every open portion of a stair, landing, or balcony that is more than 30 inches (762 mm) above the floor or grade below and is not provided with guards, or those portions in which existing guards are judged to be in danger of collapsing, shall be provided with guards.

705.10.2 Design. Guards required in accordance with Section 705.10.1 shall be designed and installed in accordance with the *International Building Code*.

SECTION 706 ACCESSIBILITY

706.1 General. A building, facility, or element that is altered shall comply with Section 605.

706.2 Stairs and escalators in existing buildings. In alterations where an escalator or stair is added where none existed previously, an accessible route shall be provided in accordance with Sections 1104.4 and 1104.5 of the *International Building Code*.

706.3 Accessible dwelling units and sleeping units. Where Group I-1, I-2, I-3, R-1, R-2 or R-4 dwelling or sleeping units are being added, the requirements of Section 1107 of the *International Building Code* for accessible units and Chapter 9 of the *International Building Code* for visible alarms apply only to the quantity of spaces being added.

706.4 Type A dwelling or sleeping units. Where more than 20 Group R-2 dwelling or sleeping units are being added, the requirements of Section 1107 of the *International Building Code* for Type A units and Chapter 9 of the *International Building Code* for visible alarms apply only to the quantity of the spaces being added.

706.5 Type B dwelling or sleeping units. Where four or more Group I-1, I-2, R-1, R-2, R-3 or R-4 dwelling or sleeping units are being added, the requirements of Section 1107 of the *International Building Code* for Type B units and Chapter 9 of the *International Building Code* for visible alarms apply only to the quantity of the spaces being added.

SECTION 707 STRUCTURAL

707.1 General. Structural elements and systems within buildings undergoing Level 2 alterations shall comply with this section.

707.2 New structural elements. New structural elements in alterations, including connections and anchorage, shall comply with the *International Building Code*.

707.3 Minimum design loads. The minimum design loads on existing elements of a structure that do not support additional loads as a result of an *alteration* shall be the loads applicable at the time the building was constructed.

707.4 Existing structural elements carrying gravity loads. Alterations shall not reduce the capacity of existing gravity load-carrying structural elements unless it is demonstrated that the elements have the capacity to carry the applicable design gravity loads required by the *International Building Code*. Existing structural elements supporting any additional gravity loads as a result of the alterations, including the effects of snow drift, shall comply with the *International Building Code*.

Exceptions:

1. Structural elements whose stress is not increased by more than 5 percent.
2. Buildings of Group R occupancy with not more than five dwelling or sleeping units used solely for residential purposes where the *existing building* and its *alteration* comply with the conventional light-frame construction methods of the *International Building Code* or the provisions of the *International Residential Code*.

707.5 Existing structural elements resisting lateral loads. Any existing lateral load-resisting structural element whose demand-capacity ratio with the *alteration* considered is more than 10 percent greater than its demand-capacity ratio with the *alteration* ignored shall comply with the structural requirements specified in Section 807.4. For purposes of calculating demand-capacity ratios, the demand shall consider applicable load combinations with design lateral loads or forces in accordance with Sections 1609 and 1613 of the *International Building Code*. For purposes of this section, comparisons of demand-capacity ratios and calculation of design lateral loads, forces and capacities shall account for the cumulative effects of *additions* and alterations since original construction.

707.6 Voluntary improvement of the seismic force-resisting system. Alterations to existing structural elements or additions of new structural elements that are not otherwise required by this chapter and are initiated for the purpose of improving the performance of the seismic force-resisting system of an existing structure or the performance of seismic bracing or anchorage of existing nonstructural elements shall be permitted, provided that an engineering analysis is submitted demonstrating the following:

1. The altered structure and the altered nonstructural elements are no less conforming with the provisions of this code with respect to earthquake design than they were prior to the alteration.
2. New structural elements are detailed and connected to the existing structural elements as required by Chapter 16 of the *International Building Code*.
3. New or relocated nonstructural elements are detailed and connected to existing or new structural elements as required by Chapter 16 of the *International Building Code*.
4. The alterations do not create a structural irregularity as defined in ASCE 7 or make an existing structural irregularity more severe.

Voluntary alterations to the seismic force-resisting system in accordance with the applicable chapters of Appendix A of this code shall be permitted.

SECTION 708 ELECTRICAL

708.1 New installations. All newly installed electrical equipment and wiring relating to work done in any *work area* shall comply with the materials and methods requirements of Chapter 6.

Exception: Electrical equipment and wiring in newly installed partitions and ceilings shall comply with all applicable requirements of NFPA 70.

708.2 Existing installations. Existing wiring in all work areas in Group A-1, A-2, A-5, H, and I occupancies shall be upgraded to meet the materials and methods requirements of Chapter 6.

708.3 Residential occupancies. In Group R-2, R-3, and R-4 occupancies and buildings regulated by the *International Residential Code*, the requirements of Sections 708.3.1 through 708.3.7 shall be applicable only to work areas located within a dwelling unit.

708.3.1 Enclosed areas. All enclosed areas, other than closets, kitchens, basements, garages, hallways, laundry areas, utility areas, storage areas, and bathrooms shall have a minimum of two duplex receptacle outlets or one duplex receptacle outlet and one ceiling or wall-type lighting outlet.

708.3.2 Kitchens. Kitchen areas shall have a minimum of two duplex receptacle outlets.

708.3.3 Laundry areas. Laundry areas shall have a minimum of one duplex receptacle outlet located near the laundry equipment and installed on an independent circuit.

708.3.4 Ground fault circuit interruption. Newly installed receptacle outlets shall be provided with ground fault circuit interruption as required by NFPA 70.

708.3.5 Minimum lighting outlets. At least one lighting outlet shall be provided in every bathroom, hallway, stairway, attached garage, and detached garage with electric power, and to illuminate outdoor entrances and exits.

708.3.6 Utility rooms and basements. At least one lighting outlet shall be provided in utility rooms and basements where such spaces are used for storage or contain equipment requiring service.

708.3.7 Clearance for equipment. Clearance for electrical service equipment shall be provided in accordance with the NFPA 70.

SECTION 709 MECHANICAL

709.1 Reconfigured or converted spaces. All reconfigured spaces intended for occupancy and all spaces converted to habitable or occupiable space in any *work area* shall be provided with natural or mechanical ventilation in accordance with the *International Mechanical Code.*

Exception: Existing mechanical ventilation systems shall comply with the requirements of Section 709.2.

709.2 Altered existing systems. In mechanically ventilated spaces, existing mechanical ventilation systems that are altered, reconfigured, or extended shall provide not less than 5 cubic feet per minute (cfm) (0.0024 m^3/s) per person of outdoor air and not less than 15 cfm (0.0071 m^3/s) of ventilation air per person; or not less than the amount of ventilation air determined by the Indoor Air Quality Procedure of ASHRAE 62.

709.3 Local exhaust. All newly introduced devices, equipment, or operations that produce airborne particulate matter, odors, fumes, vapor, combustion products, gaseous contaminants, pathogenic and allergenic organisms, and microbial contaminants in such quantities as to affect adversely or impair health or cause discomfort to occupants shall be provided with local exhaust.

SECTION 710 PLUMBING

710.1 Minimum fixtures. Where the occupant load of the story is increased by more than 20 percent, plumbing fixtures for the story shall be provided in quantities specified in the *International Plumbing Code* based on the increased occupant load.

SECTION 711 ENERGY CONSERVATION

711.1 Minimum requirements. Level 2 *alterations* to existing buildings or structures are permitted without requiring the entire building or structure to comply with the energy requirements of the *International Energy Conservation Code* or *International Residential Code.* The *alterations* shall conform to the energy requirements of the *International Energy Conservation Code* or *International Residential Code* as they relate to new construction only.

CHAPTER 8
ALTERATIONS—LEVEL 3

SECTION 801
GENERAL

801.1 Scope. Level 3 alterations as described in Section 405 shall comply with the requirements of this chapter.

801.2 Compliance. In addition to the provisions of this chapter, work shall comply with all of the requirements of Chapters 6 and 7. The requirements of Sections 703, 704, and 705 shall apply within all work areas whether or not they include exits and corridors shared by more than one tenant and regardless of the occupant load.

Exception: Buildings in which the reconfiguration of space affecting exits or shared egress access is exclusively the result of compliance with the accessibility requirements of Section 605.2 shall not be required to comply with this chapter.

SECTION 802
SPECIAL USE AND OCCUPANCY

802.1 High-rise buildings. Any building having occupied floors more than 75 feet (22 860 mm) above the lowest level of fire department vehicle access shall comply with the requirements of Sections 802.1.1 and 802.1.2.

802.1.1 Recirculating air or exhaust systems. When a floor is served by a recirculating air or exhaust system with a capacity greater than 15,000 cubic feet per minute (701 m^3/s), that system shall be equipped with approved smoke and heat detection devices installed in accordance with the *International Mechanical Code*.

802.1.2 Elevators. Where there is an elevator or elevators for public use, at least one elevator serving the *work area* shall comply with this section. Existing elevators with a travel distance of 25 feet (7620 mm) or more above or below the main floor or other level of a building and intended to serve the needs of emergency personnel for fire-fighting or rescue purposes shall be provided with emergency operation in accordance with ASME A17.3. New elevators shall be provided with Phase I emergency recall operation and Phase II emergency in-car operation in accordance with ASME A17.1.

802.2 Boiler and furnace equipment rooms. Boiler and furnace equipment rooms adjacent to or within the following facilities shall be enclosed by 1-hour fire-resistance-rated construction: day nurseries, children's shelter facilities, residential childcare facilities, and similar facilities with children below the age of $2^1/_2$ years or that are classified as Group I-2 occupancies, shelter facilities, residences for the developmentally disabled, group homes, teaching family homes, transitional living homes, rooming and boarding houses, hotels, and multiple dwellings.

Exceptions:

1. Furnace and boiler equipment of low-pressure type, operating at pressures of 15 pounds per square inch gauge (psig) (103.4 KPa) or less for steam equipment or 170 psig (1171 KPa) or less for hot water equipment, when installed in accordance with manufacturer recommendations.
2. Furnace and boiler equipment of residential R-3 type with 200,000 British thermal units (Btu) (2.11 × 108 J) per hour input rating or less is not required to be enclosed.
3. Furnace rooms protected with automatic sprinkler protection.

802.2.1 Emergency controls. Emergency controls for boilers and furnace equipment shall be provided in accordance with the *International Mechanical Code* in all buildings classified as day nurseries, children's shelter facilities, residential childcare facilities, and similar facilities with children below the age of $2^1/_2$ years or that are classified as Group I-2 occupancies, and in group homes, teaching family homes, and supervised transitional living homes in accordance with the following:

1. Emergency shutoff switches for furnaces and boilers in basements shall be located at the top of the stairs leading to the basement; and
2. Emergency shutoff switches for furnaces and boilers in other enclosed rooms shall be located outside of such room.

SECTION 803
BUILDING ELEMENTS AND MATERIALS

803.1 Existing shafts and vertical openings. Existing stairways that are part of the means of egress shall be enclosed in accordance with Section 703.2.1 from the highest *work area* floor to, and including, the level of exit discharge and all floors below.

803.2 Fire partitions in Group R-3. Fire separation in Group R-3 occupancies shall be in accordance with Section 803.2.1.

803.2.1 Separation required. Where the *work area* is in any attached dwelling unit in Group R-3 or any multiple single family dwelling (townhouse), walls separating the dwelling-units that are not continuous from the foundation to the underside of the roof sheathing shall be constructed to provide a continuous fire separation using construction materials consistent with the existing wall or complying with the requirements for new structures. All work shall be performed on the side of the dwelling unit wall that is part of the *work area.*

Exception: Where *alterations* or *repairs* do not result in the removal of wall or ceiling finishes exposing the structure, walls are not required to be continuous through concealed floor spaces.

803.3 Interior finish. Interior finish in exits serving the *work area* shall comply with Section 703.4 between the highest floor on which there is a *work area* to the floor of exit discharge.

SECTION 804 FIRE PROTECTION

804.1 Automatic sprinkler systems. Automatic sprinkler systems shall be provided in all work areas when required by Section 704.2 or this section.

804.1.1 High-rise buildings. In high-rise buildings, work areas shall be provided with automatic sprinkler protection where the building has a sufficient municipal water supply system to the site. Where the *work area* exceeds 50 percent of floor area, sprinklers shall be provided in the specified areas where sufficient municipal water supply for design and installation of a fire sprinkler system is available at the site.

804.1.2 Rubbish and linen chutes. Rubbish and linen chutes located in the *work area* shall be provided with sprinklered protection or an approved fire suppression system where protection of the rubbish and linen chute would be required under the provisions of the *International Building Code* for new construction.

804.2 Fire alarm and detection systems. Fire alarm and detection systems complying with Sections 704.4.1 and 704.4.3 shall be provided throughout the building in accordance with the *International Building Code.*

804.2.1 Manual fire alarm systems. Where required by the *International Building Code,* a manual fire alarm system shall be provided throughout the *work area.* Alarm notification appliances shall be provided on such floors and shall be automatically activated as required by the *International Building Code.*

Exceptions:

1. Alarm-initiating and notification appliances shall not be required to be installed in tenant spaces outside of the *work area.*
2. Visual alarm notification appliances are not required, except where an existing alarm system is upgraded or replaced or where a new fire alarm system is installed.

804.2.2 Automatic fire detection. Where required by the *International Building Code* for new buildings, automatic fire detection systems shall be provided throughout the *work area.*

SECTION 805 MEANS OF EGRESS

805.1 General. The means of egress shall comply with the requirements of Section 705 except as specifically required in Sections 805.2 and 805.3.

805.2 Means-of-egress lighting. Means of egress from the highest *work area* floor to the floor of exit discharge shall be provided with artificial lighting within the exit enclosure in accordance with the requirements of the *International Building Code.*

805.3 Exit signs. Means of egress from the highest *work area* floor to the floor of exit discharge shall be provided with exit signs in accordance with the requirements of the *International Building Code.*

SECTION 806 ACCESSIBILITY

806.1 General. A building, facility or element that is altered shall comply with Sections 605 and 706.

SECTION 807 STRUCTURAL

807.1 General. Where buildings are undergoing Level 3 alterations including structural alterations, the provisions of this section shall apply.

807.2 New structural elements. New structural elements shall comply with Section 707.2.

807.3 Existing structural elements carrying gravity loads. Existing structural elements carrying gravity loads shall comply with Section 707.4.

807.4 Structural alterations. All structural elements of the lateral-force-resisting system in buildings undergoing Level 3 structural alterations or buildings undergoing Level 2 alterations as triggered by Section 707.5 shall comply with this section.

Exceptions:

1. Buildings of Group R occupancy with no more than five dwelling or sleeping units used solely for residential purposes that are altered based on the conventional light-frame construction methods of the *International Building Code* or in compliance with the provisions of the *International Residential Code.*
2. Where such *alterations* involve only the lowest story of a building and the *change of occupancy* provisions of Chapter 9 do not apply, only the lateral-force-resisting components in and below that story need comply with this section.

807.4.1 Evaluation and analysis. An engineering evaluation and analysis that establishes the structural adequacy of the altered structure shall be prepared by a registered design professional and submitted to the *code official.*

807.4.2 Substantial structural alteration. Where more than 30 percent of the total floor and roof areas of the building or structure have been or are proposed to be involved in structural *alteration* within a 12-month period, the evaluation and analysis shall demonstrate that the altered building or structure complies with the *International Building Code* for wind loading and with reduced *International Building Code* level seismic forces as specified in Section 101.5.4.2 for seismic loading. For seismic considerations, the analysis shall be based on one of the procedures specified in Section 101.5.4. The areas to be counted toward the 30 percent shall be those areas tributary to the vertical load-carrying components, such as joists, beams, columns, walls and other structural components that have been or will be removed, added or altered, as well as areas such as mezzanines, penthouses, roof structures and in-filled courts and shafts.

807.4.3 Limited structural alteration. Where not more than 30 percent of the total floor and roof areas of the building are involved in structural *alteration* within a 12-month period, the evaluation and analysis shall demonstrate that the altered building or structure complies with the loads applicable at the time of the original construction or of the most recent substantial structural *alteration* as defined by Section 807.4.2. Any existing structural element whose seismic demand-capacity ratio with the *alteration* considered is more than 10 percent greater than its demand-capacity ratio with the *alteration* ignored shall comply with the reduced *International Building Code* level seismic forces as specified in Section 101.5.4.2.

SECTION 808 ENERGY CONSERVATION

808.1 Minimum requirements. Level 3 alterations to existing buildings or structures are permitted without requiring the entire building or structure to comply with the energy requirements of the *International Energy Conservation Code* or *International Residential Code.* The alterations shall conform to the energy requirements of the *International Energy Conservation Code* or *International Residential Code* as they relate to new construction only.

CHAPTER 9
CHANGE OF OCCUPANCY

SECTION 901
GENERAL

901.1 Scope. The provisions of this chapter shall apply where a *change of occupancy* occurs, as defined in Section 202, including:

1. Where the occupancy classification is not changed, or
2. Where there is a change in occupancy classification or the occupancy group designation changes.

901.2 Change in occupancy with no change of occupancy classification. A change in occupancy, as defined in Section 202, with no *change of occupancy* classification shall not be made to any structure that will subject the structure to any special provisions of the applicable *International Codes*, including the provisions of Sections 902 through 911, without the approval of the *code official*. A certificate of occupancy shall be issued where it has been determined that the requirements for the change in occupancy have been met.

901.2.1 Repair and alteration with no change of occupancy classification. Any *repair* or *alteration* work undertaken in connection with a *change of occupancy* that does not involve a *change of occupancy* classification shall conform to the applicable requirements for the work as classified in Chapter 4 and to the requirements of Sections 902 through 911.

Exception: As modified in Section 1105 for historic buildings.

901.3 Change of occupancy classification. Where the occupancy classification of a building changes, the provisions of Sections 902 through 912 shall apply. This includes a *change of occupancy* classification within a group as well as a *change of occupancy* classification from one group to a different group.

901.3.1 Partial change of occupancy classification. Where a portion of an *existing building* is changed to a new occupancy classification, Section 912 shall apply.

901.4 Certificate of occupancy required. A certificate of occupancy shall be issued where a *change of occupancy* occurs that results in a different occupancy classification as determined by the *International Building Code*.

SECTION 902
SPECIAL USE AND OCCUPANCY

902.1 Compliance with the building code. Where the character or use of an *existing building* or part of an *existing building* is changed to one of the following special use or occupancy categories as defined in the *International Building Code*, the building shall comply with all of the applicable requirements of the *International Building Code*:

1. Covered mall buildings.
2. Atriums.
3. Motor vehicle-related occupancies.
4. Aircraft-related occupancies.
5. Motion picture projection rooms.
6. Stages and platforms.
7. Special amusement buildings.
8. Incidental use areas.
9. Hazardous materials.

902.2 Underground buildings. An underground building in which there is a change of use shall comply with the requirements of the *International Building Code* applicable to underground structures.

SECTION 903
BUILDING ELEMENTS AND MATERIALS

903.1 General. Building elements and materials in portions of buildings undergoing a *change of occupancy* classification shall comply with Section 912.

SECTION 904
FIRE PROTECTION

904.1 General. Fire protection requirements of Section 912 shall apply where a building or portions thereof undergo a *change of occupancy* classification.

SECTION 905
MEANS OF EGRESS

905.1 General. Means of egress in portions of buildings undergoing a *change of occupancy* classification shall comply with Section 912.

SECTION 906
ACCESSIBILITY

906.1 General. Accessibility in portions of buildings undergoing a *change of occupancy* classification shall comply with Section 912.8.

SECTION 907
STRUCTURAL

907.1 Gravity loads. Buildings or portions thereof subject to a *change of occupancy* where such change in the nature of occupancy results in higher uniform or concentrated loads based on Tables 1607.1 and 1607.6 of the *International Building Code* shall comply with the gravity load provisions of the *International Building Code*.

Exception: Structural elements whose stress is not increased by more than 5 percent.

907.2 Snow and wind loads. Buildings and structures subject to a *change of occupancy* where such change in the nature of occupancy results in higher wind or snow occupancy categories based on Table 1604.5 of the *International Building Code* shall be analyzed and shall comply with the applicable wind or snow load provisions of the *International Building Code*.

Exception: Where the new occupancy with a higher importance factor is less than or equal to 10 percent of the total building floor area. The cumulative effect of the area of occupancy changes shall be considered for the purposes of this exception.

907.3 Seismic loads. Existing buildings with a *change of occupancy* shall comply with the seismic provisions of Sections 907.3.1 and 907.3.2.

907.3.1 Compliance with the International Building Code level seismic forces. Where a building or portion thereof is subject to a *change of occupancy* that results in the building being assigned to a higher occupancy category based on Table 1604.5 of the *International Building Code*; or where such *change of occupancy* results in a reclassification of a building to a higher hazard category as shown in Table 912.4; or where a change of a Group M occupancy to a Group A, E, I-1, R-1, R-2 or R-4 occupancy with two-thirds or more of the floors involved in Level 3 *alteration* work, the building shall comply with the requirements for *International Building Code* level seismic forces as specified in Section 101.5.4.1 for the new occupancy category.

Exceptions:

1. Group M occupancies being changed to Group A, E, I-1, R-1, R-2 or R-4 occupancies for buildings less than six stories in height and in Seismic Design Category A, B or C.
2. Where approved by the *code official*, specific detailing provisions required for a new structure are not required to be met where it can be shown that an equivalent level of performance and seismic safety is obtained for the applicable occupancy category based on the provision for reduced *International Building Code* level seismic forces as specified in Section 101.5.4.2.
3. Where the area of the new occupancy with a higher hazard category is less than or equal to 10 percent of the total building floor area and the new occupancy is not classified as Occupancy Category IV. For the purposes of this exception, buildings occupied by two or more occupancies not included in the same occupancy category, shall be subject to the provisions of Section 1604.5.1 of the *International Building Code*. The cumulative effect of the area of occupancy changes shall be considered for the purposes of this exception.
4. Unreinforced masonry bearing wall buildings in Occupancy Category III when assigned to Seismic Design Category A or B shall be allowed to be strengthened to meet the requirements of Appendix Chapter A1 of this code [Guidelines for the Seismic Retrofit of Existing Buildings (GSREB)].

907.3.2 Access to Occupancy Category IV. Where a *change of occupancy* is such that compliance with Section 907.3.1 is required and the building is assigned to Occupancy Category IV, the operational access to the building shall not be through an adjacent structure, unless that structure conforms to the requirements for Occupancy Category IV structures. Where operational access is less than 10 feet (3048 mm) from either an interior lot line or from another structure, access protection from potential falling debris shall be provided by the owner of the Occupancy Category IV structure.

SECTION 908 ELECTRICAL

908.1 Special occupancies. Where the occupancy of an *existing building* or part of an *existing building* is changed to one of the following special occupancies as described in NFPA 70, the electrical wiring and equipment of the building or portion thereof that contains the proposed occupancy shall comply with the applicable requirements of NFPA 70 whether or not a *change of occupancy* group is involved:

1. Hazardous locations.
2. Commercial garages, *repair*, and storage.
3. Aircraft hangars.
4. Gasoline dispensing and service stations.
5. Bulk storage plants.
6. Spray application, dipping, and coating processes.
7. Health care facilities.
8. Places of assembly.
9. Theaters, audience areas of motion picture and television studios, and similar locations.
10. Motion picture and television studios and similar locations.
11. Motion picture projectors.
12. Agricultural buildings.

908.2 Unsafe conditions. Where the occupancy of an *existing building* or part of an *existing building* is changed, all unsafe conditions shall be corrected without requiring that all parts of the electrical system comply with NFPA 70.

908.3 Service upgrade. Where the occupancy of an *existing building* or part of an *existing building* is changed, electrical service shall be upgraded to meet the requirements of NFPA 70 for the new occupancy.

908.4 Number of electrical outlets. Where the occupancy of an *existing building* or part of an *existing building* is changed, the number of electrical outlets shall comply with NFPA 70 for the new occupancy.

SECTION 909
MECHANICAL

909.1 Mechanical requirements. Where the occupancy of an *existing building* or part of an *existing building* is changed such that the new occupancy is subject to different kitchen exhaust requirements or to increased mechanical ventilation requirements in accordance with the *International Mechanical Code*, the new occupancy shall comply with the intent of the respective *International Mechanical Code* provisions.

SECTION 910
PLUMBING

910.1 Increased demand. Where the occupancy of an *existing building* or part of an *existing building* is changed such that the new occupancy is subject to increased or different plumbing fixture requirements or to increased water supply requirements in accordance with the *International Plumbing Code*, the new occupancy shall comply with the intent of the respective *International Plumbing Code* provisions.

910.2 Food-handling occupancies. If the new occupancy is a food-handling establishment, all existing sanitary waste lines above the food or drink preparation or storage areas shall be panned or otherwise protected to prevent leaking pipes or condensation on pipes from contaminating food or drink. New drainage lines shall not be installed above such areas and shall be protected in accordance with the *International Plumbing Code.*

910.3 Interceptor required. If the new occupancy will produce grease or oil-laden wastes, interceptors shall be provided as required in the *International Plumbing Code.*

910.4 Chemical wastes. If the new occupancy will produce chemical wastes, the following shall apply:

1. If the existing piping is not compatible with the chemical waste, the waste shall be neutralized prior to entering the drainage system, or the piping shall be changed to a compatible material.
2. No chemical waste shall discharge to a public sewer system without the approval of the sewage authority.

910.5 Group I-2. If the occupancy group is changed to Group I-2, the plumbing system shall comply with the applicable requirements of the *International Plumbing Code.*

SECTION 911
OTHER REQUIREMENTS

911.1 Light and ventilation. Light and ventilation shall comply with the requirements of the *International Building Code* for the new occupancy.

SECTION 912
CHANGE OF OCCUPANCY CLASSIFICATION

912.1 General. The provisions of this section shall apply to buildings or portions thereof undergoing a change of occupancy classification. This includes a change of occupancy classification within a group as well as a change of occupancy classification from one group to a different group. Such buildings shall also comply with Sections 902 through 911. The application of requirements for the change of occupancy shall be as set forth in Sections 912.1.1 through 912.1.4. A change of occupancy, as defined in Section 202, without a corresponding change of occupancy classification shall comply with Section 901.2.

912.1.1 Compliance with Chapter 8. The requirements of Chapter 8 shall be applicable throughout the building for the new occupancy classification based on the separation conditions set forth in Sections 912.1.1.1 and 912.1.1.2.

912.1.1.1 Change of occupancy classification without separation. Where a portion of an *existing building* is changed to a new occupancy classification and that portion is not separated from the remainder of the building with fire barriers having a fire-resistance rating as required in the *International Building Code* for the separate occupancy, the entire building shall comply with all of the requirements of Chapter 8 applied throughout the building for the most restrictive occupancy classification in the building and with the requirements of this chapter.

912.1.1.2 Change of occupancy classification with separation. Where a portion of an *existing building* that is changed to a new occupancy classification and that portion is separated from the remainder of the building with fire barriers having a fire-resistance rating as required in the *International Building Code* for the separate occupancy, that portion shall comply with all the requirements of Chapter 8 for the new occupancy classification and with the requirements of this chapter.

912.1.2 Fire protection and interior finish. The provisions of Sections 912.2 and 912.3 for fire protection and interior finish, respectively, shall apply to all buildings undergoing a change of occupancy classification.

912.1.3 Change of occupancy classification based on hazard category. The relative degree of hazard between different occupancy classifications shall be determined in accordance with the category specified in Tables 912.4, 912.5 and 912.6. Such a determination shall be the basis for the application of Sections 912.4 through 912.7.

912.1.4 Accessibility. All buildings undergoing a *change of occupancy* classification shall comply with Section 912.8.

912.2 Fire protection systems. Fire protection systems shall be provided in accordance with Sections 912.2.1 and 912.2.2.

912.2.1 Fire sprinkler system. Where a change in occupancy classification occurs that requires an automatic fire sprinkler system to be provided based on the new occupancy in accordance with Chapter 9 of the *International Building Code*, such system shall be provided throughout the area where the *change of occupancy* occurs.

912.2.2 Fire alarm and detection system. Where a change in occupancy classification occurs that requires a fire alarm and detection system to be provided based on the new occupancy in accordance with Chapter 9 of the *International Building Code*, such system shall be provided throughout the area where the *change of occupancy* occurs. Existing alarm notification appliances shall be automatically activated throughout the building. Where the building is not equipped with a fire alarm system, alarm notification appliances shall be provided throughout the area where the *change of occupancy* occurs and shall be automatically activated.

912.3 Interior finish. In areas of the building undergoing the change of occupancy classification, the interior finish of walls and ceilings shall comply with the requirements of the *International Building Code* for the new occupancy classification.

912.4 Means of egress, general. Hazard categories in regard to life safety and means of egress shall be in accordance with Table 912.4.

TABLE 912.4
MEANS OF EGRESS HAZARD CATEGORIES

RELATIVE HAZARD	OCCUPANCY CLASSIFICATIONS
1 (Highest Hazard)	H
2	I-2, I-3, I-4
3	A, E, I-1, M, R-1, R-2, R-4
4	B, F-1, R-3, S-1
5 (Lowest Hazard)	F-2, S-2, U

912.4.1 Means of egress for change to higher hazard category. When a change of occupancy classification is made to a higher hazard category (lower number) as shown in Table 912.4, the means of egress shall comply with the requirements of Chapter 10 of the *International Building Code.*

Exceptions:

1. Stairways shall be enclosed in compliance with the applicable provisions of Section 803.1.
2. Existing stairways including handrails and guards complying with the requirements of Chapter 8 shall be permitted for continued use subject to approval of the *code official*.
3. Any stairway replacing an existing stairway within a space where the pitch or slope cannot be reduced because of existing construction shall not be required to comply with the maximum riser height and minimum tread depth requirements.
4. Existing corridor walls constructed on both sides of wood lath and plaster in good condition or $^1/_2$-inch-thick (12.7 mm) gypsum wallboard shall be permitted. Such walls shall either terminate at the underside of a ceiling of equivalent construction or extend to the underside of the floor or roof next above.
5. Existing corridor doorways, transoms and other corridor openings shall comply with the requirements in Sections 705.5.1, 705.5.2 and 705.5.3.
6. Existing dead-end corridors shall comply with the requirements in Section 705.6.
7. An existing operable window with clear opening area no less than 4 square feet (0.38 m^2) and minimum opening height and width of 22 inches (559 mm) and 20 inches (508 mm), respectively, shall be accepted as an emergency escape and rescue opening.

912.4.2 Means of egress for change of use to equal or lower hazard category. When a change of occupancy classification is made to an equal or lesser hazard category (higher number) as shown in Table 912.4, existing elements of the means of egress shall comply with the requirements of Section 805 for the new occupancy classification. Newly constructed or configured means of egress shall comply with the requirements of Chapter 10 of the *International Building Code.*

Exception: Any stairway replacing an existing stairway within a space where the pitch or slope cannot be reduced because of existing construction shall not be required to comply with the maximum riser height and minimum tread depth requirements.

912.4.3 Egress capacity. Egress capacity shall meet or exceed the occupant load as specified in the *International Building Code* for the new occupancy.

912.4.4 Handrails. Existing stairways shall comply with the handrail requirements of Section 705.9 in the area of the change of occupancy classification.

912.4.5 Guards. Existing guards shall comply with the requirements in Section 705.10 in the area of the change of occupancy classification.

912.5 Heights and areas. Hazard categories in regard to height and area shall be in accordance with Table 912.5.

TABLE 912.5
HEIGHTS AND AREAS HAZARD CATEGORIES

RELATIVE HAZARD	OCCUPANCY CLASSIFICATIONS
1 (Highest Hazard)	H
2	A-1, A-2, A-3, A-4, I, R-1, R-2, R-4
3	E, F-1, S-1, M
4 (Lowest Hazard)	B, F-2, S-2, A-5, R-3, U

912.5.1 Height and area for change to higher hazard category. When a change of occupancy classification is made to a higher hazard category as shown in Table 912.5, heights and areas of buildings and structures shall comply with the requirements of Chapter 5 of the *International Building Code* for the new occupancy classification.

Exception: In other than Groups H, F-1 and S-1, in lieu of fire walls, use of fire barriers having a fire-resistance rating of not less than that specified in Table 706.4 of the *International Building Code*, constructed in accordance with Section 707 of the *International Building Code*, shall be permitted to meet area limitations required for the new occupancy in buildings protected throughout with an automatic sprinkler system in accordance with Section 903.3.1.1 of the *International Fire Code.*

912.5.2 Height and area for change to equal or lesser hazard category. When a change of occupancy classification is made to an equal or lesser hazard category as shown in Table 912.5, the height and area of the *existing building* shall be deemed acceptable.

912.5.3 Fire barriers. When a change of occupancy classification is made to a higher hazard category as shown in Table 912.5, fire barriers in separated mixed-use buildings shall comply with the fire resistance requirements of the *International Building Code.*

Exception: Where the fire barriers are required to have a 1-hour fire-resistance rating, existing wood lath and plaster in good condition or existing $^1/_2$-inch-thick (12.7 mm) gypsum wallboard shall be permitted.

912.6 Exterior wall fire-resistance ratings. Hazard categories in regard to fire-resistance ratings of exterior walls shall be in accordance with Table 912.6.

TABLE 912.6
EXPOSURE OF EXTERIOR WALLS HAZARD CATEGORIES

RELATIVE HAZARD	OCCUPANCY CLASSIFICATION
1 (Highest Hazard)	H
2	F-1, M, S-1
3	A, B, E, I, R
4 (Lowest Hazard)	F-2, S-2, U

912.6.1 Exterior wall rating for change of occupancy classification to a higher hazard category. When a change of occupancy classification is made to a higher hazard category as shown in Table 912.6, exterior walls shall have fire resistance and exterior opening protectives as required by the *International Building Code.*

Exception: A 2-hour fire-resistance rating shall be allowed where the building does not exceed three stories in height and is classified as one of the following groups: A-2 and A-3 with an occupant load of less than 300, B, F, M or S.

912.6.2 Exterior wall rating for change of occupancy classification to an equal or lesser hazard category. When a change of occupancy classification is made to an equal or lesser hazard category as shown in Table 912.6, existing exterior walls, including openings, shall be accepted.

912.6.3 Opening protectives. Openings in exterior walls shall be protected as required by the *International Building Code.* Where openings in the exterior walls are required to be protected because of their distance from the property line, the sum of the area of such openings shall not exceed 50 percent of the total area of the wall in each story.

Exceptions:

1. Where the *International Building Code* permits openings in excess of 50 percent.
2. Protected openings shall not be required in buildings of Group R occupancy that do not exceed three stories in height and that are located not less than 3 feet (914 mm) from the property line.
3. Where exterior opening protectives are required, an automatic sprinkler system throughout may be substituted for opening protection.
4. Exterior opening protectives are not required when the change of occupancy group is to an equal or lower hazard classification in accordance with Table 912.6

912.7 Enclosure of vertical shafts. Enclosure of vertical shafts shall be in accordance with Sections 912.7.1 through 912.7.4.

912.7.1 Minimum requirements. Vertical shafts shall be designed to meet the *International Building Code* requirements for atriums or the requirements of this section.

912.7.2 Stairways. When a change of occupancy classificiation is made to a higher hazard category as shown in Table 912.4, interior stairways shall be enclosed as required by the *International Building Code.*

Exceptions:

1. In other than Group I occupancies, an enclosure shall not be required for openings serving only one adjacent floor and that are not connected with corridors or stairways serving other floors.
2. Unenclosed existing stairways need not be enclosed in a continuous vertical shaft if each story is separated from other stories by 1-hour fire-resistance-rated construction or approved wired glass set in steel frames and all exit corridors are sprinklered. The openings between the corridor and the occupant space shall have at least one sprinkler head above the openings on the tenant side. The sprinkler system shall be permitted to be supplied from the domestic water-supply systems, provided the system is of adequate pressure, capacity, and sizing for the combined domestic and sprinkler requirements.
3. Existing penetrations of stairway enclosures shall be accepted if they are protected in accordance with the *International Building Code.*

912.7.3 Other vertical shafts. Interior vertical shafts other than stairways, including but not limited to elevator hoistways and service and utility shafts, shall be enclosed as required by the *International Building Code* when there is a change of use to a higher hazard category as specified in Table 912.4.

Exceptions:

1. Existing 1-hour interior shaft enclosures shall be accepted where a higher rating is required.
2. Vertical openings, other than stairways, in buildings of other than Group I occupancy and connecting less than six stories shall not be required to be enclosed if the entire building is provided with an approved automatic sprinkler system.

912.7.4 Openings. All openings into existing vertical shaft enclosures shall be protected by fire assemblies having a fire-protection rating of not less than 1 hour and shall be maintained self-closing or shall be automatic closing by actuation of a smoke detector. All other openings shall be fire protected in an approved manner. Existing fusible link-type automatic door-closing devices shall be permitted in all shafts except stairways if the fusible link rating does not exceed 135°F (57°C).

912.8 Accessibility. Existing buildings that undergo a change of group or occupancy classification shall comply with this section.

912.8.1 Partial change in occupancy. Where a portion of the building is changed to a new occupancy classification, any alterations shall comply with Sections 605 and 706, as applicable.

912.8.2 Complete change of occupancy. Where an entire building undergoes a *change of occupancy*, it shall comply with Section 912.8.1 and shall have all of the following accessible features:

1. At least one accessible building entrance.
2. At least one accessible route from an accessible building entrance to *primary function* areas.
3. Signage complying with Section 1110 of the *International Building Code*.
4. Accessible parking, where parking is provided.
5. At least one accessible passenger loading zone, where loading zones are provided.
6. At least one accessible route connecting accessible parking and accessible passenger loading zones to an accessible entrance.

Where it is *technically infeasible* to comply with the new construction standards for any of these requirements for a change of group or occupancy, the above items shall conform to the requirements to the maximum extent technically feasible.

CHAPTER 10
ADDITIONS

SECTION 1001 GENERAL

1001.1 Scope. An *addition* to a building or structure shall comply with the *International Codes* as adopted for new construction without requiring the *existing building* or structure to comply with any requirements of those codes or of these provisions, except as required by this chapter. Where an *addition* impacts the *existing building* or structure, that portion shall comply with this code.

1001.2 Creation or extension of nonconformity. An *addition* shall not create or extend any nonconformity in the *existing building* to which the *addition* is being made with regard to accessibility, structural strength, fire safety, means of egress, or the capacity of mechanical, plumbing, or electrical systems.

1001.3 Other work. Any *repair* or *alteration* work within an *existing building* to which an *addition* is being made shall comply with the applicable requirements for the work as classified in Chapter 4.

SECTION 1002 HEIGHTS AND AREAS

1002.1 Height limitations. No *addition* shall increase the height of an *existing building* beyond that permitted under the applicable provisions of Chapter 5 of the *International Building Code* for new buildings.

1002.2 Area limitations. No *addition* shall increase the area of an *existing building* beyond that permitted under the applicable provisions of Chapter 5 of the *International Building Code* for new buildings unless fire separation as required by the *International Building Code* is provided.

Exception: In-filling of floor openings and nonoccupiable appendages such as elevator and exit stair shafts shall be permitted beyond that permitted by the *International Building Code.*

1002.3 Fire protection systems. Existing fire areas increased by the *addition* shall comply with Chapter 9 of the *International Building Code.*

SECTION 1003 STRUCTURAL

1003.1 Compliance with the International Building Code. *Additions* to existing buildings or structures are new construction and shall comply with the *International Building Code.*

1003.2 Additional gravity loads. Existing structural elements supporting any additional gravity loads as a result of additions shall comply with the *International Building Code.*

Exceptions:

1. Structural elements whose stress is not increased by more than 5 percent.
2. Buildings of Group R occupancy with no more than five dwelling units or sleeping units used solely for residential purposes where the *existing building* and the *addition* comply with the conventional light-frame construction methods of the *International Building Code* or the provisions of the *International Residential Code.*

1003.3 Lateral-force-resisting system. The lateral-force-resisting system of existing buildings to which additions are made shall comply with Sections 1003.3.1, 1003.3.2 and 1003.3.3.

Exceptions:

1. Buildings of Group R occupancy with no more than five dwelling or sleeping units used solely for residential purposes where the *existing building* and the *addition* comply with the conventional light-frame construction methods of the *International Building Code* or the provisions of the *International Residential Code.*
2. In other existing buildings where the lateral-force story shear in any story is not increased by more than 10 percent cumulative.

1003.3.1 Vertical addition. Any element of the lateral-force-resisting system of an *existing building* subjected to an increase in vertical or lateral loads from the vertical *addition* shall comply with the *International Building Code* wind provisions and the *International Building Code* level seismic forces specified in Section 101.5.4.1 of this code.

1003.3.2 Horizontal addition. Where horizontal *additions* are structurally connected to an existing structure, all lateral-force-resisting elements of the existing structure affected by such *addition* shall comply with the *International Building Code* wind provisions and the *International Building Code* level seismic forces specified in Section 101.5.4.1 of this code.

1003.3.3 Voluntary addition of structural elements to improve the lateral-force-resisting system. Voluntary *addition* of structural elements to improve the lateral-force-resisting system of an *existing building* shall comply with Section 707.6.

1003.4 Snow drift loads. Any structural element of an *existing building* subjected to additional loads from the effects of snow drift as a result of an *addition* shall comply with the *International Building Code.*

Exceptions:

1. Structural elements whose stress is not increased by more than 5 percent.
2. Buildings of Group R occupancy with no more than five dwelling units or sleeping units used solely for residential purposes where the *existing building* and

the *addition* comply with the conventional light-frame construction methods of the *International Building Code* or the provisions of the *International Residential Code*.

1003.5 Flood hazard areas. Additions and foundations in flood hazard areas shall comply with the following requirements:

1. For horizontal additions that are structurally interconnected to the *existing building*:

   1.1. If the *addition* and all other proposed work, when combined, constitute *substantial improvement*, the *existing building* and the *addition* shall comply with Section 1612 of the *International Building Code*.

   1.2. If the *addition* constitutes *substantial improvement*, the *existing building* and the *addition* shall comply with Section 1612 of the *International Building Code*.

2. For horizontal additions that are not structurally interconnected to the *existing building*:

   2.1. The *addition* shall comply with Section 1612 of the *International Building Code*.

   2.2. If the *addition* and all other proposed work, when combined, constitute *substantial improvement*, the *existing building* and the *addition* shall comply with Section 1612 of the *International Building Code*.

3. For vertical additions and all other proposed work that, when combined, constitute *substantial improvement*, the *existing building* shall comply with Section 1612 of the *International Building Code*.

4. For a new, replacement, raised, or extended foundation, if the foundation work and all other proposed work, when combined, constitute *substantial improvement*, the *existing building* shall comply with Section 1612 of the *International Building Code*.

SECTION 1004 SMOKE ALARMS IN OCCUPANCY GROUPS R-3 AND R-4

1004.1 Smoke alarms in existing portions of a building. Whenever an *addition* is made to a building or structure of a Group R-3 or R-4 occupancy, the *existing building* shall be provided with smoke alarms as required by the *International Building Code* or *International Residential Code* as applicable.

SECTION 1005 ACCESSIBILITY

1005.1 Minimum requirements. Accessibility provisions for new construction shall apply to additions. An *addition* that affects the accessibility to, or contains an area of, *primary function* shall comply with the requirements of Sections 605 and 706, as applicable.

CHAPTER 11
HISTORIC BUILDINGS

SECTION 1101
GENERAL

1101.1 Scope. It is the intent of this chapter to provide means for the preservation of historic buildings. Historical buildings shall comply with the provisions of this chapter relating to their *repair*, *alteration*, relocation and *change of occupancy*.

1101.2 Report. A *historic building* undergoing *repair*, *alteration* or *change of occupancy* shall be investigated and evaluated. If it is intended that the building meet the requirements of this chapter, a written report shall be prepared and filed with the *code official* by a registered design professional when such a report is necessary in the opinion of the *code official*. Such report shall be in accordance with Chapter 1 and shall identify each required safety feature that is in compliance with this chapter and where compliance with other chapters of these provisions would be damaging to the contributing historic features. For buildings assigned to Seismic Design Category D, E or F, a structural evaluation describing, at minimum, a complete load path and other earthquake-resistant features shall be prepared. Additionally, the report shall describe each feature that is not in compliance with these provisions and shall demonstrate how the intent of these provisions is complied with in providing an equivalent level of safety.

1101.3 Special occupancy exceptions—museums. When a building in Group R-3 is also used for Group A, B, or M purposes such as museum tours, exhibits, and other public assembly activities, or for museums less than 3,000 square feet (279 m^2), the *code official* may determine that the occupancy is Group B when life-safety conditions can be demonstrated in accordance with Section 1101.2. Adequate means of egress in such buildings, which may include a means of maintaining doors in an open position to permit egress, a limit on building occupancy to an occupant load permitted by the means of egress capacity, a limit on occupancy of certain areas or floors, or supervision by a person knowledgeable in the emergency exiting procedures, shall be provided.

1101.4 Flood hazard areas. In flood hazard areas, if all proposed work, including repairs, work required because of a *change of occupancy*, and alterations, constitutes *substantial improvement*, then the *existing building* shall comply with Section 1612 of the *International Building Code*.

> **Exception:** If an *historic building* will continue to be an *historic building* after the proposed work is completed, then the proposed work is not considered a *substantial improvement*. For the purposes of this exception, an *historic building* is:
>
> 1. Listed or preliminarily determined to be eligible for listing in the National Register of Historic Places;
> 2. Determined by the Secretary of the U.S. Department of Interior to contribute to the historical significance of a registered historic district or a district preliminarily determined to qualify as a historic district; or
> 3. Designated as historic under a state or local historic preservation program that is approved by the Department of Interior.

SECTION 1102
REPAIRS

1102.1 General. Repairs to any portion of an *historic building* or structure shall be permitted with original or like materials and original methods of construction, subject to the provisions of this chapter.

1102.2 Dangerous buildings. When an *historic building* is determined to be *dangerous*, no work shall be required except as necessary to correct identified unsafe conditions.

1102.3 Relocated buildings. Foundations of relocated historic buildings and structures shall comply with the *International Building Code*. Relocated historic buildings shall otherwise be considered an *historic building* for the purposes of this code. Relocated historic buildings and structures shall be sited so that exterior wall and opening requirements comply with the *International Building Code* or with the compliance alternatives of this code.

1102.4 Chapter 5 compliance. Historic buildings undergoing repairs shall comply with all of the applicable requirements of Chapter 5, except as specifically permitted in this chapter.

1102.5 Replacement. Replacement of existing or missing features using original materials shall be permitted. Partial replacement for repairs that match the original in configuration, height, and size shall be permitted. Such replacements shall not be required to meet the materials and methods requirements of Section 501.2.

> **Exception:** Replacement glazing in hazardous locations shall comply with the safety glazing requirements of Chapter 24 of the *International Building Code*.

SECTION 1103
FIRE SAFETY

1103.1 Scope. Historic buildings undergoing alterations, changes of occupancy, or that are moved shall comply with Section 1103.

1103.2 General. Every *historic building* that does not conform to the construction requirements specified in this code for the occupancy or use and that constitutes a distinct fire hazard as defined herein shall be provided with an approved automatic fire-extinguishing system as determined appropriate by the *code official*. However, an automatic fire-extinguishing system shall not be used to substitute for, or act as an alternative to, the required number of exits from any facility.

1103.3 Means of egress. Existing door openings and corridor and stairway widths less than those specified elsewhere in this code may be approved, provided that, in the opinion of the *code official*, there is sufficient width and height for a person to pass through the opening or traverse the means of egress. When approved by the *code official*, the front or main exit doors need not swing in the direction of the path of exit travel, provided that other approved means of egress having sufficient capacity to serve the total occupant load are provided.

1103.4 Transoms. In fully sprinklered buildings of Group R-1, R-2 or R-3 occupancy, existing transoms in corridors and other fire-resistance-rated walls may be maintained if fixed in the closed position. A sprinkler shall be installed on each side of the transom.

1103.5 Interior finishes. The existing finishes of walls and ceilings shall be accepted when it is demonstrated that they are the historic finishes.

1103.6 Stairway enclosure. In buildings of three stories or less, exit enclosure construction shall limit the spread of smoke by the use of tight-fitting doors and solid elements. Such elements are not required to have a fire-resistance rating.

1103.7 One-hour fire-resistant assemblies. Where 1-hour fire-resistance-rated construction is required by these provisions, it need not be provided, regardless of construction or occupancy, where the existing wall and ceiling finish is wood or metal lath and plaster.

1103.8 Glazing in fire-resistance-rated systems. Historic glazing materials are permitted in interior walls required to have a 1-hour fire-resistance rating where the opening is provided with approved smoke seals and the area affected is provided with an automatic sprinkler system.

1103.9 Stairway railings. Grand stairways shall be accepted without complying with the handrail and guard requirements. Existing handrails and guards at all stairs shall be permitted to remain, provided they are not structurally *dangerous*.

1103.10 Guards. Guards shall comply with Sections 1103.10.1 and 1103.10.2.

1103.10.1 Height. Existing guards shall comply with the requirements of Section 505.

1103.10.2 Guard openings. The spacing between existing intermediate railings or openings in existing ornamental patterns shall be accepted. Missing elements or members of a guard may be replaced in a manner that will preserve the historic appearance of the building or structure.

1103.11 Exit signs. Where exit sign or egress path marking location would damage the historic character of the building, alternative exit signs are permitted with approval of the *code official*. Alternative signs shall identify the exits and egress path.

1103.12 Automatic fire-extinguishing systems. Every historical building that cannot be made to conform to the construction requirements specified in the *International Building Code* for the occupancy or use and that constitutes a distinct fire hazard shall be deemed to be in compliance if provided with an approved automatic fire-extinguishing system.

Exception: When the *code official* approves an alternative life-safety system.

SECTION 1104 ALTERATIONS

1104.1 Accessibility requirements. The provisions of 605 and 706, as applicable, shall apply to buildings and facilities designated as historic structures that undergo alterations, unless *technically infeasible*. Where compliance with the requirements for accessible routes, entrances or toilet facilities would threaten or destroy the historic significance of the building or facility, as determined by the *code official*, the alternative requirements of Sections 1104.1.1 through 1104.1.4 for that element shall be permitted.

1104.1.1 Site arrival points. At least one main entrance shall be accessible.

1104.1.2 Multilevel buildings and facilities. An accessible route from an accessible entrance to public spaces on the level of the accessible entrance shall be provided.

1104.1.3 Entrances. At least one main entrance shall be accessible.

Exceptions:

1. If a main entrance cannot be made accessible, an accessible nonpublic entrance that is unlocked while the building is occupied shall be provided; or
2. If a main entrance cannot be made accessible, a locked accessible entrance with a notification system or remote monitoring shall be provided.

1104.1.4 Toilet and bathing facilities. Where toilet rooms are provided, at least one accessible family or assisted-use toilet room complying with Section 1109.2.1 of the *International Building Code* shall be provided.

SECTION 1105 CHANGE OF OCCUPANCY

1105.1 General. Historic buildings undergoing a *change of occupancy* shall comply with the applicable provisions of Chapter 9, except as specifically permitted in this chapter. When Chapter 9 requires compliance with specific requirements of Chapter 5, Chapter 6, or Chapter 7 and when those requirements are subject to the exceptions in Section 1102, the same exceptions shall apply to this section.

1105.2 Building area. The allowable floor area for historic buildings undergoing a *change of occupancy* shall be permitted to exceed by 20 percent the allowable areas specified in Chapter 5 of the *International Building Code*.

1105.3 Location on property. Historic structures undergoing a change of use to a higher hazard category in accordance with Section 912.6 may use alternative methods to comply with the fire-resistance and exterior opening protective requirements. Such alternatives shall comply with Section 1101.2.

1105.4 Occupancy separation. Required occupancy separations of 1 hour may be omitted when the building is provided with an approved automatic sprinkler system throughout.

1105.5 Roof covering. Regardless of occupancy or use group, roof-covering materials not less than Class C shall be permitted where a fire-retardant roof covering is required.

1105.6 Means of egress. Existing door openings and corridor and stairway widths less than those that would be acceptable for nonhistoric buildings under these provisions shall be approved, provided that, in the opinion of the *code official*, there is sufficient width and height for a person to pass through the opening or traverse the exit and that the capacity of the exit system is adequate for the occupant load, or where other operational controls to limit occupancy are approved by the *code official.*

1105.7 Door swing. When approved by the *code official*, existing front doors need not swing in the direction of exit travel, provided that other approved exits having sufficient capacity to serve the total occupant load are provided.

1105.8 Transoms. In corridor walls required by these provisions to be fire-resistance rated, existing transoms may be maintained if fixed in the closed position, and fixed wired glass set in a steel frame or other approved glazing shall be installed on one side of the transom.

> **Exception:** Transoms conforming to Section 1103.4 shall be accepted.

1105.9 Finishes. Where interior finish materials are required to have a flame spread index of Class C or better, existing nonconforming materials shall be surfaced with approved fire-retardant paint or finish.

> **Exception:** Existing nonconforming materials need not be surfaced with an approved fire-retardant paint or finish where the building is equipped throughout with an automatic fire-suppression system installed in accordance with the *International Building Code* and the nonconforming materials can be substantiated as being historic in character.

1105.10 One-hour fire-resistant assemblies. Where 1-hour fire-resistance-rated construction is required by these provisions, it need not be provided, regardless of construction or occupancy, where the existing wall and ceiling finish is wood lath and plaster.

1105.11 Stairs and railings. Existing stairways shall comply with the requirements of these provisions. The *code official* shall grant alternatives for stairways and railings if alternative stairways are found to be acceptable or are judged to meet the intent of these provisions. Existing stairways shall comply with Section 1103.

> **Exception:** For buildings less than 3,000 square feet (279 m^2), existing conditions are permitted to remain at all stairs and rails.

1105.12 Exit signs. The *code official* may accept alternative exit sign locations where such signs would damage the historic character of the building or structure. Such signs shall identify the exits and exit path.

1105.13 Exit stair live load. Existing historic stairways in buildings changed to a Group R-1 or R-2 occupancy shall be accepted where it can be shown that the stairway can support a 75-pounds-per-square-foot (366 kg/m^2) live load.

1105.14 Natural light. When it is determined by the *code official* that compliance with the natural light requirements of Section 911.1 will lead to loss of historic character or historic materials in the building, the existing level of natural lighting shall be considered acceptable.

1105.15 Accessibility requirements. The provisions of Section 912.8 shall apply to buildings and facilities designated as historic structures that undergo a *change of occupancy*, unless *technically infeasible*. Where compliance with the requirements for accessible routes, ramps, entrances, or toilet facilities would threaten or destroy the historic significance of the building or facility, as determined by the authority having jurisdiction, the alternative requirements of Sections 1104.1.1 through 1104.1.4 for those elements shall be permitted.

SECTION 1106
STRUCTURAL

1106.1 General. Historic buildings shall comply with the applicable structural provisions for the work as classified in Chapter 4.

> **Exception:** The *code official* shall be authorized to accept existing floors and approve operational controls that limit the live load on any such floor.

1106.2 Unsafe structural elements. Where the *code official* determines that a component or a portion of a building or structure is *dangerous* as defined in this code and is in need of *repair*, strengthening, or replacement by provisions of this code, only that specific component or portion shall be required to be repaired, strengthened or replaced.

CHAPTER 12

RELOCATED OR MOVED BUILDINGS

SECTION 1201
GENERAL

1201.1 Scope. This chapter provides requirements for relocated or moved structures.

1201.2 Conformance. The building shall be safe for human occupancy as determined by the *International Fire Code* and the *International Property Maintenance Code*. Any *repair*, *alteration*, or *change of occupancy* undertaken within the moved structure shall comply with the requirements of this code applicable to the work being performed. Any field-fabricated elements shall comply with the requirements of the *International Building Code* or the *International Residential Code* as applicable.

SECTION 1202
REQUIREMENTS

1202.1 Location on the lot. The building shall be located on the lot in accordance with the requirements of the *International Building Code* or the *International Residential Code* as applicable.

1202.2 Foundation. The foundation system of relocated buildings shall comply with the *International Building Code* or the *International Residential Code* as applicable.

1202.2.1 Connection to the foundation. The connection of the relocated building to the foundation shall comply with the *International Building Code* or the *International Residential Code* as applicable.

1202.3 Wind loads. Buildings shall comply with *International Building Code* or *International Residential Code* wind provisions as applicable.

Exceptions:

1. Detached one- and two-family dwellings and Group U occupancies where wind loads at the new location are not higher than those at the previous location.
2. Structural elements whose stress is not increased by more than 5 percent.

1202.4 Seismic loads. Buildings shall comply with *International Building Code* or *International Residential Code* seismic provisions at the new location as applicable.

Exceptions:

1. Structures in Seismic Design Categories A and B and detached one- and two-family dwellings in Seismic Design Categories A, B, and C where the seismic loads at the new location are not higher than those at the previous location.
2. Stuctural elements whose stress is not increased by more than 5 percent.

1202.5 Snow loads. Structures shall comply with *International Building Code* or *International Residential Code* snow loads as applicable where snow loads at the new location are higher than those at the previous location.

Exception: Structural elements whose stress is not increased by more than 5 percent.

1202.6 Flood hazard areas. If relocated or moved into a *flood hazard area*, structures shall comply with Section 1612 of the *International Building Code*.

1202.7 Required inspection and repairs. The code official shall be authorized to inspect, or to require approved professionals to inspect at the expense of the owner, the various structural parts of a relocated building to verify that structural components and connections have not sustained structural damage. Any repairs required by the code official as a result of such inspection shall be made prior to the final approval.

CHAPTER 13
PERFORMANCE COMPLIANCE METHODS

SECTION 1301
GENERAL

1301.1 Scope. The provisions of this chapter shall apply to the *alteration*, *repair*, *addition* and *change of occupancy* of existing structures, including historic and moved structures, as referenced in Section 101.5.3. The provisions of this chapter are intended to maintain or increase the current degree of public safety, health and general welfare in existing buildings while permitting *repair*, *alteration*, *addition* and *change of occupancy* without requiring full compliance with Chapters 4 through 12, except where compliance with other provisions of this code is specifically required in this chapter.

1301.1.1 Compliance with other methods. *Alterations*, *repairs*, *additions* and *changes of occupancy* to existing structures shall comply with the provisions of this chapter or with one of the methods provided in Section 101.5.

[B] 1301.2 Applicability. Structures existing prior to [DATE TO BE INSERTED BY THE JURISDICTION. Note: it is recommended that this date coincide with the effective date of building codes within the jurisdiction], in which there is work involving *additions*, *alterations* or *changes of occupancy* shall be made to conform to the requirements of this chapter or the provisions of Chapters 4 through 12. The provisions of Sections 1301.2.1 through 1301.2.5 shall apply to existing occupancies that will continue to be, or are proposed to be, in Groups A, B, E, F, M, R, and S. These provisions shall not apply to buildings with occupancies in Group H or I.

[B] 1301.2.1 Change in occupancy. Where an *existing building* is changed to a new occupancy classification and this section is applicable, the provisions of this section for the new occupancy shall be used to determine compliance with this code.

[B] 1301.2.2 Partial change in occupancy. Where a portion of the building is changed to a new occupancy classification and that portion is separated from the remainder of the building with fire barriers or horizontal assemblies having a fire-resistance rating as required by Table 508.4 of the *International Building Code* or Section R317 of the *International Residential Code* for the separate occupancies, or with approved compliance alternatives, the portion changed shall be made to conform to the provisions of this section.

Where a portion of the building is changed to a new occupancy classification and that portion is not separated from the remainder of the building with fire barriers or horizontal assemblies having a fire-resistance rating as required by Table 508.4 of the *International Building Code* or Section R317 of the *International Residential Code* for the separate occupancies, or with approved compliance alternatives, the provisions of this section which apply to each occupancy shall apply to the entire building. Where there are conflicting provisions, those requirements which secure the greater public safety shall apply to the entire building or structure.

[B] 1301.2.3 Additions. Additions to existing buildings shall comply with the requirements of the *International Building Code*, *International Residential Code*, and this code for new construction. The combined height and area of the *existing building* and the new *addition* shall not exceed the height and area allowed by Chapter 5 of the *International Building Code*. Where a fire wall that complies with Section 706 of the *International Building Code* is provided between the *addition* and the *existing building*, the *addition* shall be considered a separate building.

[B] 1301.2.4 Alterations and repairs. An *existing building* or portion thereof that does not comply with the requirements of this code for new construction shall not be altered or repaired in such a manner that results in the building being less safe or sanitary than such building is currently. If, in the *alteration* or *repair*, the current level of safety or sanitation is to be reduced, the portion altered or repaired shall conform to the requirements of Chapters 2 through 12 and Chapters 14 through 33 of the *International Building Code*.

[B] 1301.2.5 Accessibility requirements. All portions of the buildings proposed for *change of occupancy* shall conform to the accessibility provisions of Section 310.

[B] 1301.3 Acceptance. For repairs, alterations, additions, and changes of occupancy to existing buildings that are evaluated in accordance with this section, compliance with this section shall be accepted by the *code official*.

[B] 1301.3.1 Hazards. Where the *code official* determines that an unsafe condition exists as provided for in Section 116, such unsafe condition shall be abated in accordance with Section 116.

[B] 1301.3.2 Compliance with other codes. Buildings that are evaluated in accordance with this section shall comply with the *International Fire Code* and *International Property Maintenance Code*.

[B] 1301.3.3 Compliance with flood hazard provisions. In flood hazard areas, buildings that are evaluated in accordance with this section shall comply with Section 1612 of the *International Building Code* if the work covered by this section constitutes *substantial improvement*.

[B] 1301.4 Investigation and evaluation. For proposed work covered by this chapter, the building owner shall cause the *existing building* to be investigated and evaluated in accordance with the provisions of Sections 1301.4 through 1301.9.

[B] 1301.4.1 Structural analysis. The owner shall have a structural analysis of the *existing building* made to determine adequacy of structural systems for the proposed *alteration*, *addition* or *change of occupancy*. The analysis shall demonstrate that the building with the work completed is capable of resisting the loads specified in Chapter 16 of the *International Building Code*.

[B] 1301.4.2 Submittal. The results of the investigation and evaluation as required in Section 1301.4, along with proposed compliance alternatives, shall be submitted to the *code official.*

[B] 1301.4.3 Determination of compliance. The *code official* shall determine whether the *existing building*, with the proposed *addition*, *alteration*, or *change of occupancy*, complies with the provisions of this section in accordance with the evaluation process in Sections 1301.5 through 1301.9.

[B] 1301.5 Evaluation. The evaluation shall be comprised of three categories: fire safety, means of egress, and general safety, as defined in Sections 1301.5.1 through 1301.5.3.

[B] 1301.5.1 Fire safety. Included within the fire safety category are the structural fire resistance, automatic fire detection, fire alarm, and fire-suppression system features of the facility.

[B] 1301.5.2 Means of egress. Included within the means of egress category are the configuration, characteristics, and support features for means of egress in the facility.

[B] 1301.5.3 General safety. Included within the general safety category are the fire safety parameters and the means-of-egress parameters.

[B] 1301.6 Evaluation process. The evaluation process specified herein shall be followed in its entirety to evaluate existing buildings. Table 1301.7 shall be utilized for tabulating the results of the evaluation. References to other sections of this code indicate that compliance with those sections is required in order to gain credit in the evaluation herein outlined. In applying this section to a building with mixed occupancies, where the separation between the mixed occupancies does not qualify for any category indicated in Section 1301.6.16, the score for each occupancy shall be determined, and the lower score determined for each section of the evaluation process shall apply to the entire building.

Where the separation between the mixed occupancies qualifies for any category indicated in Section 1301.6.16, the score for each occupancy shall apply to each portion of the building based on the occupancy of the space.

[B] 1301.6.1 Building height. The value for building height shall be the lesser value determined by the formula in Section 1301.6.1.1. Chapter 5 of the *International Building Code*, including allowable increases due to automatic sprinklers as provided for in Section 504.2 of the *International Building Code*, shall be used to determine the allowable height of the building. Subtract the actual building height from the allowable height and divide by $12^{1}/_{2}$ feet (3810 mm). Enter the height value and its sign (positive or negative) in Table 1301.7 under Safety Parameter 1301.6.1, Building Height, for fire safety, means of egress, and general safety. The maximum score for a building shall be 10.

[B] 1301.6.1.1 Height formula. The following formulas shall be used in computing the building height value.

$$\text{Height value, feet} = \frac{(AH)-(EBH)}{12.5} \times CF$$

(Equation (13-1)

$$\text{Height value, stories} = (AS - EBS) \times CF$$

(Equation 13-2)

where:

AH = Allowable height in feet (mm) from Table 503 of the *International Building Code*.

EBH = Existing building height in feet (mm).

AS = Allowable height in stories from Table 503 of the *International Building Code*.

EBS = Existing building height in stories.

CF = 1 if *(AH)* – *(EBH)* is positive.

CF = Construction-type factor shown in Table 1301.6.6(2) if *(AH)* – *(EBH)* is negative.

Note: Where mixed occupancies are separated and individually evaluated as indicated in Section 1301.6, the values *AH, AS, EBH,* and *EBS* shall be based on the height of the occupancy being evaluated.

[B] 1301.6.2 Building area. The value for building area shall be determined by the formula in Section 1301.6.2.2. Section 503 of the *International Building Code* and the formula in Section 1301.6.2.1 shall be used to determine the allowable area of the building. This shall include any allowable increases due to frontage and automatic sprinklers as provided for in Section 506 of the *International Building Code*. Subtract the actual building area from the allowable area and divide by 1,200 square feet (112 m^2). Enter the area value and its sign (positive or negative) in Table 1301.7 under Safety Parameter 1301.6.2, Building Area, for fire safety, means of egress and general safety. In determining the area value, the maximum permitted positive value for area is 50 percent of the fire safety score as listed in Table 1301.8, Mandatory Safety Scores.

[B] 1301.6.2.1 Allowable area formula. The following formula shall be used in computing allowable area:

$$A_a = (1 + I_f + I_s) \times A_t$$

(Equation 13-3)

where:

A_a = Allowable area.

A_t = Tabular area per story in accordance with Table 503 (square feet) of the *International Building Code*.

I_s = Area increase factor for sprinklers (Section 506.3 of the *International Building Code*).

I_f = Area increase factor for frontage (Section 506.2 of the *International Building Code*).

[B] 1301.6.2.2 Area formula. The following formula shall be used in computing the area value. Determine the area value for each occupancy floor area on a floor-by-floor basis. For each occupancy, choose the minimum area value of the set of values obtained for the particular occupancy.

$$\text{Area value}_i = \frac{\text{Allowable area}_i}{1200\text{ square feet}}\left[1 - \left(\frac{\text{Actual area}_i}{\text{Allowable area}_i} + \ldots + \frac{\text{Actual area}_n}{\text{Allowable area}_n}\right)\right]$$

(Equation 13-4)

where:

i = Value for an individual separated occupancy on a floor.

n = Number of separated occupancies on a floor.

[B] 1301.6.3 Compartmentation. Evaluate the compartments created by fire barriers or horizontal assemblies which comply with Sections 1301.6.3.1 and 1301.6.3.2 and which are exclusive of the wall elements considered under Sections 1301.6.4 and 1301.6.5. Conforming compartments shall be figured as the net area and do not include shafts, chases, stairways, walls, or columns. Using Table 1301.6.3, determine the appropriate compartmentation value (CV) and enter that value into Table 1301.7 under Safety Parameter 1301.6.3, Compartmentation, for fire safety, means of egress, and general safety.

[B] 1301.6.3.1 Wall construction. A wall used to create separate compartments shall be a fire barrier conforming to Section 707 of the *International Building Code* with a fire-resistance rating of not less than 2 hours. Where the building is not divided into more than one compartment, the compartment size shall be taken as the total floor area on all floors. Where there is more than one compartment within a story, each compartmented area on such story shall be provided with a horizontal exit conforming to Section 1025 of the *International Building Code*. The fire door serving as the horizontal exit between compartments shall be so installed, fitted, and gasketed that such fire door will provide a substantial barrier to the passage of smoke.

[B] 1301.6.3.2 Floor/ceiling construction. A floor/ceiling assembly used to create compartments shall conform to Section 712 of the *International Building Code* and shall have a fire-resistance rating of not less than 2 hours.

[B] 1301.6.4 Tenant and dwelling unit separations. Evaluate the fire-resistance rating of floors and walls separating tenants, including dwelling units, and not evaluated under Sections 1301.6.3 and 1301.6.5. Under the categories and occupancies in Table 1301.6.4, determine the appropriate value and enter that value in Table 1301.7 under Safety Parameter 1301.6.4, Tenant and Dwelling Unit Separation, for fire safety, means of egress, and general safety.

TABLE 1301.6.4
SEPARATION VALUES

OCCUPANCY	CATEGORIES				
	a	b	c	d	e
A-1	0	0	0	0	1
A-2	-5	-3	0	1	3
R	-4	-2	0	2	4
A-3, A-4, B, E, F, M, S-1	-4	-3	0	2	4
S-2	-5	-2	0	2	4

[B] 1301.6.4.1 Categories. The categories for tenant and dwelling unit separations are:

1. Category a—No fire partitions; incomplete fire partitions; no doors; doors not self-closing or automatic closing.
2. Category b—Fire partitions or floor assembly less than 1-hour fire-resistance rating or not constructed in accordance with Section 709 or 712 of the *International Building Code*, respectively.
3. Category c—Fire partitions with 1-hour or greater fire-resistance rating constructed in accordance with Section 709 of the *International Building Code* and floor assemblies with 1-hour but less than 2-hour fire-resistance rating constructed in accordance with Section 712 of the *International Building Code* or with only one tenant within the floor area.
4. Category d—Fire barriers with 1-hour but less than 2-hour fire-resistance rating constructed in accordance with Section 707 of the *International Building Code* and floor assemblies with 2-hour or greater fire-resistance rating constructed in accor-

TABLE 1301.6.3
COMPARTMENTATION VALUES

OCCUPANCY	CATEGORIES				
	a Compartment size equal to or greater than 15,000 square feet	b Compartment size of 10,000 square feet	c Compartment size of 7,500 square feet	d Compartment size of 5,000 square feet	e Compartment size of 2,500 square feet or less
A-1, A-3	0	6	10	14	18
A-2	0	4	10	14	18
A-4, B, E, S-2	0	5	10	15	20
F, M, R, S-1	0	4	10	16	22

For SI: 1 square foot = 0.0929 m^2.

dance with Section 712 of the *International Building Code*.

5. Category e—Fire barriers and floor assemblies with 2-hour or greater fire-resistance rating and constructed in accordance with Sections 707 and 712 of the *International Building Code*, respectively.

[B] 1301.6.5 Corridor walls. Evaluate the fire-resistance rating and degree of completeness of walls which create corridors serving the floor and that are constructed in accordance with Section 1018 of the *International Building Code*. This evaluation shall not include the wall elements considered under Sections 1301.6.3 and 1301.6.4. Under the categories and groups in Table 1301.6.5, determine the appropriate value and enter that value into Table 1301.7 under Safety Parameter 1301.6.5, Corridor Walls, for fire safety, means of egress, and general safety.

TABLE 1301.6.5
CORRIDOR WALL VALUES

OCCUPANCY	CATEGORIES			
	a	b	c[a]	d[a]
A-1	-10	-4	0	2
A-2	-30	-12	0	2
A-3, F, M, R, S-1	-7	-3	0	2
A-4, B, E, S-2	-5	-2	0	5

a. Corridors not providing at least one-half the travel distance for all occupants on a floor shall use Category b.

[B] 1301.6.5.1 Categories. The categories for corridor walls are:

1. Category a—No fire partitions; incomplete fire partitions; no doors; or doors not self-closing.
2. Category b—Less than 1-hour fire-resistance rating or not constructed in accordance with Section 709.4 of the *International Building Code*.
3. Category c—1-hour to less than 2-hour fire-resistance rating, with doors conforming to Section 715 of the *International Building Code* or without corridors as permitted by Section 1018 of the *International Building Code*.
4. Category d—2-hour or greater fire-resistance rating, with doors conforming to Section 715 of the *International Building Code*.

[B] 1301.6.6 Vertical openings. Evaluate the fire-resistance rating of exit enclosures, hoistways, escalator openings, and other shaft enclosures within the building, and openings between two or more floors. Table 1301.6.6(1) contains the appropriate protection values. Multiply that value by the construction type factor found in Table 1301.6.6(2). Enter the vertical opening value and its sign (positive or negative) in Table 1301.7 under Safety Parameter 1301.6.6, Vertical Openings, for fire safety, means of egress, and general safety. If the structure is a one-story building or if all the unenclosed vertical openings within the building conform to the requirements of Section 708 of the *International Building Code*, enter a value of 2. The maximum positive value for this requirement shall be 2.

TABLE 1301.6.6(1)
VERTICAL OPENING PROTECTION VALUE

PROTECTION	VALUE
None (unprotected opening)	-2 times number of floors connected
Less than 1 hour	-1 times number of floors connected
1 to less than 2 hours	1
2 hours or more	2

TABLE 1301.6.6(2)
CONSTRUCTION-TYPE FACTOR

FACTOR	TYPE OF CONSTRUCTION								
	IA	IB	IIA	IIB	IIIA	IIIB	IV	VA	VB
	1.2	1.5	2.2	3.5	2.5	3.5	2.3	3.3	7

[B] 1301.6.6.1 Vertical opening formula. The following formula shall be used in computing vertical opening value.

$$VO = PV \times CF$$ **(Equation 13-5)**

where:

VO = Vertical opening value.

PV = Protection value from Table 1301.6.6.(1).

CF = Construction type factor from Table 1301.6.6.(2).

[B] 1301.6.7 HVAC systems. Evaluate the ability of the HVAC system to resist the movement of smoke and fire beyond the point of origin. Under the categories in Section 1301.6.7.1, determine the appropriate value and enter that value into Table 1301.7 under Safety Parameter 1301.6.7, HVAC Systems, for fire safety, means of egress, and general safety.

[B] 1301.6.7.1 Categories. The categories for HVAC systems are:

1. Category a—Plenums not in accordance with Section 602 of the *International Mechanical Code*. -10 points.
2. Category b—Air movement in egress elements not in accordance with Section 1018.5 of the *International Building Code*. -5 points.
3. Category c—Both Categories a and b are applicable. -15 points.
4. Category d—Compliance of the HVAC system with Section 1018.5 of the *International Building Code* and Section 602 of the *International Mechanical Code*. 0 points.
5. Category e—Systems serving one story; or a central boiler/chiller system without ductwork connecting two or more stories. +5 points.

[B] 1301.6.8 Automatic fire detection. Evaluate the smoke detection capability based on the location and operation of automatic fire detectors in accordance with Section 907 of the *International Building Code* and the *International Mechanical Code*. Under the categories and occupancies in Table 1301.6.8, determine the appropriate value and enter that value into Table 1301.7 under Safety Parameter 1301.6.8, Automatic Fire Detection, for fire safety, means of egress, and general safety.

TABLE 1301.6.8
AUTOMATIC FIRE DETECTION VALUES

OCCUPANCY	CATEGORIES				
	a	b	c	d	e
A-1, A-3, F, M, R, S-1	-10	-5	0	2	6
A-2	-25	-5	0	5	9
A-4, B, E, S-2	-4	-2	0	4	8

[B] 1301.6.8.1 Categories. The categories for automatic fire detection are:

1. Category a—None.
2. Category b—Existing smoke detectors in HVAC systems and maintained in accordance with the *International Fire Code*.
3. Category c—Smoke detectors in HVAC systems. The detectors are installed in accordance with the requirements for new buildings in the *International Mechanical Code*.
4. Category d—Smoke detectors throughout all floor areas other than individual sleeping units, tenant spaces and dwelling units.
5. Category e—Smoke detectors installed throughout the floor area.

[B] 1301.6.9 Fire alarm systems. Evaluate the capability of the fire alarm system in accordance with Section 907 of the *International Building Code*. Under the categories and occupancies in Table 1301.6.9, determine the appropriate value and enter that value into Table 1301.7 under Safety Parameter 1301.6.9, Fire Alarm System, for fire safety, means of egress, and general safety.

TABLE 1301.6.9
FIRE ALARM SYSTEM VALUES

OCCUPANCY	CATEGORIES			
	a	b[a]	c	d
A-1, A-2, A-3, A-4, B, E, R	-10	-5	0	5
F, M, S	0	5	10	15

a. For buildings equipped throughout with an automatic sprinkler system, add 2 points for activation by a sprinkler water-flow device.

[B] 1301.6.9.1 Categories. The categories for fire alarm systems are:

1. Category a—None.
2. Category b—Fire alarm system with manual fire alarm boxes in accordance with Section 907.3 of the *International Building Code* and alarm notification appliances in accordance with Section 907.5.2 of the *International Building Code*.
3. Category c—Fire alarm system in accordance with Section 907 of the *International Building Code*.
4. Category d—Category c plus a required emergency voice/alarm communications system and a fire command station that conforms to Section 403.4.5 of the *International Building Code* and contains the emergency voice/alarm communications system controls, fire department communication system controls, and any other controls specified in Section 911 of the *International Building Code* where those systems are provided.

[B] 1301.6.10 Smoke control. Evaluate the ability of a natural or mechanical venting, exhaust, or pressurization system to control the movement of smoke from a fire. Under the categories and occupancies in Table 1301.6.10, determine the appropriate value and enter that value into Table 1301.7 under Safety Parameter 1301.6.10, Smoke Control, for means of egress and general safety.

TABLE 1301.6.10
SMOKE CONTROL VALUES

OCCUPANCY	CATEGORIES					
	a	b	c	d	e	f
A-1, A-2, A-3	0	1	2	3	6	6
A-4, E	0	0	0	1	3	5
B, M, R	0	2[a]	3[a]	3[a]	3[a]	4[a]
F, S	0	2[a]	2[a]	3[a]	3[a]	3[a]

a. This value shall be 0 if compliance with Category d or e in Section 1301.6.8.1 has not been obtained.

[B] 1301.6.10.1 Categories. The categories for smoke control are:

1. Category a—None.
2. Category b—The building is equipped throughout with an automatic sprinkler system. Openings are provided in exterior walls at the rate of 20 square feet (1.86 m^2) per 50 linear feet (15 240 mm) of exterior wall in each story and distributed around the building perimeter at intervals not exceeding 50 feet (15 240 mm). Such openings shall be readily openable from the inside without a key or separate tool and shall be provided with ready access thereto. In lieu of operable openings, clearly and permanently marked tempered glass panels shall be used.
3. Category c—One enclosed exit stairway, with ready access thereto, from each occupied floor of the building. The stairway has operable exterior windows, and the building has openings in accordance with Category b.
4. Category d—One smokeproof enclosure and the building has openings in accordance with Category b.
5. Category e—The building is equipped throughout with an automatic sprinkler system. Each floor area is provided with a mechanical air-handling

system designed to accomplish smoke containment. Return and exhaust air shall be moved directly to the outside without recirculation to other floor areas of the building under fire conditions. The system shall exhaust not less than six air changes per hour from the floor area. Supply air by mechanical means to the floor area is not required. Containment of smoke shall be considered as confining smoke to the floor area involved without migration to other floor areas. Any other tested and approved design that will adequately accomplish smoke containment is permitted.

6. Category f—Each stairway shall be one of the following: a smokeproof enclosure in accordance with Section 1022.9 of the *International Building Code*; pressurized in accordance with Section 909.20.5 of the *International Building Code*; or shall have operable exterior windows.

[B] 1301.6.11 Means-of-egress capacity and number. Evaluate the means-of-egress capacity and the number of exits available to the building occupants. In applying this section, the means of egress are required to conform to the following sections of the *International Building Code*: 1003.7, 1004, 1005.1, 1014.2, 1014.3, 1015.2, 1021, 1025.1, 1027.2, 1027.6, 1028.2, 1028.3, 1028.4 and 1029. [except that the minimum width required by this section shall be determined solely by the width for the required capacity in accordance with Table 1301.6.11(1)]. The number of exits credited is the number that is available to each occupant of the area being evaluated. Existing fire escapes shall be accepted as a component in the means of egress when conforming to Section 705.3.1.2. Under the categories and occupancies in Table 1301.6.11(2), determine the appropriate value and enter that value into Table 1301.7 under Safety Parameter 1301.6.11, Means-of-Egress Capacity, for means of egress and general safety.

[B] 1301.6.11.1 Categories. The categories for means-of-egress capacity and number of exits are:

1. Category a—Compliance with the minimum required means-of-egress capacity or number of exits is achieved through the use of a fire escape in accordance with Section 305.

2. Category b—Capacity of the means of egress complies with Section 1004 of the *International Building Code*, and the number of exits complies with the minimum number required by Section 1021 of the *International Building Code*.

3. Category c—Capacity of the means of egress is equal to or exceeds 125 percent of the required means-of-egress capacity, the means of egress complies with the minimum required width dimensions specified in the *International Building Code*, and the number of exits complies with the minimum number required by Section 1021of the *International Building Code*.

4. Category d—The number of exits provided exceeds the number of exits required by Section 1021 of the *International Building Code*. Exits shall be located a distance apart from each other equal to not less than that specified in Section 1015.2 of the *International Building Code*.

5. Category e—The area being evaluated meets both Categories c and d.

TABLE 1301.6.11(1)
EGRESS WIDTH PER OCCUPANT SERVED

OCCUPANCY	WITHOUT SPRINKLER SYSTEM		WITH SPRINKLER SYSTEM[a]	
	Stairways (inches per occupancy)	Other egress components (inches per occupant)	Stairways (inches per occupant)	Other egress components (inches per occupant)
Occupancies other than those listed below	0.3	0.2	0.2	0.15
Hazardous: H-1, H-2, H-3, H-4	Not permitted	Not permitted	0.3	0.2
Institutional: I-2	Not permitted	Not permitted	0.3	0.2

For SI: 1 inch = 25.4 mm.

a. Buildings equipped throughout with an automatic sprinkler system in accordance with Section 903.3.1.1 or 903.3.1.2 of the *International Building Code*.

TABLE 1301.6.11(2)
MEANS OF EGRESS VALUES

OCCUPANCY	CATEGORIES				
	a[a]	b	c	d	e
A-1, A-2, A-3, A-4, E	-10	0	2	8	10
M	-3	0	1	2	4
B, F, S	-1	0	0	0	0
R	-3	0	0	0	0

a. The values indicated are for buildings six stories or less in height. For buildings over six stories above grade plane, add an additional -10 points.

[B] 1301.6.12 Dead ends. In spaces required to be served by more than one means of egress, evaluate the length of the exit access travel path in which the building occupants are confined to a single path of travel. Under the categories and occupancies in Table 1301.6.12, determine the appropriate value and enter that value into Table 1301.7 under Safety Parameter 1301.6.12, Dead Ends, for means of egress and general safety.

TABLE 1301.6.12
DEAD-END VALUES

OCCUPANCY	CATEGORIES[a]		
	a	b	c
A-1, A-3, A-4, B, F, M, R, S	-2	0	2
A-2, E	-2	0	2

a. For dead-end distances between categories, the dead end value shall be obtained by linear interpolation.

[B] 1301.6.12.1 Categories. The categories for dead ends are:

1. Category a—Dead end of 35 feet (10 670 mm) in nonsprinklered buildings or 70 feet (21 340 mm) in sprinklered buildings.

2. Category b—Dead end of 20 feet (6096 mm); or 50 feet (15 240 mm) in Group B in accordance with Section 1018.4, Exception 2 of the *International Building Code*.

3. Category c—No dead ends; or ratio of length to width (*l/w*) is less than 2.5:1.

[B] 1301.6.13 Maximum exit access travel distance to an exit. Evaluate the length of exit access travel to an approved exit. Determine the appropriate points in accordance with the following equation and enter that value into Table 1301.7 under Safety Parameter 1301.6.13, Maximum Exit Access Travel Distance for means of egress and general safety. The maximum allowable exit access travel distance shall be determined in accordance with Section 1016.1 of the *International Building Code*.

$$\text{Points} = 20 \times \frac{\text{Maximum allowable travel distance} - \text{Maximum actual travel distance}}{\text{Maximum allowable travel distance}}$$

(Equation 13-6)

[B] 1301.6.14 Elevator control. Evaluate the passenger elevator equipment and controls that are available to the fire department to reach all occupied floors. Elevator recall controls shall be provided in accordance with the *International Fire Code*. Under the categories an occupancies in Table 1301.6.14, determine the appropriate value and enter that value into Table 1301.7 under Safety Parameter 1301.6.14, Elevator Control, for fire safety, means of egress, and general safety. The values shall be zero for a single story building.

TABLE 1301.6.14
ELEVATOR CONTROL VALUES

ELEVATOR TRAVEL	CATEGORIES			
	a	b	c	d
Less than 25 feet of travel above or below the primary level of elevator access for emergency fire-fighting or rescue personnel	-2	0	0	+2
Travel of 25 feet or more above or below the primary level of elevator access for emergency fire-fighting or rescue personnel	-4	NP	0	+4

For SI: 1 foot = 304.8 mm.
NP = Not permitted.

[B] 1301.6.14.1 Categories. The categories for elevator controls are:

1. Category a—No elevator.
2. Category b—Any elevator without Phase I and II recall.
3. Category c—All elevators with Phase I and II recall as required by the *International Fire Code*.
4. Category d—All meet Category c; or Category b where permitted to be without recall; and at least one elevator that complies with new construction requirements serves all occupied floors.

[B] 1301.6.15 Means-of-egress emergency lighting. Evaluate the presence of and reliability of means-of-egress emergency lighting. Under the categories and occupancies in Table 1301.6.15, determine the appropriate value and enter that value into Table 1301.7 under Safety Parameter 1301.6.15, Means-of-Egress Emergency Lighting, for means of egress and general safety.

TABLE 1301.6.15
MEANS-OF-EGRESS EMERGENCY LIGHTING VALUES

NUMBER OF EXITS REQUIRED BY SECTION 1015 OF THE *INTERNATIONAL BUILDING CODE*	CATEGORIES		
	a	b	c
Two or more exits	NP	0	4
Minimum of one exit	0	1	1

NP = Not permitted.

[B] 1301.6.15.1 Categories. The categories for means-of-egress emergency lighting are:

1. Category a—Means-of-egress lighting and exit signs not provided with emergency power in accordance with Section 2702 of the *International Building Code*.
2. Category b—Means-of-egress lighting and exit signs provided with emergency power in accordance with Section 2702 of the *International Building Code*.
3. Category c—Emergency power provided to means-of- egress lighting and exit signs, which provides protection in the event of power failure to the site or building.

[B] 1301.6.16 Mixed occupancies. Where a building has two or more occupancies that are not in the same occupancy classification, the separation between the mixed occupancies shall be evaluated in accordance with this section. Where there is no separation between the mixed occupancies or the separation between mixed occupancies does not qualify for any of the categories indicated in Section 1301.6.16.1, the building shall be evaluated as indicated in Section 1301.6, and the value for mixed occupancies shall be zero. Under the categories and occupancies in Table 1301.6.16, determine the appropriate value and enter that value into Table 1301.7 under Safety Parameter 1301.6.16, Mixed Occupancies, for fire safety and general safety. For buildings without mixed occupancies, the value shall be zero.

TABLE 1301.6.16
MIXED OCCUPANCY VALUES[a]

OCCUPANCY	CATEGORIES		
	a	b	c
A-1, A-2, R	-10	0	10
A-3, A-4, B, E, F, M, S	-5	0	5

a. For fire-resistance ratings between categories, the value shall be obtained by linear interpolation.

[B] 1301.6.16.1 Categories. The categories for mixed occupancies are:

1. Category a—Occupancies separated by minimum 1-hour fire barriers or minimum 1-hour horizontal assemblies, or both.

2. Category b—Separations between occupancies in accordance with Section 508.4 of the *International Building Code*.

3. Category c—Separations between occupancies having a fire-resistance rating of not less than twice that required by Section 508.4 of the *International Building Code*.

[B] 1301.6.17 Automatic sprinklers. Evaluate the ability to suppress a fire based on the installation of an automatic sprinkler system in accordance with Section 903.3.1.1 of the *International Building Code*. "Required sprinklers" shall be based on the requirements of this code. Under the categories and occupancies in Table 1301.6.17, determine the appropriate value and enter that value into Table 1301.7 under Safety Parameter 1301.6.17, Automatic Sprinklers, for fire safety, means of egress divided by 2, and general safety. High-rise buildings defined in Section 403.1 of the *International Building Code* that undergo a *change of occupancy* to Group R shall be equipped throughout with an automatic sprinkler system in accordance with Section 403.2 of the *International Building Code* and Chapter 9 of the *International Building Code*.

TABLE 1301.6.17
SPRINKLER SYSTEM VALUES

OCCUPANCY	CATEGORIES					
	a[a]	b[a]	c	d	e	f
A-1, A-3, F, M, R, S-1	-6	-3	0	2	4	6
A-2	-4	-2	0	1	2	4
A-4, B, E, S-2	-12	-6	0	3	6	12

a. These options cannot be taken if Category a in Section 1301.6.18 is used.

[B] 1301.6.17.1 Categories. The categories for automatic sprinkler system protection are:

1. Category a—Sprinklers are required throughout; sprinkler protection is not provided or the sprinkler system design is not adequate for the hazard protected in accordance with Section 903 of the *International Building Code*.

2. Category b—Sprinklers are required in a portion of the building; sprinkler protection is not provided or the sprinkler system design is not adequate for the hazard protected in accordance with Section 903 of the *International Building Code*.

3. Category c—Sprinklers are not required; none are provided.

4. Category d—Sprinklers are required in a portion of the building; sprinklers are provided in such portion; the system is one that complied with the code at the time of installation and is maintained and supervised in accordance with Section 903 of the *International Building Code*.

5. Category e—Sprinklers are required throughout; sprinklers are provided throughout in accordance with Chapter 9 of the *International Building Code*.

6. Category f—Sprinklers are not required throughout; sprinklers are provided throughout in accordance with Chapter 9 of the *International Building Code*.

[B] 1301.6.18 Standpipes. Evaluate the ability to initiate attack on a fire by making a supply of water available readily through the installation of standpipes in accordance with Section 905 of the *International Building Code*. "Required Standpipes" shall be based on the requirements of the *International Building Code*. Under the categories and occupancies in Table 1301.6.18, determine the appropriate value and enter that value into Table 1301.7 under Safety Parameter 1301.6.18, Standpipes, for fire safety, means of egress, and general safety.

[B] 1301.6.18.1 Standpipe catagories. The categories for standpipe systems are:

1. Category a—Standpipes are required; standpipe is not provided or the standpipe system design is not in compliance with Section 905.3 of the *International Building Code*.

2. Category b—Standpipes are not required; none are provided.

3. Category c—Standpipes are required; standpipes are provided in accordance with Section 905 of the *International Building Code*.

4. Category d—Standpipes are not required; standpipes are provided in accordance with Section 905 of the *International Building Code*.

TABLE 1301.6.18
STANDPIPE SYSTEM VALUES

OCCUPANCY	CATEGORIES			
	a[a]	b	c	d
A-1, A-3, F, M, R, S-1	-6	0	4	6
A-2	-4	0	2	4
A-4, B, E, S-2	-12	0	6	12

a. This option cannot be taken if Category a or Category b in Section 1301.6.17 is used.

[B] 1301.6.19 Incidental accessory occupancy. Evaluate the protection of incidental accessory occupancies in accordance with Section 508.2.5 of the *International Building Code*. Do not include those where this code requires suppression throughout the building including covered mall buildings, high-rise buildings, public garages and unlimited area buildings. Assign the lowest score from Table 1301.6.19 for the building or floor area being evaluated and enter that value into Table 1301.7 under Safety Parameter 1301.6.19, Incidental Accessory Occupancy, for fire safety, means of egress and general safety. If there are no specific occupancy areas in the building or floor area being evaluated, the value shall be zero.

[B] 1301.7 Building score. After determining the appropriate data from Section 1301.6, enter those data in Table 1301.7 and total the building score.

[B] 1301.8 Safety scores. The values in Table 1301.8 are the required mandatory safety scores for the evaluation process listed in Section 1301.6.

[B] 1301.9 Evaluation of building safety. The mandatory safety score in Table 1301.8 shall be subtracted from the building score in Table 1301.7 for each category. Where the final score for any category equals zero or more, the building is in compliance with the requirements of this section for that category. Where the final score for any category is less than zero, the building is not in compliance with the requirements of this section.

[B] 1301.9.1 Mixed occupancies. For mixed occupancies, the following provisions shall apply:

1. Where the separation between mixed occupancies does not qualify for any category indicated in Section 1301.6.16, the mandatory safety scores for the occupancy with the lowest general safety score in Table 1301.8 shall be utilized. (See Section 1301.6.)
2. Where the separation between mixed occupancies qualifies for any category indicated in Section 1301.6.16, the mandatory safety scores for each occupancy shall be placed against the evaluation scores for the appropriate occupancy.

TABLE 1301.6.19
INCIDENTAL ACCESSORY OCCUPANCY VALUES[a]

PROTECTION REQUIRED BY TABLE 508.2.5 OF THE *INTERNATIONAL BUILDING CODE*	PROTECTION PROVIDED						
	None	1 hour	AFSS	AFSS with SP	1 hour and AFSS	2 hours	2 hours and AFSS
2 hours and AFSS	-4	-3	-2	-2	-1	-2	0
2 hours, or 1 hour and AFSS	-3	-2	-1	-1	0	0	0
1 hour and AFSS	-3	-2	-1	-1	0	-1	0
1 hour	-1	0	-1	-1	0	0	
1 hour, or AFSS with SP	-1	0	-1	-1	0	0	0
AFSS with SP	-1	-1	-1	-1	0	-1	0
1 hour or AFSS	-1	0	0	0	0	0	0

a. AFSS = Automatic fire suppression system; SP = Smoke partitions (See IBC Section 508.2.5).
Note: For Table 1301.7, see page 68.

TABLE 1301.7
SUMMARY SHEET—BUILDING CODE

Existing occupancy ______________________ Proposed occupancy ______________________

Year building was constructed ______________________ Number of stories ________ Height in feet ________

Type of construction ______________________ Area per floor ______________________

Percentage of open perimeter increase _______%

Completely suppressed: Yes _______ No _______ Corridor wall rating ______________________

Compartmentation: Yes _______ No _______ Required door closers: Yes _______ No _______

Fire-resistance rating of vertical opening enclosures ______________________

Type of HVAC system ______________________, serving number of floors ______________________

Automatic fire detection: Yes _______ No _______, Type and location ______________________

Fire alarm system: Yes _______ No _______, Type ______________________

Smoke control: Yes _______ No _______, Type ______________________

Adequate exit routes: Yes _______ No _______ Dead ends: ______________ Yes _______ No _______

Maximum exit access travel distance ______________________ Elevator controls: Yes _______ No _______

Means of egress emergency lighting: Yes _______ No _______ Mixed occupancies: Yes _______ No _______

SAFETY PARAMETERS	FIRE SAFETY (FS)	MEANS OF EGRESS (ME)	GENERAL SAFETY (GS)
1301.6.1 Building Height 1301.6.2 Building Area 1301.6.3 Compartmentation			
1301.6.4 Tenant and Dwelling Unit Separations 1301.6.5 Corridor Walls 1301.6.6 Vertical Openings			
1301.6.7 HVAC Systems 1301.6.8 Automatic Fire Detection 1301.6.9 Fire Alarm System			
1301.6.10 Smoke control 1301.6.11 Means of Egress 1301.6.12 Dead ends	* * * * * * * * * * * *		
1301.6.13 Maximum Exit Access Travel Distance 1301.6.14 Elevator Control 1301.6.15 Means of Egress Emergency Lighting	* * * * * * * *		
1301.6.16 Mixed Occupancies 1301.6.17 Automatic Sprinklers 1301.6.18 Standpipes 1301.6.19 Incidental Accessory Occupancy		* * * * ÷ 2 =	
Building score — total value			

* * * *No applicable value to be inserted.

TABLE 1301.8
MANDATORY SAFETY SCORES[a]

OCCUPANCY	FIRE SAFETY (MFS)	MEANS OF EGRESS (MME)	GENERAL SAFETY (MGS)
A-1	20	31	31
A-2	21	32	32
A-3	22	33	33
A-4, E	29	40	40
B	30	40	40
F	24	34	34
M	23	40	40
R	21	38	38
S-1	19	29	29
S-2	29	39	39

a. MFS = Mandatory Fire Safety
MME = Mandatory Means of Egress
MGS = Mandatory General Safety

TABLE 1301.9
EVALUATION FORMULAS[a]

FORMULA	T1201.7	T1201.8		SCORE	PASS	FAIL
FS - MFS > 0	______ (FS) -	______ (MFS)	=	______	______	______
ME - MME ≥ 0	______ (ME) -	______ (MME)	=	______	______	______
GS - MGS ≥ 0	______ (GS -	______ (MGS)	=	______	______	______

a. FS = Fire Safety; MFS = Mandatory Fire Safety
ME = Means of Egress; MME = Mandatory Means of Egress
GS = General Safety; MGS = Mandatory General Safety

CHAPTER 14
CONSTRUCTION SAFEGUARDS

SECTION 1401
GENERAL

[B] 1401.1 Scope. The provisions of this chapter shall govern safety during construction that is under the jurisdiction of this code and the protection of adjacent public and private properties.

[B] 1401.2 Storage and placement. Construction equipment and materials shall be stored and placed so as not to endanger the public, the workers or adjoining property for the duration of the construction project.

1401.3 Alterations, repairs, and additions. Required exits, existing structural elements, fire protection devices, and sanitary safeguards shall be maintained at all times during alterations, repairs, or additions to any building or structure.

Exceptions:

1. When such required elements or devices are being altered or repaired, adequate substitute provisions shall be made.
2. When the existing building is not occupied.

[B] 1401.4 Manner of removal. Waste materials shall be removed in a manner which prevents injury or damage to persons, adjoining properties, and public rights-of-way.

[B] 1401.5 Facilities required. Sanitary facilities shall be provided during construction or demolition activities in accordance with the *International Plumbing Code*.

[B] 1401.6 Protection of pedestrians. Pedestrians shall be protected during construction and demolition activities as required by Sections 1401.6.1 through 1401.6.7 and Table 1401.6. Signs shall be provided to direct pedestrian traffic.

[B] 1401.6.1 Walkways. A walkway shall be provided for pedestrian travel in front of every construction and demolition site unless the applicable governing authority authorizes the sidewalk to be fenced or closed. Walkways shall be of sufficient width to accommodate the pedestrian traffic, but in no case shall they be less than 4 feet (1219 mm) in width. Walkways shall be provided with a durable walking surface. Walkways shall be accessible in accordance with Chapter 11 of the *International Building Code* and shall be designed to support all imposed loads and in no case shall the design live load be less than 150 pounds per square foot (psf) (7.2 kN/m^2).

[B] 1401.6.2 Directional barricades. Pedestrian traffic shall be protected by a directional barricade where the walkway extends into the street. The directional barricade shall be of sufficient size and construction to direct vehicular traffic away from the pedestrian path.

[B] 1401.6.3 Construction railings. Construction railings shall be at least 42 inches (1067 mm) in height and shall be sufficient to direct pedestrians around construction areas.

[B] 1401.6.4 Barriers. Barriers shall be a minimum of 8 feet (2438 mm) in height and shall be placed on the side of the walkway nearest the construction. Barriers shall extend the entire length of the construction site. Openings in such barriers shall be protected by doors which are normally kept closed.

[B] 1401.6.4.1 Barrier design. Barriers shall be designed to resist loads required in Chapter 16 of the *International Building Code* unless constructed as follows:

1. Barriers shall be provided with 2 × 4 top and bottom plates.
2. The barrier material shall be a minimum of $^3/_4$ inch (19.1 mm) inch boards or $^1/_4$ inch (6.4 mm) wood structural use panels.

[B] TABLE 1401.6
PROTECTION OF PEDESTRIANS

HEIGHT OF CONSTRUCTION	DISTANCE OF CONSTRUCTION TO LOTLINE	TYPE OF PROTECTION REQUIRED
8 feet or less	Less than 5 feet	Construction railings
	5 feet or more	None
More than 8 feet	Less than 5 feet	Barrier and covered walkway
	5 feet or more, but not more than one-fourth the height of construction	Barrier and covered walkway
	5 feet or more, but between one-fourth and one-half the height of construction	Barrier
	5 feet or more, but exceeding one-half the height of construction	None

For SI: 1 foot = 304.8 mm.

3. Wood structural use panels shall be bonded with an adhesive identical to that for exterior wood structural use panels.
4. Wood structural use panels $^1/_4$ inch (6.4 mm) or $^1/_{16}$ inch (23.8 mm) in thickness shall have studs spaced not more than 2 feet (610 mm) on center.
5. Wood structural use panels $^3/_8$ inch (9.5 mm) or $^1/_2$ inch (12.7 mm) in thickness shall have studs spaced not more than 4 feet (1219 mm) on center, provided a 2 inch by 4 inch (51 mm by 102 mm) stiffener is placed horizontally at the mid height where the stud spacing exceeds 2 feet (610 mm) on center.
6. Wood structural use panels $^5/_8$ inch (15.9 mm) or thicker shall not span over 8 feet (2438 mm).

[B] 1401.6.5 Covered walkways. Covered walkways shall have a minimum clear height of 8 feet (2438 mm) as measured from the floor surface to the canopy overhead. Adequate lighting shall be provided at all times. Covered walkways shall be designed to support all imposed loads. In no case shall the design live load be less than 150 psf (7.2 kN/m^2) for the entire structure.

Exception: Roofs and supporting structures of covered walkways for new, light-frame construction not exceeding two stories above grade plane are permitted to be designed for a live load of 75 psf (3.6 kN/m^2) or the loads imposed on them, whichever is greater. In lieu of such designs, the roof and supporting structure of a covered walkway are permitted to be constructed as follows:

1. Footings shall be continuous 2 × 6 members.
2. Posts not less than 4 × 6 shall be provided on both sides of the roof and spaced not more than 12 feet (3658 mm) on center.
3. Stringers not less than 4 × 12 shall be placed on edge upon the posts.
4. Joists resting on the stringers shall be at least 2 × 8 and shall be spaced not more than 2 feet (610 mm) on center.
5. The deck shall be planks at least 2 inches (51 mm) thick or wood structural panels with an exterior exposure durability classification at least $^{23}/_{32}$ inch (18.3 mm) thick nailed to the joists.
6. Each post shall be knee-braced to joists and stringers by 2 × 4 minimum members 4 feet (1219 mm) long.
7. A 2 × 4 minimum curb shall be set on edge along the outside edge of the deck.

[B] 1401.6.6 Repair, maintenance and removal. Pedestrian protection required by Section 1401.6 shall be maintained in place and kept in good order for the entire length of time pedestrians may be endangered. The owner or the owner's agent, upon the completion of the construction activity, shall immediately remove walkways, debris and other obstructions and leave such public property in as good a condition as it was before such work was commenced.

[B] 1401.6.7 Adjacent to excavations. Every excavation on a site located 5 feet (1524 mm) or less from the street lot line shall be enclosed with a barrier not less than 6 feet (1829 mm) high. Where located more than 5 feet (1524 mm) from the street lot line, a barrier shall be erected when required by the *code official*. Barriers shall be of adequate strength to resist wind pressure as specified in Chapter 16 of the *International Building Code*.

[B] SECTION 1402
PROTECTION OF ADJOINING PROPERTY

1402.1 Protection required. Adjoining public and private property shall be protected from damage during construction and demolition work. Protection must be provided for footings, foundations, party walls, chimneys, skylights and roofs. Provisions shall be made to control water run-off and erosion during construction or demolition activities. The person making or causing an excavation to be made shall provide written notice to the owners of adjoining buildings advising them that the excavation is to be made and that the adjoining buildings should be protected. Said notification shall be delivered not less than 10 days prior to the scheduled starting date of the excavation.

[B] SECTION 1403
TEMPORARY USE OF STREETS, ALLEYS AND PUBLIC PROPERTY

1403.1 Storage and handling of materials. The temporary use of streets or public property for the storage or handling of materials or equipment required for construction or demolition, and the protection provided to the public shall comply with the provisions of the applicable governing authority and this chapter.

1403.2 Obstructions. Construction materials and equipment shall not be placed or stored so as to obstruct access to fire hydrants, standpipes, fire or police alarm boxes, catch basins or manholes, nor shall such material or equipment be located within 20 feet (6.1 m) of a street intersection, or placed so as to obstruct normal observations of traffic signals or to hinder the use of public transit loading platforms.

1403.3 Utility fixtures. Building materials, fences, sheds or any obstruction of any kind shall not be placed so as to obstruct free approach to any fire hydrant, fire department connection, utility pole, manhole, fire alarm box, or catch basin, or so as to interfere with the passage of water in the gutter. Protection against damage shall be provided to such utility fixtures during the progress of the work, but sight of them shall not be obstructed.

SECTION 1404
FIRE EXTINGUISHERS

[F] 1404.1 Where required. All structures under construction, *alteration*, or demolition shall be provided with not less than one approved portable fire extinguisher in accordance with Section 906 of the *International Fire Code* and sized for not less than ordinary hazard as follows:

1. At each stairway on all floor levels where combustible materials have accumulated.
2. In every storage and construction shed.
3. Additional portable fire extinguishers shall be provided where special hazards exist including, but not limited to, the storage and use of flammable and combustible liquids.

[B] 1404.2 Fire hazards. The provisions of this code and of the *International Fire Code* shall be strictly observed to safeguard against all fire hazards attendant upon construction operations.

SECTION 1405
MEANS OF EGRESS

[B] 1405.1 Stairways required. Where a building has been constructed to a building height of 50 feet (15 240 mm) or four stories, or where an *existing building* exceeding 50 feet (15 240 mm) in building height is altered, at least one temporary lighted stairway shall be provided unless one or more of the permanent stairways are erected as the construction progresses.

[B] 1405.2 Maintenance of means of egress. Required means of egress shall be maintained at all times during construction, demolition, remodeling or *alterations* and *additions* to any building.

Exception: Approved temporary means of egress systems and facilities.

[F] SECTION 1406
STANDPIPE SYSTEMS

1406.1 Where required. In buildings required to have standpipes by Section 905.3.1, not less than one standpipe shall be provided for use during construction. Such standpipes shall be installed where the progress of construction is not more than 40 feet (12 192 mm) in height above the lowest level of fire department vehicle access. Such standpipe shall be provided with fire department hose connections at accessible locations adjacent to usable stairs. Such standpipes shall be extended as construction progresses to within one floor of the highest point of construction having secured decking or flooring.

1406.2 Buildings being demolished. Where a building or portion of a building is being demolished and a standpipe is existing within such a building, such standpipe shall be maintained in an operable condition so as to be available for use by the fire department. Such standpipe shall be demolished with the building but shall not be demolished more than one floor below the floor being demolished.

1406.3 Detailed requirements. Standpipes shall be installed in accordance with the provisions of Chapter 9 of the *International Building Code.*

Exception: Standpipes shall be either temporary or permanent in nature, and with or without a water supply, provided that such standpipes conform to the requirements of Section 905 of the *International Building Code* as to capacity, outlets and materials.

[F] SECTION 1407
AUTOMATIC SPRINKLER SYSTEM

1407.1 Completion before occupancy. In portions of a building where an automatic sprinkler system is required by this code, it shall be unlawful to occupy those portions of the building until the automatic sprinkler system installation has been tested and approved, except as provided in Section 110.3.

1407.2 Operation of valves. Operation of sprinkler control valves shall be permitted only by properly authorized personnel and shall be accompanied by notification of duly designated parties. When the sprinkler protection is being regularly turned off and on to facilitate connection of newly completed segments, the sprinkler control valves shall be checked at the end of each work period to ascertain that protection is in service.

SECTION 1408
ACCESSIBILITY

1408.1 Construction sites. Structures, sites, and equipment directly associated with the actual process of construction, including but not limited to scaffolding, bridging, material hoists, material storage, or construction trailers are not required to be accessible.

[F] SECTION 1409
WATER SUPPLY FOR FIRE PROTECTION

1409.1 When required. An approved water supply for fire protection, either temporary or permanent, shall be made available as soon as combustible material arrives on the site.

CHAPTER 15
REFERENCED STANDARDS

This chapter lists the standards that are referenced in various sections of this document. The standards are listed herein by the promulgating agency of the standard, the standard identification, the effective date and title, and the section or sections of this document that reference the standard. The application of the referenced standards shall be as specified in Section 102.4.

ASCE

American Society of Civil Engineers Structural Engineering Institute
1801 Alexander Bell Drive
Reston, VA 20191-4400

Standard reference number	Title	Referenced in code section number
7—05	Minimum Design Loads for Buildings and Other Structures with Supplement No. 1	101.5.4.1, A104, A506.1, A507.1
31—03	Seismic Evaluation of Existing Buildings	101.5.4, Table 101.5.4.1, 101.5.4.2
41—06	Seismic Rehabilitation of Existing Buildings	Table 101.5.4.1, 101.5.4.2

ASHRAE

American Society of Heating, Refrigerating and Air Conditioning Engineers
1791 Tullie Circle, NE
Atlanta, GA 30329

Standard reference number	Title	Referenced in code section number
62—04	Ventilation for Acceptable Indoor Air Quality	709.2

ASME

American Society of Mechanical Engineers
3 Park Avenue
New York, NY 10016

Standard reference number	Title	Referenced in code section number
A17.1/CSA B44—2007	Safety Code for Elevators and Escalators	310.8.2, 605.1.2, 802.1.2
A17.3—2002	Safety Code for Existing Elevators and Escalators	802.1.2
A18.1—2005	Safety Standard for Platform Lifts and Stairway Chair Lifts	310.8.3, 605.1.3

ASTM

ASTM International
100 Barr Harbor Drive
West Conshohocken, PA 19428-2959

Standard reference number	Title	Referenced in code section number
C 90—03	Standard Specification for Load-bearing Concrete Masonry Units	A505.2.3
C 496—96	Standard Test Method for Splitting Tensile Strength of Cylindrical Concrete Specimens	A104, A106.3.3.2
E 519—00e1	Standard Test Method for Diagonal Tension (Shear) in Masonry Assemblages	A104, A106.3.3.2

DOC

United States Department of Commerce
1401 Constitution Avenue, NW
Washington, DC 20230

Standard reference number	Title	Referenced in code section number
PS 1—07	Structural Plywood	A302
PS 2—04	Performance Standard for Wood-based Structural-use Panels	A302

ICC

International Code Council, Inc.
500 New Jersey Avenue, NW, 6th Floor
Washington, DC 20001

Standard reference number	Title	Referenced in code section number
IBC—09	International Building Code®	101, 106, 109, 110, 202, 302, 303, 304, 307, 310, 401, 501, 502, 506, 601, 602, 605, 606, 701, 702, 703, 704, 705, 706, 707, 804, 805, 807, 901, 902, 907, 911, 912, 1002, 1003, 1004, 1101, 1102, 1103, 1104, 1105, 1201, 1202, 1301, 1401, 1406
ICC A117.1—03	Guidelines for Accessible and Usable Buildings and Facilities	310.6, 310.8.2, 605.1, 605.1.2, 605.1.3,
IECC—09	International Energy Conservation Code®	602.4, 607.1, 711.1, 808.1,
IFC—09	International Fire Code®	101.4.2, 101.5.1, 703.2.1, 703.2.3, 704.4.1.1, 704.4.1.2, 704.4.1.3, 704.4.1.4, 704.4.1.5, 704.4.1.6, 70, 4.4.1.7, 704.4.3, 1201.2, 1301.3.2, 1301.6.8.1, 1301.6.14, 1301.6.14.1, 1404.1, 1404.2,
IFGC—09	International Fuel Gas Code®	307.7, 602.4.1
IMC—09	International Mechanical Code®	307.8, 602.4, 709.1, 802.1.1, 802.2.1, 909.1, 1301.6.7.1, 1301.6.8, 1301.6.8.1
IPC—09	International Plumbing Code®	307.9, 509.1, 602.4, 710.1, 910.2, 910.3, 910.5, 1401.5
IPMC—09	International Property Maintenance Code®	101.4.2, 1201.2, 1301.3.2
IRC—09	International Residential Code®	101.4.1, 502.3, 506, 606.2.1, 607.1, 707.4, 708.3, 711.1 807.4, 807.4.2, 808.1, 1003.2, 1003.3, 1003.4, 1004.1, 1202.1, 1202.2, 1202.2.1, 1202.3, 1202.4, 1202.5, 1301.2.2, 1301.2.3

NFPA

National Fire Protection Agency
1 Batterymarch Park
Quincy, MA 02269-9101

Standard reference number	Title	Referenced in code section number
NFPA 13R—07	Installation of Sprinkler Systems in Residential Occupancies up to and Including Four Stories in Height	704.2.5
NFPA 70—05	National Electrical Code	507.1.1, 507.1.2, 507.1.3, 507.1.4, 507.1.5
NFPA 72—07	National Fire Alarm Code	704.2.5, 704.4
NFPA 99—05	Health Care Facilities	507.1.4
NFPA 101—06	Life Safety Code	705.2

Appendix A: Guidelines for the Seismic Retrofit of Existing Buildings
CHAPTER A1
SEISMIC STRENGTHENING PROVISIONS FOR UNREINFORCED MASONRY BEARING WALL BUILDINGS

SECTION A101 PURPOSE

The purpose of this chapter is to promote public safety and welfare by reducing the risk of death or injury that may result from the effects of earthquakes on existing unreinforced masonry bearing wall buildings.

The provisions of this chapter are intended as minimum standards for structural seismic resistance, and are established primarily to reduce the risk of life loss or injury. Compliance with these provisions will not nec essarily prevent loss of life or injury, or prevent earthquake damage to rehabilitated buildings.

SECTION A102 SCOPE

A102.1 General. The provisions of this chapter shall apply to all existing buildings having at least one unreinforced masonry bearing wall. The elements regulated by this chapter shall be determined in accordance with Table A1-A. Except as provided herein, other structural provisions of the building code shall apply. This chapter does not apply to the *alteration* of existing electrical, plumbing, mechanical or fire safety systems.

A102.2 Essential and hazardous facilities. The provisions of this chapter shall not apply to the strengthening of buildings or structures in Occupancy Category III when assigned to Seismic Design Category C, D, or E or buildings or structures in Occupancy Category IV. Such buildings or structures shall be strengthened to meet the requirements of the *International Building Code* for new buildings of the same occupancy category or other such criteria that have been established by the jurisdiction.

SECTION A103 DEFINITIONS

For the purpose of this chapter, the applicable definitions in the building code shall also apply.

BUILDING CODE. The code currently adopted by the jurisdiction for new buildings.

COLLAR JOINT. The vertical space between adjacent wythes. A collar joint may contain mortar or grout.

CROSSWALL. A new or existing wall that meets the requirements of Section A111.3 and the definition of Section A111.3. A crosswall is not a shear wall.

CROSSWALL SHEAR CAPACITY. The unit shear value times the length of the crosswall, $v_c L_c$.

DIAPHRAGM EDGE. The intersection of the horizontal diaphragm and a shear wall.

DIAPHRAGM SHEAR CAPACITY. The unit shear value times the depth of the diaphragm, $v_u D$.

INTERNATIONAL BUILDING CODE. The *2009 International Building Code* (IBC).

NORMAL WALL. A wall perpendicular to the direction of seismic forces.

OPEN FRONT. An exterior building wall line without vertical elements of the lateral-force-resisting system in one or more stories.

POINTING. The partial reconstruction of the bed joints of an unreinforced masonry wall as defined in UBC Standard 21-8.

RIGID DIAPHRAGM. A diaphragm of reinforced concrete construction supported by concrete beams and columns or by structural steel beams and columns.

UNREINFORCED MASONRY. Includes burned clay, concrete or sand-lime brick; hollow clay or concrete block; plain concrete; and hollow clay tile. These materials shall comply with the requirements of Section A106 as applicable.

UNREINFORCED MASONRY BEARING WALL. A URM wall that provides the vertical support for the reaction of floor or roof-framing members.

UNREINFORCED MASONRY (URM) WALL. A masonry wall that relies on the tensile strength of masonry units, mortar and grout in resisting design loads, and in which the area of reinforcement is less than 25 percent of the minimum ratio required by the building code for reinforced masonry.

YIELD STORY DRIFT. The lateral displacement of one level relative to the level above or below at which yield stress is first developed in a frame member.

SECTION A104 SYMBOLS AND NOTATIONS

For the purpose of this chapter, the following notations supplement the applicable symbols and notations in the building code.

a_n = Diameter of core multiplied by its length or the area of the side of a square prism.

A = Cross-sectional area of unreinforced masonry pier or wall, square inches (10^{-6} m^2).

A_b = Total area of the bed joints above and below the test specimen for each in-place shear test, square inches (10^{-6} m^2).

D = In-plane width dimension of pier, inches (10^{-3} m), or depth of diaphragm, feet (m).

DCR = Demand-capacity ratio specified in Section A111.4.2.

f'_m = Compressive strength of masonry.

f_{sp} = Tensile-splitting strength of masonry.

F_{wx} = Force applied to a wall at level x, pounds (N).

H = Least clear height of opening on either side of a pier, inches (10^{-3} m).

h/t = Height-to-thickness ratio of URM wall. Height, h, is measured between wall anchorage levels and/or slab-on-grade.

L = Span of diaphragm between shear walls, or span between shear wall and open front, feet (m).

L_c = Length of crosswall, feet (m).

L_i = Effective span for an open-front building specified in Section A111.8, feet (m).

P = Applied force as determined by standard test method of ASTM C 496 or ASTM E 519, pounds (N).

P_D = Superimposed dead load at the location under consideration, pounds (kN). For determination of the rocking shear capacity, dead load at the top of the pier under consideration shall be used.

p_{D+L} = Press resulting from the dead plus actual live load in place at the time of testing, pounds per square inch (kPa).

P_w = Weight of wall, pounds (N).

R = Response modification factor for Ordinary plain masonry shear walls in Bearing Wall System from Table 12.2-1 of ASCE 7, where $R = 1.5$.

S_{DS} = Design spectral acceleration at short period, in g units.

S_{D1} = Design spectral acceleration at 1-second period, in g units.

v_a = The shear strength of any URM pier, $v_m A/1.5$ pounds (N).

v_c = Unit shear capacity value for a crosswall sheathed with any of the materials given in Table A1-D or A1-E, pounds per foot (N/m).

v_m = Shear strength of unreinforced masonry, pounds per square inch (kPa).

V_a = The shear strength of any URM pier or wall, pounds (N).

V_{ca} = Total shear capacity of crosswalls in the direction of analysis immediately above the diaphragm level being investigated, $v_c L_c$, pounds (N).

V_{cb} = Total shear capacity of crosswalls in the direction of analysis immediately below the diaphragm level being investigated, $v_c L_c$, pounds (N).

V_p = Shear force assigned to a pier on the basis of its relative shear rigidity, pounds (N).

V_r = Pier rocking shear capacity of any URM wall or wall pier, pounds (N).

v_t = Mortar shear strength as specified in Section A106.3.3.5, pounds per square inch (kPa).

V_{test} = Load at incipient cracking for each in-place shear test per UBC Standard 21-6, pounds (kN).

v_{to} = Mortar shear test values as specified in Section A106.3.3.5, pounds per square inch (kPa).

v_u = Unit shear capacity value for a diaphragm sheathed with any of the materials given in Table A1-D or A1-E, pounds per foot (N/m).

V_{wx} = Total shear force resisted by a shear wall at the level under consideration, pounds (N).

W = Total seismic dead load as defined in the building code, pounds (N).

W_d = Total dead load tributary to a diaphragm level, pounds (N).

W_w = Total dead load of a URM wall above the level under consideration or above an open-front building, pounds (N).

W_{wx} = Dead load of a URM wall assigned to level x halfway above and below the level under consideration, pounds (N).

$\Sigma v_u D$ = Sum of diaphragm shear capacities of both ends of the diaphragm, pounds (N).

$\Sigma\Sigma v_u D$ = For diaphragms coupled with crosswalls, $v_u D$ includes the sum of shear capacities of both ends of diaphragms coupled at and above the level under consideration, pounds (N).

ΣW_d = Total dead load of all the diaphragms at and above the level under consideration, pounds (N).

SECTION A105 GENERAL REQUIREMENTS

A105.1 General. The seismic-force-resisting system specified in this chapter shall comply with the building code, except as modified herein.

A105.2 Alterations and repairs. Alterations and repairs required to meet the provisions of this chapter shall comply with applicable structural requirements of the building code unless specifically provided for in this chapter.

A105.3 Requirements for plans. The following construction information shall be included in the plans required by this chapter:

1. Dimensioned floor and roof plans showing existing walls and the size and spacing of floor and roof-framing members and sheathing materials. The plans shall indicate all existing and new crosswalls and shear walls and their materials of construction. The location of these walls and their openings shall be fully dimensioned and drawn to scale on the plans.
2. Dimensioned wall elevations showing openings, piers, wall classes as defined in Section A106.3.3.8, thickness,

heights, wall shear test locations, cracks or damaged portions requiring repairs, the general condition of the mortar joints, and if and where pointing is required. Where the exterior face is veneer, the type of veneer, its thickness and its bonding and/or ties to the structural wall masonry shall also be noted.

3. The type of interior wall and ceiling materials, and framing.
4. The extent and type of existing wall anchorage to floors and roof when used in the design.
5. The extent and type of parapet corrections that were previously performed, if any.
6. *Repair* details, if any, of cracked or damaged unreinforced masonry walls required to resist forces specified in this chapter.
7. All other plans, sections and details necessary to delineate required retrofit construction.
8. The design procedure used shall be stated on both the plans and the permit application.
9. Details of the anchor prequalification program required by UBC Standard 21-7, if used, including location and results of all tests.

A105.4 Structural observation, testing and inspection. Structural observation, in accordance with Section 1709 of the *International Building Code*, shall be required for all structures in which seismic retrofit is being performed in accordance with this chapter. Structural observation shall include visual oservation of work for conformance with the approved construction documents and confirmation of existing conditions assumed during design.

Structural testing and inspection for new construction materials shall be in accordance with the building code, except as modified by this chapter.

SECTION A106 MATERIALS REQUIREMENTS

A106.1 General. Materials permitted by this chapter, including their appropriate strength design values and those existing configurations of materials specified herein, may be used to meet the requirements of this chapter.

A106.2 Existing materials. Existing materials used as part of the required vertical-load-carrying or lateral-force-resisting system shall be in sound condition, or shall be repaired or removed and replaced with new materials. All other unreinforced masonry materials shall comply with the following requirements:

1. The lay-up of the masonry units shall comply with Section A106.3.2, and the quality of bond between the units has been verified to the satisfaction of the building official;
2. Concrete masonry units are verified to be load-bearing units complying with UBC Standard 21-4 or such other standard as is acceptable to the building official; and
3. The compressive strength of plain concrete walls shall be determined based on cores taken from each class of concrete wall. The location and number of tests shall be the same as those prescribed for tensile-splitting strength tests in Sections A106.3.3.3 and A106.3.3.4, or in Section A108.1.

The use of materials not specified herein or in Section A108.1 shall be based on substantiating research data or engineering judgment, with the approval of the building official.

A106.3 Existing unreinforced masonry.

A106.3.1 General. Unreinforced masonry walls used to carry vertical loads or seismic forces parallel and perpendicular to the wall plane shall be tested as specified in this section. All masonry that does not meet the minimum standards established by this chapter shall be removed and replaced with new materials, or alternatively, shall have its structural functions replaced with new materials and shall be anchored to supporting elements.

A106.3.2 Lay-up of walls.

A106.3.2.1 Multiwythe solid brick. The facing and backing shall be bonded so that not less than 10 percent of the exposed face area is composed of solid headers extending not less than 4 inches (102 mm) into the backing. The clear distance between adjacent full-length headers shall not exceed 24 inches (610 mm) vertically or horizontally. Where the backing consists of two or more wythes, the headers shall extend not less than 4 inches (102 mm) into the most distant wythe, or the backing wythes shall be bonded together with separate headers with their area and spacing conforming to the foregoing. Wythes of walls not bonded as described above shall be considered veneer. Veneer wythes shall not be included in the effective thickness used in calculating the height-to-thickness ratio and the shear capacity of the wall.

Exception: Veneer wythes anchored as specified in the building code and made composite with backup masonry may be used for calculation of the effective thickness, where S_{D1} exceeds 0.3.

A106.3.2.2 Grouted or ungrouted hollow concrete or clay block and structural hollow clay tile. Grouted or ungrouted hollow concrete or clay block and structural hollow clay tile shall be laid in a running bond pattern.

A106.3.2.3 Other lay-up patterns. Lay-up patterns other than those specified in Sections A106.3.2.1 and A106.3.2.2 above are allowed if their performance can be justified.

A106.3.3 Testing of masonry.

A106.3.3.1 Mortar tests. The quality of mortar in all masonry walls shall be determined by performing in-place shear tests in accordance with the following:

1. The bed joints of the outer wythe of the masonry should be tested in shear by laterally displacing a single brick relative to the adjacent bricks in the same wythe. The head joint opposite the loaded end of the test brick should be carefully excavated

and cleared. The brick adjacent to the loaded end of the test brick should be carefully removed by sawing or drilling and excavating to provide space for a hydraulic ram and steel loading blocks. Steel blocks, the size of the end of the brick, should be used on each end of the ram to distribute the load to the brick. The blocks should not contact the mortar joints. The load should be applied horizontally, in the plane of the wythe. The load recorded at first movement of the test brick as indicated by spalling of the face of the mortar bed joints is V_{test} in Equation (A1-3).

2. Alternative procedures for testing shall be used where in-place testing is not practical because of crushing or other failure mode of the masonry unit (see Section A106.3.3.2).

A106.3.3.2 Alternative procedures for testing masonry. The tensile-splitting strength of existing masonry, f_{sp}, or the prism strength of existing masonry, f'_m may be determined in accordance with one of the following procedures:

1. Wythes of solid masonry units shall be tested by sampling the masonry by drilled cores of not less than 8 inches (203 mm) in diameter. A bed joint intersection with a head joint shall be in the center of the core. The tensile-splitting strength of these cores should be determined by the standard test method of ASTM C 496. The core should be placed in the test apparatus with the bed joint 45 degrees from the horizontal. The tensile-splitting strength should be determined by the following equation:

$$f_{sp} = \frac{2P}{\pi a_n}$$ **(Equation A1-1)**

2. Hollow unit masonry constructed of through-the-wall units shall be tested by sampling the masonry by a sawn square prism of not less than 18 inches square (11 613 mm^2). The tensile-splitting strength should be determined by the standardtest method of ASTM E 519. The diagonal of the prism should be placed in a vertical position. The tensile-splitting strength should be determined by the following equation:

$$f_{sp} = \frac{0.494P}{a_n}$$ **(Equation A1-2)**

3. An alternative to material testing is estimation of the f'_m of the existing masonry. This alternative should be limited to recently constructed masonry. The determination of f'_m requires that the unit correspond to a specification of the unit by an ASTM standard and classification of the mortar by type.

A106.3.3.3 Location of tests. The shear tests shall be taken at locations representative of the mortar conditions throughout the entire building, taking into account variations in workmanship at different building height levels, variations in weathering of the exterior surfaces, and variations in the condition of the interior surfaces due to deterioration caused by leaks and condensation of water and/or by the deleterious effects of other substances contained within the building. The exact test locations shall be determined at the building site by the engineer or architect in responsible charge of the structural design work. An accurate record of all such tests and their locations in the building shall be recorded, and these results shall be submitted to the building department for approval as part of the structural analysis.

A106.3.3.4 Number of tests. The minimum number of tests per class shall be as follows:

1. At each of both the first and top stories, not less than two tests per wall or line of wall elements providing a common line of resistance to lateral forces.
2. At each of all other stories, not less than one test per wall or line of wall elements providing a common line of resistance to lateral forces.
3. In any case, not less than one test per 1,500 square feet (139.4 m^2) of wall surface and not less than a total of eight tests.

A106.3.3.5 Minimum quality of mortar.

1. Mortar shear test values, v_{to}, in pounds per square inch (kPa) shall be obtained for each in-place shear test in accordance with the following equation:

$$v_{to} = (V_{test}/A_b) - p_{D+L}$$ **(Equation A1-3)**

2. Individual unreinforced masonry walls with v_{to} consistently less than 30 pounds per square inch (207 kPa) shall be entirely pointed prior to retesting.
3. The mortar shear strength, v_t, is the value in pounds per square inch (kPa) that is exceeded by 80 percent of the mortar shear test values, v_{to}.
4. Unreinforced masonry with mortar shear strength, v_t, less than 30 pounds per square inch (207 kPa) shall be removed, pointed and retested or shall have its structural function replaced, and shall be anchored to supporting elements in accordance with Sections A106.3.1 and A113.8. When existing mortar in any wythe is pointed to increase its shear strength and is retested, the condition of the mortar in the adjacent bed joints of the inner wythe or wythes and the opposite outer wythe shall be examined for extent of deterioration. The shear strength of any wall class shall be no greater than that of the weakest wythe of that class.

A106.3.3.6 Minimum quality of masonry.

1. The minimum average value of tensile-splitting strength determined by Equation (A1-1) or (A1-2) shall be 50 pounds per square inch (344.7 kPa). The minimum value of f'_m determined by categorization of the masonry units and mortar should be 1,000 pounds per square inch (6895 kPa).

2. Individual unreinforced masonry walls with average tensile-splitting strength of less than 50 pounds per square inch (344.7 kPa) shall be entirely pointed prior to retesting.
3. Hollow unit unreinforced masonry walls with estimated prism compressive strength of less than 1,000 pounds per square inch (6895 kPa) shall be grouted to increase the average net area compressive strength.

A106.3.3.7 Collar joints. The collar joints shall be inspected at the test locations during each in-place shear test, and estimates of the percentage of adjacent wythe surfaces that are covered with mortar shall be reported along with the results of the in-place shear tests.

A106.3.3.8 Unreinforced masonry classes. Existing unreinforced masonry shall be categorized into one or more classes based on shear strength, quality of construction, state of *repair*, deterioration and weathering. A class shall be characterized by the allowable masonry shear stress determined in accordance with Section A108.2. Classes shall be defined for whole walls, not for small areas of masonry within a wall.

A106.3.3.9 Pointing. Deteriorated mortar joints in unreinforced masonry walls shall be pointed according to UBC Standard 21-8. Nothing shall prevent pointing of any deteriorated masonry wall joints before the tests are made, except as required in Section A107.1.

SECTION A107
QUALITY CONTROL

A107.1 Pointing. Preparation and mortar pointing shall be performed with special inspection.

Exception: At the discretion of the building official, incidental pointing may be performed without special inspection.

A107.2 Masonry shear tests. In-place masonry shear tests shall comply with Section A106.3.3.1. Testing of masonry for determination of tensile-splitting strength shall comply with Section A106.3.3.2.

A107.3 Existing wall anchors. Existing wall anchors used as all or part of the required tension anchors shall be tested in pull-out according to UBC Standard 21-7. The minimum number of anchors tested shall be four per floor, with two tests at walls with joists framing into the wall and two tests at walls with joists parallel to the wall, but not less than 10 percent of the total number of existing tension anchors at each level.

A107.4 New bolts. All new embedded bolts shall be subject to periodic special inspection in accordance with the building code, prior to placement of the bolt and grout or adhesive in the drilled hole. Five percent of all bolts that do not extend through the wall shall be subject to a direct-tension test, and an additional 20 percent shall be tested using a calibrated torque wrench. Testing shall be performed in accordance with UBC Standard 21-7. New bolts that extend through the wall with steel plates on the far side of the wall need not be tested.

Exception: Special inspection in accordance with the building code may be provided during installation of new anchors in lieu of testing.

All new embedded bolts resisting tension forces or a combination of tension and shear forces shall be subject to periodic special inspection in accordance with the building code, prior to placement of the bolt and grout or adhesive in the drilled hole. Five percent of all bolts resisting tension forces shall be subject to a direct-tension test, and an additional 20 percent shall be tested using a calibrated torque wrench. Testing shall be performed in accordance with UBC Standard 21-7. New through-bolts need not be tested.

SECTION A108
DESIGN STRENGTHS

A108.1 Values.

1. Strength values for existing materials are given in Table A1-D and for new materials in Table A1-E.
2. Capacity reduction factors need not be used.
3. The use of new materials not specified herein shall be based on substantiating research data or engineering judgment, with the approval of the building official.

A108.2 Masonry shear strength. The unreinforced masonry shear strength, v_m, shall be determined for each masonry class from one of the following equations:

1. The unreinforced masonry shear strength, v_m, shall be determined by Equation (A1-4) when the mortar shear strength has been determined by Section A106.3.3.1.

$$v_m = 0.56 v_t + \frac{0.75 P_D}{A} \quad \textbf{(Equation A1-4)}$$

The mortar shear strength values, v_t, shall be determined in accordance with Section A106.3.3.5 and shall not exceed 100 pounds per square inch (689.5 kPa) for the determination of v_m.

2. The unreinforced masonry shear, v_m, shall be determined by Equation (A1-5) when tensile-splitting strength has been determined in accordance with Section A106.3.3.2, Item 1 or 2.

$$v_m = 0.8 f_{sp} + 0.5 \frac{P_D}{A} \quad \textbf{(Equation A1-5)}$$

3. When f'_m has been estimated by categorization of the units and mortar in accordance with Section 2105.2.2.1 of the *International Building Code*, the unreinforced masonry shear strength, v_m, shall not exceed 200 pounds per square inch (1380 kPa) or the lesser of the following:

a) $2.5\sqrt{f'_m}$ or

b) 200 psi or

c) $v + 0.75 \frac{P_D}{A}$ **(Equation A1-6)**

For SI: 1 psi = 6.895 kPa.

where:

v = 62.5 psi (430 kPa) for running bond masonry not grouted solid.

v = 100 psi (690 kPa) for running bond masonry grouted solid.

v = 25 psi (170 kPa) for stack bond grouted solid.

A108.3 Masonry compression. Where any increase in dead plus live compression stress occurs, the compression stress in unreinforced masonry shall not exceed 300 pounds per square inch (2070 kPa).

A108.4 Masonry tension. Unreinforced masonry shall be assumed to have no tensile capacity.

A108.5 Existing tension anchors. The resistance values of the existing anchors shall be the average of the tension tests of existing anchors having the same wall thickness and joist orientation.

A108.6 Foundations. For existing foundations, new total dead loads may be increased over the existing dead load by 25 percent. New total dead load plus live load plus seismic forces may be increased over the existing dead load plus live load by 50 percent. Higher values may be justified only in conjunction with a geotechnical investigation.

SECTION A109 ANALYSIS AND DESIGN PROCEDURE

A109.1 General. The elements of buildings hereby required to be analyzed are specified in Table A1-A.

A109.2 Selection of procedure. Buildings with rigid diaphragms shall be analyzed by the general procedure of Section A110, which is based on the building code. Buildings with flexible diaphragms shall be analyzed by the general procedure or, when applicable, may be analyzed by the special procedure of Section A111.

SECTION A110 GENERAL PROCEDURE

A110.1 Minimum design lateral forces. Buildings shall be analyzed to resist minimum lateral forces assumed to act nonconcurrently in the direction of each of the main axes of the structure in accordance with the following:

$$V = \frac{0.75\, S_{DS} W}{R}$$ **(Equation A1-7)**

A110.2 Lateral forces on elements of structures. Parts and portions of a structure not covered in Sections A110.3 shall be analyzed and designed per the current building code, using force levels defined in Section A110.1.

Exceptions:

1. Unreinforced masonry walls for which height-to-thickness ratios do not exceed ratios set forth in Table A1-B need not be analyzed for out-of-plane loading. Unreinforced masonry walls that exceed the allowable *h/t* ratios of Table A1-B shall be braced according to Section A113.5.
2. Parapets complying with Section A113.6 need not be analyzed for out-of-plane loading.
3. Walls shall be anchored to floor and roof diaphragms in accordance with Section A113.1.

A110.3 In-plane loading of URM shear walls and frames. Vertical lateral-load-resisting elements shall be analyzed in accordance with Section A112.

A110.4 Redundancy and overstrength factors. Any redundancy or overstrength factors contained in the building code may be taken as unity. The vertical component of earthquake load (E_v) may be taken as zero.

SECTION A111 SPECIAL PROCEDURE

A111.1 Limits for the application of this procedure. The special procedures of this section may be applied only to buildings having the following characteristics:

1. Flexible diaphragms at all levels above the base of the structure.
2. Vertical elements of the lateral-force-resisting system consisting predominantly of masonry or concrete shear walls.
3. Except for single-story buildings with an open front on one side only, a minimum of two lines of vertical elements of the lateral-force-resisting system parallel to each axis of the building (see Section A111.8 for open-front buildings).

A111.2 Lateral forces on elements of structures. With the exception of the provisions in Sections A111.4 through A111.7, elements of structures shall comply with Sections A110.2 through A110.4.

A111.3 Crosswalls. Crosswalls shall meet the requirements of this section.

A111.3.1 Crosswall definition. A crosswall is a wood-framed wall sheathed with any of the materials described in Table A1-D or A1-E or other system as defined in Section A111.3.5. Crosswalls shall be spaced no more than 40 feet (12 192 mm) on center measured perpendicular to the direction of consideration, and shall be placed in each story of the building. Crosswalls shall extend the full story height between diaphragms.

Exceptions:

1. Crosswalls need not be provided at all levels when used in accordance with Section A111.4.2, Item 4.
2. Existing crosswalls need not be continuous below a wood diaphragm at or within 4 feet (1219 mm) of grade, provided:
   - 2.1. Shear connections and anchorage requirements of Section A111.5 are satisfied at all edges of the diaphragm.

2.2. Crosswalls with total shear capacity of $0.5S_{D1}\Sigma W_d$ interconnect the diaphragm to the foundation.

2.3. The demand-capacity ratio of the diaphragm between the crosswalls that are continuous to their foundations does not exceed 2.5, calculated as follows:

$$DCR = \frac{(2.1S_{D1}W_d + V_{ca})}{2v_u D} \quad \textbf{(Equation A1-8)}$$

A111.3.2 Crosswall shear capacity. Within any 40 feet (12 192 mm) measured along the span of the diaphragm, the sum of the crosswall shear capacities shall be at least 30 percent of the diaphragm shear capacity of the strongest diaphragm at or above the level under consideration.

A111.3.3 Existing crosswalls. Existing crosswalls shall have a maximum height-to-length ratio between openings of 1.5 to 1. Existing crosswall connections to diaphragms need not be investigated as long as the crosswall extends to the framing of the diaphragms above and below.

A111.3.4 New crosswalls. New crosswall connections to the diaphragm shall develop the crosswall shear capacity. New crosswalls shall have the capacity to resist an overturning moment equal to the crosswall shear capacity times the story height. Crosswall overturning moments need not be cumulative over more than two stories.

A111.3.5 Other crosswall systems. Other systems, such as moment-resisting frames, may be used as crosswalls provided that the yield story drift does not exceed 1 inch (25.4 mm) in any story.

A111.4 Wood diaphragms.

A111.4.1 Acceptable diaphragm span. A diaphragm is acceptable if the point *(L,DCR)* on Figure A1-1 falls within Region 1, 2 or 3.

A111.4.2 Demand-capacity ratios. Demand-capacity ratios shall be calculated for the diaphragm at any level according to the following formulas:

1. For a diaphragm without qualifying crosswalls at levels immediately above or below:

$$DCR = 2.1S_{D1}W_d/\Sigma v_u D \quad \textbf{(Equation A1-9)}$$

2. For a diaphragm in a single-story building with qualifying crosswalls, or for a roof diaphragm coupled by crosswalls to the diaphragm directly below:

$$DCR = 2.1S_{D1}W_d/(\Sigma v_u D + V_{cb}) \quad \textbf{(Equation A1-10)}$$

3. For diaphragms in a multistory building with qualifying crosswalls in all levels:

$$DCR = 2.1S_{D1}\Sigma W_d/(\Sigma\Sigma v_u D + V_{cb}) \quad \textbf{(Equation A1-11)}$$

DCR shall be calculated at each level for the set of diaphragms at and above the level under consideration. In addition, the roof diaphragm shall also meet the requirements of Equation (A1-10).

4. For a roof diaphragm and the diaphragm directly below, if coupled by crosswalls:

$$DCR = 2.1S_{D1}\Sigma W_d/\Sigma\Sigma v_u D \quad \textbf{(Equation A1-12)}$$

A111.4.3 Chords. An analysis for diaphragm flexure need not be made, and chords need not be provided.

A111.4.4 Collectors. An analysis of diaphragm collector forces shall be made for the transfer of diaphragm edge shears into vertical elements of the lateral-force-resisting system. Collector forces may be resisted by new or existing elements.

A111.4.5 Diaphragm openings.

1. Diaphragm forces at corners of openings shall be investigated and shall be developed into the diaphragm by new or existing materials.
2. In addition to the demand-capacity ratios of Section A111.4.2, the demand-capacity ratio of the portion of the diaphragm adjacent to an opening shall be calculated using the opening dimension as the span.
3. Where an opening occurs in the end quarter of the diaphragm span, the calculation of $v_u D$ for the demand-capacity ratio shall be based on the net depth of the diaphragm.

A111.5 Diaphragm shear transfer. Diaphragms shall be connected to shear walls with connections capable of developing the diaphragm-loading tributary to the shear wall given by the lesser of the following formulas:

$$V = 1.2S_{D1}C_p W_d \quad \textbf{(Equation A1-13)}$$

using the C_p values in Table A1-C, or

$$V = v_u D \quad \textbf{(Equation A1-14)}$$

A111.6 Shear walls (In-plane loading).

A111.6.1 Wall story force. The wall story force distributed to a shear wall at any diaphragm level shall be the lesser value calculated as:

$$F_{wx} = 0.8S_{D1}(W_{wx} + W_d/2) \quad \textbf{(Equation A1-15)}$$

but need not exceed

$$F_{wx} = 0.8S_{D1}W_{wx} + v_u D \quad \textbf{(Equation A1-16)}$$

A111.6.2 Wall story shear. The wall story shear shall be the sum of the wall story forces at and above the level of consideration.

$$V_{wx} = \Sigma F_{wx} \quad \textbf{(Equation A1-17)}$$

A111.6.3 Shear wall analysis. Shear walls shall comply with Section A112.

A111.6.4 Moment frames. Moment frames used in place of shear walls shall be designed as required by the building code, except that the forces shall be as specified in Section A111.6.1, and the story drift ratio shall be limited to 0.015, except as further limited by Section A112.4.2.

A111.7 Out-of-plane forces—unreinforced masonry walls.

A111.7.1 Allowable unreinforced masonry wall height-to-thickness ratios. The provisions of Section A110.2 are applicable, except the allowable height-to-thickness ratios given in Table A1-B shall be determined from Figure A1-1 as follows:

1. In Region 1, height-to-thickness ratios for buildings with crosswalls may be used if qualifying crosswalls are present in all stories.
2. In Region 2, height-to-thickness ratios for buildings with crosswalls may be used whether or not qualifying crosswalls are present.
3. In Region 3, height-to-thickness ratios for "all other buildings" shall be used whether or not qualifying crosswalls are present.

A111.7.2 Walls with diaphragms in different regions. When diaphragms above and below the wall under consideration have demand-capacity ratios in different regions of Figure A1-1, the lesser height-to-thickness ratio shall be used.

A111.8 Open-front design procedure. A single-story building with an open front on one side and crosswalls parallel to the open front may be designed by the following procedure:

1. Effective diaphragm span, L_i, for use in Figure A1-1 shall be determined in accordance with the following formula:

$$L_i = 2\,[(W_w/W_d)L + L] \qquad \textbf{(Equation A1-18)}$$

2. Diaphragm demand-capacity ratio shall be calculated as:

$$DCR = 2.12S_{D1}(W_d + W_w)/[(v_uD) + V_{cb}] \qquad \textbf{(Equation A1-19)}$$

SECTION A112 ANALYSIS AND DESIGN

A112.1 General. The following requirements are applicable to both the general procedure and the special procedure for analyzing vertical elements of the lateral-force-resisting system.

A112.2 Existing unreinforced masonry walls.

A112.2.1 Flexural rigidity. Flexural components of deflection may be neglected in determining the rigidity of an unreinforced masonry wall.

A112.2.2 Shear walls with openings. Wall piers shall be analyzed according to the following procedure, which is diagramed in Figure A1-2.

1. For any pier,
   1.1. The pier shear capacity shall be calculated as:

   $$V_a = v_m A/1.5 \qquad \textbf{(Equation A1-20)}$$

   1.2. The pier rocking shear capacity shall be calculated as:

   $$V_r = 0.9P_D D/H \qquad \textbf{(Equation A1-21)}$$

2. The wall piers at any level are acceptable if they comply with one of the following modes of behavior:
   2.1. **Rocking controlled mode.** When the pier rocking shear capacity is less than the pier shear capacity, i.e., $V_r < V_a$ for each pier in a level, forces in the wall at that level, V_{wx}, shall be distributed to each pier in proportion to $P_D D/H$.

   For the wall at that level:

   $$0.7\,V_{wx} < \Sigma V_r \qquad \textbf{(Equation A1-22)}$$

   2.2. **Shear controlled mode.** Where the pier shear capacity is less than the pier rocking capacity, i.e., $V_a < V_r$ in at least one pier in a level, forces in the wall at the level, V_{wx}, shall be distributed to each pier in proportion to D/H.

   For each pier at that level:

   $$V_p < V_a \qquad \textbf{(Equation A1-23)}$$

   and

   $$V_p < V_r \qquad \textbf{(Equation A1-24)}$$

   If $V_p < V_a$ for each pier and $V_p > V_r$ for one or more piers, such piers shall be omitted from the analysis, and the procedure shall be repeated for the remaining piers, unless the wall is strengthened and reanalyzed.

3. **Masonry pier tension stress.** Unreinforced masonry wall piers need not be analyzed for tension stress.

A112.2.3 Shear walls without openings. Shear walls without openings shall be analyzed the same as for walls with openings, except that V_r shall be calculated as follows:

$$V_r = 0.9\,(P_D + 0.5P_w)\,D/H \qquad \textbf{(Equation A1-25)}$$

A112.3 Plywood-sheathed shear walls. Plywood-sheathed shear walls may be used to resist lateral forces for buildings with flexible diaphragms analyzed according to provisions of Section A111. Plywood-sheathed shear walls may not be used to share lateral forces with other materials along the same line of resistance.

A112.4 Combinations of vertical elements.

A112.4.1 Lateral-force distribution. Lateral forces shall be distributed among the vertical-resisting elements in proportion to their relative rigidities, except that moment-resisting frames shall comply with Section A112.4.2.

A112.4.2 Moment-resisting frames. Moment-resisting frames shall not be used with an unreinforced masonry wall in a single line of resistance unless the wall has piers that have adequate shear capacity to sustain rocking in accordance with Section A112.2.2. The frames shall be designed in accordance with the building code to carry 100 percent of the lateral forces tributary to that line of resistance, as determined from Equation (A1-7). The story drift ratio shall be limited to 0.0075.

SECTION A113
DETAILED SYSTEM DESIGN REQUIREMENTS

A113.1 Wall anchorage.

A113.1.1 Anchor locations. Unreinforced masonry walls shall be anchored at the roof and floor levels as required in Section A110.2. Ceilings of plaster or similar materials, when not attached directly to roof or floor framing and where abutting masonry walls, shall either be anchored to the walls at a maximum spacing of 6 feet (1829 mm), or be removed.

A113.1.2 Anchor requirements. Anchors shall consist of bolts installed through the wall as specified in Table A1-E, or an approved equivalent at a maximum anchor spacing of 6 feet (1829 mm). All wall anchors shall be secured to the joists to develop the required forces.

A113.1.3 Minimum wall anchorage. Anchorage of masonry walls to each floor or roof shall resist a minimum force determined as $0.9S_{DS}$ times the tributary weight or 200 pounds per linear foot (2920 N/m), whichever is greater, acting normal to the wall at the level of the floor or roof. Existing wall anchors, if used, must meet the requirements of this chapter or must be upgraded.

A113.1.4 Anchors at corners. At the roof and floor levels, both shear and tension anchors shall be provided within 2 feet (610 mm) horizontally from the inside of the corners of the walls.

A113.2 Diaphragm shear transfer. Bolts transmitting shear forces shall have a maximum bolt spacing of 6 feet (1829 mm) and shall have nuts installed over malleable iron or plate washers when bearing on wood, and heavy-cut washers when bearing on steel.

A113.3 Collectors. Collector elements shall be provided that are capable of transferring the seismic forces originating in other portions of the building to the element providing the resistance to those forces.

A113.4 Ties and continuity. Ties and continuity shall conform to the requirements of the building code.

A113.5 Wall bracing.

A113.5.1 General. Where a wall height-to-thickness ratio exceeds the specified limits, the wall may be laterally supported by vertical bracing members per Section A113.5.2 or by reducing the wall height by bracing per Section A113.5.3.

A113.5.2 Vertical bracing members. Vertical bracing members shall be attached to floor and roof construction for their design loads independently of required wall anchors. Horizontal spacing of vertical bracing members shall not exceed one-half of the unsupported height of the wall or 10 feet (3048 mm). Deflection of such bracing members at design loads shall not exceed one-tenth of the wall thickness.

A113.5.3 Intermediate wall bracing. The wall height may be reduced by bracing elements connected to the floor or roof. Horizontal spacing of the bracing elements and wall anchors shall be as required by design, but shall not exceed 6 feet (1829 mm) on center. Bracing elements shall be detailed to minimize the horizontal displacement of the wall by the vertical displacement of the floor or roof.

A113.6 Parapets. Parapets and exterior wall appendages not conforming to this chapter shall be removed, or stabilized or braced to ensure that the parapets and appendages remain in their original positions.

The maximum height of an unbraced unreinforced masonry parapet above the lower of either the level of tension anchors or the roof sheathing shall not exceed the height-to-thickness ratio shown in Table A1-F. If the required parapet height exceeds this maximum height, a bracing system designed for the forces determined in accordance with the building code shall support the top of the parapet. Parapet corrective work must be performed in conjunction with the installation of tension roof anchors.

The minimum height of a parapet above any wall anchor shall be 12 inches (305 mm).

Exception: If a reinforced concrete beam is provided at the top of the wall, the minimum height above the wall anchor may be 6 inches (152 mm).

A113.7 Veneer.

1. Veneer shall be anchored with approved anchor ties conforming to the required design capacity specified in the building code and shall be placed at a maximum spacing of 24 inches (610 mm) with a maximum supported area of 4 square feet ($0.372 m^2$).

   Exception: Existing anchor ties for attaching brick veneer to brick backing may be acceptable, provided the ties are in good condition and conform to the following minimum size and material requirements.

   Existing veneer anchor ties may be considered adequate if they are of corrugated galvanized iron strips not less than 1 inch (25.4 mm) in width, 8 inches (203 mm) in length and $^{1}/_{16}$ inch (1.6 mm) in thickness, or the equivalent.

2. The location and condition of existing veneer anchor ties shall be verified as follows:

   2.1. An approved testing laboratory shall verify the location and spacing of the ties and shall submit a report to the building official for approval as part of the structural analysis.

   2.2. The veneer in a selected area shall be removed to expose a representative sample of ties (not less than four) for inspection by the building official.

A113.8 Nonstructural masonry walls. Unreinforced masonry walls that carry no design vertical or lateral loads and that are not required by the design to be part of the lateral-force resisting system shall be adequately anchored to new or existing supporting elements. The anchors and elements shall be designed for the out-of-plane forces specified in the building code. The height- or length-to-thickness ratio between such supporting elements for such walls shall not exceed nine.

A113.9 Truss and beam supports. Where trusses and beams other than rafters or joists are supported on masonry, independent secondary columns shall be installed to support vertical loads of the roof or floor members.

Exception: Secondary supports are not required where S_{D1} is less than 0.3g.

A113.10 Adjacent buildings. Where elements of adjacent buildings do not have a separation of at least 5 inches (127 mm), the allowable height-to-thickness ratios for "all other buildings" per Table A1-B shall be used in the direction of consideration.

SECTION A114 WALLS OF UNBURNED CLAY, ADOBE OR STONE MASONRY

A114.1 General. Walls of unburned clay, adobe or stone masonry construction shall conform to the following:

1. Walls of unburned clay, adobe or stone masonry shall not exceed a height- or length-to-thickness ratio specified in Table A1-G.
2. Adobe may be allowed a maximum value of 9 pounds per square inch (62.1 kPa) for shear unless higher values are justified by test.
3. Mortar for repointing may be of the same soil composition and stabilization as the brick, in lieu of cement-mortar.

TABLE A1-A—ELEMENTS REGULATED BY THIS CHAPTER

BUILDING ELEMENTS	S_{D1}			
	≥ 0.067_g < 0.133_g	≥ 0.133_g < 0.20_g	≥ 0.20_g < 0.30_g	> 0.30_g
Parapets	X	X	X	X
Walls, anchorage	X	X	X	X
Walls, *h/t* ratios		X	X	X
Walls, in-plane shear		X	X	X
Diaphragms[a]			X	X
Diaphragms, shear transfer[b]		X	X	X
Diaphragms, demand-capacity ratios[b]			X	X

a. Applies only to buildings designed according to the general procedures of Section A110.

b. Applies only to buildings designed according to the special procedures of Section A111.

TABLE A1-B—ALLOWABLE VALUE OF HEIGHT-TO-THICKNESS RATIO OF UNREINFORCED MASONRY WALLS

WALL TYPES	$0.13_g \le S_{D1} < 0.25_g$	$0.25_g \le S_{D1} < 0.4_g$	$S_{D1} \ge 0.4_g$ BUILDINGS WITH CROSSWALLS[a]	$S_{D1} > 0.4_g$ ALL OTHER BUILDINGS
Walls of one-story buildings	20	16	16[b,c]	13
First-story wall of multistory building	20	18	16	15
Walls in top story of multistory building	14	14	14[b,c]	9
All other walls	20	16	16	13

a. Applies to the special procedures of Section A111 only. See Section A111.7 for other restrictions.

b. This value of height-to-thickness ratio may be used only where mortar shear tests establish a tested mortar shear strength, v_t, of not less than 100 pounds per square inch (690 kPa). This value may also be used where the tested mortar shear strength is not less than 60 pounds per square inch (414 kPa), and where a visual examination of the collar joint indicates not less than 50-percent mortar coverage.

c. Where a visual examination of the collar joint indicates not less than 50-percent mortar coverage, and the tested mortar shear strength, v_t, is greater than 30 pounds per square inch (207 kPa) but less than 60 pounds per square inch (414 kPa), the allowable height-to-thickness ratio may be determined by linear interpolation between the larger and smaller ratios in direct proportion to the tested mortar shear strength.

TABLE A1-C—HORIZONTAL FORCE FACTOR, C_p

CONFIGURATION OF MATERIALS	C_p
Roofs with straight or diagonal sheathing and roofing applied directly to the sheathing, or floors with straight tongue-and-groove sheathing.	0.50
Diaphragms with double or mulitple layers of boards with edges offset, and blocked plywood systems.	0.75
Diaphragms of metal deck without topping:	
Minimal welding or mechanical attachment.	0.6
Welded or mechanically attached for seismic resistance.	0.68

TABLE A1-D—STRENGTH VALUES FOR EXISTING MATERIALS

EXISTING MATERIALS OR CONFIGURATION OF MATERIALS[a]		STRENGTH VALUES
		× 14.594 for N/m
Horizontal diaphragms	Roofs with straight sheathing and roofing applied directly to the sheathing.	300 lbs. per ft. for seismic shear
	Roofs with diagonal sheathing and roofing applied directly to the sheathing.	750 lbs. per ft. for seismic shear
	Floors with straight tongue-and-groove sheathing.	300 lbs. per ft. for seismic shear
	Floors with straight sheathing and finished wood flooring with board edges offset or perpendicular.	1,500 lbs. per ft. for seismic shear
	Floors with diagonal sheathing and finished wood flooring.	1,800 lbs. per ft. for seismic shear
	Metal deck welded with minimal welding.[c]	1,800 lbs, per ft. for seismic shear
	Metal deck welded for seismic resistance.[d]	3,000 lbs. per ft. for seismic shear
Crosswalls[b]	Plaster on wood or metal lath.	600 lbs. per ft. for seismic shear
	Plaster on gypsum lath.	550 lbs. per ft. for seismic shear
	Gypsum wallboard, unblocked edges.	200 lbs. per ft. for seismic shear
	Gypsum wallboard, blocked edges.	400 lbs. per ft. for seismic shear
Existing footing, wood framing, structural steel, reinforcing steel	Plain concrete footings.	f'_c = 1,500 psi (10.34 MPa) unless otherwise shown by tests
	Douglas fir wood.	Same as D.F. No. 1
	Reinforcing steel.	F_y = 40,000 psi (124.1 N/mm^2) maximum
	Structural steel.	F_y = 33,000 psi (137.9 N/mm^2) maximum

a. Material must be sound and in good condition.

b. Shear values of these materials may be combined, except the total combined value should not exceed 900 pounds per foot (4380 N/m).

c. Minimum 22-gage steel deck with welds to supports satisfying the standards of the Steel Deck Institute.

d. Minimum 22-gage steel deck with $^3/_4$ ϕ plug welds at an average spacing not exceeding 8 inches (203 mm) and with sidelap welds appropriate for the deck span.

TABLE A1-E—STRENGTH VALUES OF NEW MATERIALS USED IN CONJUNCTION WITH EXISTING CONSTRUCTION

NEW MATERIALS OR CONFIGURATION OF MATERIALS		STRENGTH VALUES
Horizontal diaphragms	Plywood sheathing applied directly over existing straight sheathing with ends of plywood sheets bearing on joists or rafters and edges of plywood located on center of individual sheathing boards.	675 lbs. per ft.
Crosswalls	Plywood sheathing applied directly over wood studs; no value should be given to plywood applied over existing plaster or wood sheathing.	1.2 times the value specified in the current building code.
	Drywall or plaster applied directly over wood studs.	The value specified in the current building code.
	Drywall or plaster applied to sheathing over existing wood studs.	50 percent of the value specified in the current building code.
Tension bolts[e]	Bolts extending entirely through unreinforced masonry wall secured with bearing plates on far side of a three-wythe- minimum wall with at least 30 square inches of area.[b,c]	5,400 lbs. per bolt 2,700 lbs. for two-wythe walls
Shear bolts[e]	Bolts embedded a minimum of 8 inches into unreinforced masonry walls; bolts should be centered in $2^1/_2$-inch-diameter holes with dry-pack or nonshrink grout around the circumference of the bolt.	The value for plain masonry specified for solid masonry in the current building code; no value larger than those given for $^3/_4$-inch bolts should be used.
Combined tension and shear bolts	Through-bolts—bolts meeting the requirements for shear and for tension bolts.[b,c]	Tension—same as for tension bolts Shear—same as for shear bolts
	Embedded bolts—bolts extending to the exterior face of the wall with a $2^1/_2$-inch round plate under the head and drilled at an angle of $22^1/_2$ degrees to the horizontal; installed as specified for shear bolts.[a,b,c]	Tension—3,600 lbs. per bolt Shear—same as for shear bolts
Infilled walls	Reinforced masonry infilled openings in existing unreinforced masonry walls; provide keys or dowels to match reinforcing.	Same as values specified for unreinforced masonry walls
Reinforced masonry[d]	Masonry piers and walls reinforced per the current building code.	The value specified in the current building code for strength design.
Reinforced concrete[d]	Concrete footings, walls and piers reinforced as specified in the current building code.	The value specified in the current building code for strength design.

For SI: 1 inch = 25.4 mm, 1 square inch = 645.16 mm², 1 pound = 4.4 N.

a. Embedded bolts to be tested as specified in Section A107.4.

b. Bolts to be $^1/_2$ inch (12.7 mm) minimum in diameter.

c. Drilling for bolts and dowels shall be done with an electric rotary drill; impact tools should not be used for drilling holes or tightening anchors and shear bolt nuts.

d. No load factors or capacity reduction factor shall be used.

e. Other bolt sizes, values and installation methods may be used, provided a testing program is conducted in accordance with UBC Standard 21-7. The useable value shall be determined by multiplying the calculated allowable value, as determined by UBC Standard 21-7, by 3.0, and the useable value shall be limited to a maximum of 1.5 times the value given in the table. Bolt spacing shall not exceed 6 feet (1829 mm) on center and shall not be less than 12 inches (305 mm) on center.

TABLE A1-F—MAXIMUM ALLOWABLE HEIGHT-TO-THICKNESS RATIOS FOR PARAPETS

	S_{D1}		
	$0.13_g \le S_{D1} < 0.25_g$	$0.25_g \le S_{D1} < 0.4_g$	$S_{D1} \ge 0.4_g$
Maximum allowable height-to-thickness ratios	2.5	2.5	1.5

TABLE A1-G—MAXIMUM HEIGHT-TO-THICKNESS RATIOS FOR ADOBE OR STONE WALLS

	S_{D1}		
	$0.13_g \# S_{D1} < 0.25_g$	$0.25_g \# S_{D1} < 0.4_g$	$S_{D1} \ge 0.4_g$
One-story buildings	12	10	8
Two-story buildings			
First story	14	11	9
Second story	12	10	8

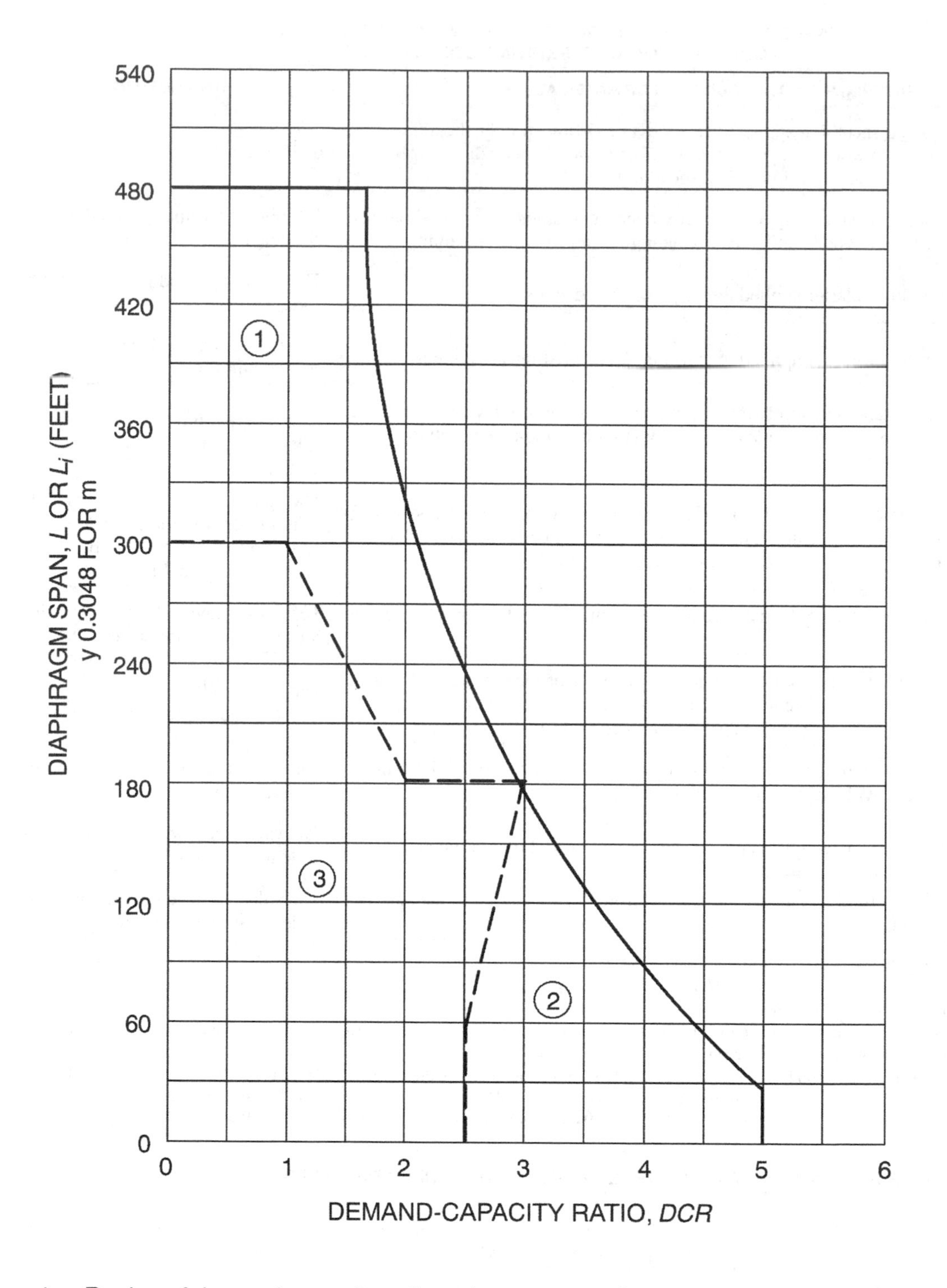

1. Region of demand-capacity ratios where crosswalls may be used to increase *h/t* ratios.
2. Region of demand-capacity ratios where *h/t* ratios of "buildings with crosswalls" may be used, whether or not crosswalls are present.
3. Region of demand-capacity ratios where *h/t* ratios of "all other buildings" shall be used, whether or not crosswalls are present.

FIGURE A1-1
ACCEPTABLE DIAPHRAGM SPAN

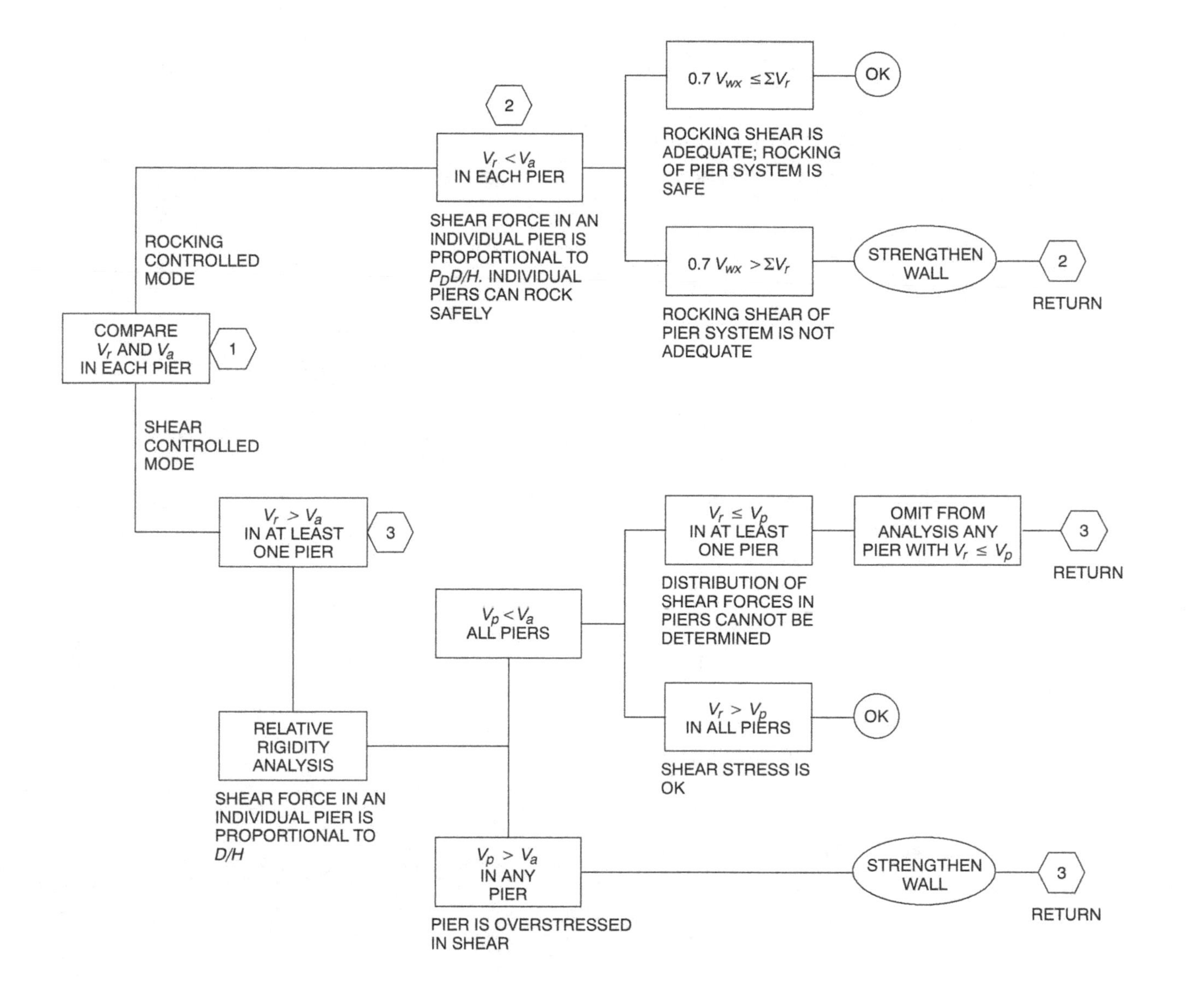

V_a = Allowable shear strength of a pier.
V_p = Shear force assigned to a pier on the basis of a relative shear rigidity analysis.
V_r = Rocking shear capacity of pier.
V_{wx} = Total shear force resisted by the wall.
ΣV_r = Rocking shear capacity of all piers in the wall.

FIGURE A1-2
ANALYSIS OF URM WALL IN-PLANE SHEAR FORCES

CHAPTER A2

EARTHQUAKE HAZARD REDUCTION IN EXISTING REINFORCED CONCRETE AND REINFORCED MASONRY WALL BUILDINGS WITH FLEXIBLE DIAPHRAGMS

SECTION A201 PURPOSE

The purpose of this chapter is to promote public safety and welfare by reducing the risk of death or injury that may result from the effects of earthquakes on reinforced concrete and reinforced masonry wall buildings with flexible diaphragms. Based on past earthquakes, these buildings have been categorized as being potentially hazardous and prone to significant damage, including possible collapse in a moderate to major earthquake. The provisions of this chapter are minimum standards for structural seismic resistance established primarily to reduce the risk of life loss or injury on both subject and adjacent properties. These provisions will not necessarily prevent loss of life or injury, or prevent earthquake damage to an *existing building* that complies with these standards.

SECTION A202 SCOPE

The provisions of this chapter shall apply to wall anchorage systems that resist out-of-plane forces and to collectors in existing reinforced concrete or reinforced masonry buildings with flexible diaphragms. Wall anchorage systems that were designed and constructed in accordance with the 1997 *Uniform Building Code*, 1999 *BOCA National Building Code*, 1999 *Standard Building Code* or the 2000 and subsequent editions of the *International Building Code* shall be deemed to comply with these provisions.

SECTION A203 DEFINITIONS

For the purpose of this chapter, the applicable definitions in Chapters 16, 19, 21, 22 and 23 of the *International Building Code* and the following shall apply:

FLEXIBLE DIAPHRAGMS. Roofs and floors including, but not limited to, those sheathed with plywood, wood decking (1-by or 2-by) or metal decks without concrete topping slabs.

SECTION A204 SYMBOLS AND NOTATIONS

For the purpose of this chapter, the applicable symbols and notations in the *International Building Code* shall apply.

SECTION A205 GENERAL REQUIREMENTS

A205.1 General. The seismic-resisting elements specified in this chapter shall comply with provisions of Section 1613 of the *International Building Code*, except as modified herein.

A205.2 Alterations and repairs. Alterations and repairs required to meet the provisions of this chapter shall comply with applicable structural requirements of the building code unless specifically modified in this chapter.

A205.3 Requirements for plans. The plans shall accurately reflect the results of the engineering investigation and design and shall show all pertinent dimensions and sizes for plan review and construction. The following shall be provided:

1. Floor plans and roof plans shall show existing framing construction, diaphragm construction, proposed wall anchors, cross-ties and collectors. Existing nailing, anchors, cross-ties and collectors shall also be shown on the plans if they are considered part of the lateral-force-resisting systems.
2. At elevations where there are alterations or damage, details shall show roof and floor heights, dimensions of openings, location and extent of existing damage and proposed *repair*.
3. Typical wall panel details and sections with panel thickness, height, pilasters and location of anchors shall be provided.
4. Details shall include existing and new anchors and the method of developing anchor forces into the diaphragm framing, existing and/or new cross-ties, and existing and/or new or improved support of roof and floor girders at pilasters or walls.
5. The basis for design and the building code used for the design shall be stated on the plans.

A205.4 Structural observation, testing and inspection. Structural observation, in accordance with Section 1709 of the *International Building Code*, shall be required for all structures in which seismic retrofit is being performed in accordance with this chapter. Structural observation shall include visual observation of work for conformance with the approved construction documents and confirmation of existing conditions assumed during design.

Structural testing and inspection for new construction materials shall be in accordance with the building code, except as modified by this chapter.

SECTION A206
ANALYSIS AND DESIGN

A206.1 Reinforced concrete and reinforced masonry wall anchorage. Concrete and masonry walls shall be anchored to all floors and roofs that provide lateral support for the wall. The anchorage shall provide a positive direct connection between the wall and floor or roof construction capable of resisting 75 percent of the horizontal forces specified in Section 1613 of the *International Building Code.*

A206.2 Special requirements for wall anchorage systems. The steel elements of the wall anchorage system shall be designed in accordance with the building code without the use of the 1.33 short duration allowable stress increase when using allowable stress design.

Wall anchors shall be provided to resist out-of-plane forces, independent of existing shear anchors.

> **Exception:** Existing cast-in-place shear anchors are allowed to be used as wall anchors if the tie element can be readily attached to the anchors, and if the engineer or architect can establish tension values for the existing anchors through the use of approved as-built plans or testing and through analysis showing that the bolts are capable of resisting the total shear load (including dead load) while being acted upon by the maximum tension force due to an earthquake. Criteria for analysis and testing shall be determined by the building official.

Expansion anchors are only allowed with special inspection and approved testing for seismic loading.

Attaching the edge of plywood sheathing to steel ledgers is not considered compliant with the positive anchoring requirements of this chapter. Attaching the edge of steel decks to steel ledgers is not considered as providing the positive anchorage of this chapter unless testing and/or analysis are performed to establish shear values for the attachment perpendicular to the edge of the deck. Where steel decking is used as a wall anchor system, the existing connections shall be subject to field verification and the new connections shall be subject to special inspection.

A206.3 Development of anchor loads into the diaphragm. Development of anchor loads into roof and floor diaphragms shall comply with Section 1613 of the *International Building Code* using horizontal forces that are 75 percent of those used for new construction.

> **Exception:** If continuously tied girders are present, the maximum spacing of the continuity ties is the greater of the girder spacing or 24 feet (7315 mm).

In wood diaphragms, anchorage shall not be accomplished by use of toenails or nails subject to withdrawal. Wood ledgers, top plates or framing shall not be used in cross-grain bending or cross-grain tension. The continuous ties required in Section 1613 of the *International Building Code* shall be in addition to the diaphragm sheathing.

Lengths of development of anchor loads in wood diaphragms shall be based on existing field nailing of the sheathing unless existing edge nailing is positively identified on the original construction plans or at the site.

A206.4 Anchorage at pilasters. Anchorage at pilasters shall be designed for the tributary wall-anchoring load per Section A206.1, considering the wall as a two-way slab. The edges of the two-way slab shall be considered fixed when there is continuity at pilasters and shall be considered pinned at roof and floor. The pilasters or the walls immediately adjacent to the pilasters shall be anchored directly to the roof framing such that the existing vertical anchor bolts at the top of the pilasters are bypassed without permitting tension or shear failure at the top of the pilasters.

> **Exception:** If existing vertical anchor bolts at the top of the pilasters are used for the anchorage, additional exterior confinement shall be provided as required to resist the total anchorage force.

The minimum anchorage force at a floor or roof between the pilasters shall be that specified in Section A206.1.

A206.5 Symmetry. Symmetry of wall anchorage and continuity connectors about the minor axis of the framing member is required.

> **Exception:** Eccentricity may be allowed when it can be shown that all components of forces are positively resisted. The resistance must be supported by calculations or tests.

A206.6 Minimum member size. Wood members used to develop anchorage forces to the diaphragm must be at least 3-inch (76 mm) nominal members for new construction and replacement. All such members must be checked for gravity and earthquake loading as part of the wall-anchorage system.

> **Exception:** Existing 2-inch (51 mm) nominal members may be doubled and internailed to meet the strength requirement.

A206.7 Combination of anchor types. New anchors used in combination on a single framing member shall be of compatible behavior and stiffness.

A206.8 Anchorage at interior walls. Existing interior reinforced concrete or reinforced masonry walls that extend to the floor above or to the roof diaphragm shall be anchored for out-of-plane forces per Sections A206.1 and A206.3.Walls extending through the roof diaphragm shall be anchored for out-of-plane forces on both sides, and continuity ties shall be spliced across or continuous through the interior wall to provide diaphragm continuity.

A206.9 Collectors. If collectors are not present at reentrant corners or interior shear walls, they shall be provided. Existing or new collectors shall be designed for the capacity required to develop into the diaphragm a force equal to the lesser of the rocking or shear capacity of the reentrant wall or the tributary shear based on 75 percent of the horizontal forces specified in Chapter 16 of the *International Building Code*. The capacity of the collector need not exceed the capacity of the diaphragm to deliver loads to the collector. A connection shall be provided from the collector to the reentrant wall to transfer the full collector force (load). If a truss or beam other than a rafter or purlin is supported by the reentrant wall or by a column integral with the reentrant wall, then an independent secondary column is required to support the roof or floor members whenever rocking or shear capacity of the reentrant wall is less than the tributary shear.

A206.10 Mezzanines. Existing mezzanines relying on reinforced concrete or reinforced masonry walls for vertical and/or lateral support shall be anchored to the walls for the tributary mezzanine load. Walls depending on the mezzanine for lateral support shall be anchored per Sections A206.1, A206.2 and A206.3.

Exception: Existing mezzanines that have independent lateral and vertical support need not be anchored to the walls.

SECTION A207 MATERIALS OF CONSTRUCTION

All materials permitted by the building code, including their appropriate strength or allowable stresses, may be used to meet the requirements of this chapter.

CHAPTER A3

PRESCRIPTIVE PROVISIONS FOR SEISMIC STRENGTHENING OF CRIPPLE WALLS AND SILL PLATE ANCHORAGE OF LIGHT, WOOD-FRAME RESIDENTIAL BUILDINGS

SECTION A301 GENERAL

A301.1 Purpose. The provisions of this chapter are intended to promote public safety and welfare by reducing the risk of earthquake-induced damage to existing wood-frame residential buildings. The requirements contained in this chapter are prescriptive minimum standards intended to improve the seismic performance of residential buildings; however, they will not necessarily prevent earthquake damage.

This chapter sets standards for strengthening that may be approved by the building official without requiring plans or calculations prepared by an architect or engineer. The provisions of this chapter are not intended to prevent the use of any material or method of construction not prescribed herein. The building official may require that construction documents for strengthening using alternative materials or methods be prepared by an architect or engineer.

A301.2 Scope. The provisions of this chapter apply to residential buildings of light-frame wood construction assigned to Seismic Design Category C, D or E of the *International Building Code* containing one or more of the structural weaknesses specified in Section A303.

Exception: The provisions of this chapter do not apply to the buildings, or elements thereof, listed below. These buildings or elements require analysis by an engineer or architect in accordance with Section A301.3 to determine appropriate strengthening:

1. Group R-1, R-2 or R-4 occupancies with more than four dwelling units.
2. Buildings with a lateral-force-resisting system using poles or columns embedded in the ground.
3. Cripple walls that exceed 4 feet (1219 mm) in height.
4. Buildings exceeding three stories in height and any three-story building with cripple wall studs exceeding 14 inches (356 mm) in height.
5. Buildings where the building official determines that conditions exist that are beyond the scope of the prescriptive requirements of this chapter.
6. Buildings or portions thereof constructed on concrete slabs on grade.

The details and prescriptive provisions herein are not intended to be the only acceptable strengthening methods permitted. Alternative details and methods may be used when approved by the building official. Approval of alternatives shall be based on test data showing that the method or material used is at least equivalent in terms of strength, deflection and capacity to that provided by the prescriptive methods and materials.

The provisions of this chapter may be used to strengthen historic structures, provided they are not in conflict with other related provisions and requirements that may apply.

A301.3 Alternative design procedures. When analysis by an engineer or architect is required in accordance with Section A301.2, such analysis shall be in accordance with all requirements of the building code, except that the base shear may be taken as 75 percent of the horizontal forces specified in the building code.

SECTION A302 DEFINITIONS

For the purpose of this chapter, in addition to the applicable definitions in the building code, certain additional terms are defined as follows:

CHEMICAL ANCHOR. An assembly consisting of a threaded rod, washer, nut and chemical adhesive approved by the building official for installation in existing concrete or masonry.

COMPOSITE PANEL. A wood structural panel product composed of a combination of wood veneer and wood-based material, and bonded with waterproof adhesive.

CRIPPLE WALL. A wood-frame stud wall extending from the top of the foundation to the underside of the lowest floor framing.

EXPANSION BOLT. A single assembly approved by the building official for installation in existing concrete or masonry. For the purpose of this chapter, expansion bolts shall contain a base designed to expand when properly set, wedging the bolt in the pre-drilled hole. Assembly shall also include appropriate washer and nut.

ORIENTED STRAND BOARD (OSB). A mat-formed wood structural panel product composed of thin rectangular wood strands or wafers arranged in oriented layers and bonded with waterproof adhesive.

PERIMETER FOUNDATION. A foundation system that is located under the exterior walls of a building.

PLYWOOD. A wood structural panel product composed of sheets of wood veneer bonded together with the grain of adjacent layers oriented at right angles to one another.

SNUG-TIGHT. As tight as an individual can torque a nut on a bolt by hand, using a wrench with a 10-inch-long (254 mm) handle, and the point at which the full surface of the plate washer is contacting the wood member and slightly indenting the wood surface.

WAFERBOARD. A mat-formed wood structural panel product composed of thin rectangular wood wafers arranged in random layers and bonded with waterproof adhesive.

WOOD STRUCTURAL PANEL. A structural panel product composed primarily of wood and meeting the requirements of United States Voluntary Product Standard PS 1 and United States Voluntary Product Standard PS 2. Wood structural panels include all-veneer plywood, composite panels containing a combination of veneer and wood-based material, and mat-formed panels such as oriented strand board and waferboard.

SECTION A303 STRUCTURAL WEAKNESSES

For the purpose of this chapter, structural weaknesses shall be as specified below.

1. Sill plates or floor framing that are supported directly on the ground without an approved foundation system.
2. A perimeter foundation system that is constructed only of wood posts supported on isolated pad footings.
3. Perimeter foundation systems that are not continuous.

   Exceptions:

   1. Existing single-story exterior walls not exceeding 10 feet (3048 mm) in length, forming an extension of floor area beyond the line of an existing continuous perimeter foundation.
   2. Porches, storage rooms and similar spaces not containing fuel-burning appliances.
4. A perimeter foundation system that is constructed of unreinforced masonry or stone.
5. Sill plates that are not connected to the foundation or that are connected with less than what is required by the building code.

   Exception: When approved by the building official, connections of a sill plate to the foundation made with other than sill bolts may be accepted if the capacity of the connection is equivalent to that required by the building code.
6. Cripple walls that are not braced in accordance with the requirements of Section A304.4 and Table A3-A, or cripple walls not braced with diagonal sheathing or wood structural panels in accordance with the building code.

SECTION A304 STRENGTHENING REQUIREMENTS

A304.1 General.

A304.1.1 Scope. The structural weaknesses noted in Section A303 shall be strengthened in accordance with the requirements of this section. Strengthening work may include both new construction and *alteration* of existing construction. Except as provided herein, all strengthening work and materials shall comply with the applicable provisions of the building code. Alternative methods of strengthening may be used, provided such systems are designed by an engineer or architect and are approved by the building official.

A304.1.2 Condition of existing wood materials. All existing wood materials that will be a part of the strengthening work (sills, studs, sheathing, etc.) shall be in a sound condition and free from defects that substantially reduce the capacity of the member. Any wood material found to contain fungus infection shall be removed and replaced with new material. Any wood material found to be infested with insects or to have been infested with insects shall be strengthened or replaced with new materials to provide a net dimension of sound wood at least equal to its undamaged original dimension.

A304.1.3 Floor joists not parallel to foundations. Floor joists framed perpendicular or at an angle to perimeter foundations shall be restrained either by an existing nominal 2-inch-wide (51 mm) continuous rim joist or by a nominal 2-inch-wide (51 mm) full-depth blocking between alternate joists in one-and two-story buildings, and between each joist in three-story buildings. Existing blocking for multistory buildings must occur at each joist space above a braced cripple wall panel.

Existing connections at the top and bottom edges of an existing rim joist or blocking need not be verified in one-story buildings. In multistory buildings, the existing top edge connection need not be verified; however, the bottom edge connection to either the foundation sill plate or the top plate of a cripple wall shall be verified. The minimum existing bottom edge connection shall consist of 8d toenails spaced 6 inches (152 mm) apart for a continuous rim joist, or three 8d toenails per block. When this minimum bottom edge-connection is not present or cannot be verified, a supplemental connection installed as shown in Figure A3-8 shall beprovided.

Where an existing continuous rim joist or the minimum existing blocking does not occur, new $^3/_4$-inch (19 mm) wood structural panel blocking installed tightly between floor joists and nailed as shown in Figure A3-8 shall be provided at the inside face of the cripple wall. In lieu of $^3/_4$-inch (19 mm) wood structural panel blocking, tightfitting, full-depth 2-inch (51 mm) blocking may be used. New blocking may be omitted where it will interfere with vents or plumbing that penetrates the wall.

A304.1.4 Floor joists parallel to foundations. Where existing floor joists are parallel to the perimeter foundations, the end joist shall be located over the foundation and, except for required ventilation openings, shall be continuous and in continuous contact with the foundation sill plate or the top plate of the cripple wall. Existing connections at the top and bottom edges of the end joist need not be verified in one-story buildings. In multistory buildings, the existing top edge connection of the end joist need not be verified; however, the bottom edge connection to either the foundation sill plate or the top plate of a cripple wall shall be verified. The minimum bottom edge connection shall be 8d toenails spaced 6 inches (152 mm) apart. If this minimum bottom edge connection is not present or cannot be verified, a supplemental connection installed as shown in Figure A3-9 shall be provided.

A304.2 Foundations.

A304.2.1 New perimeter foundations. New perimeter foundations shall be provided for structures with the structural weaknesses noted in Items 1 and 2 of Section A303. Soil investigations or geotechnical studies are not required for this work unless the building is located in a special study zone as designated by the jurisdiction or other public agency.

A304.2.2 Foundation evaluation by an engineer or architect. Partial perimeter foundations or unreinforced masonry foundations shall be evaluated by an engineer or architect for the force levels noted in Section A301.3. Test reports or other substantiating data to determine existing foundation material strengths shall be submitted for review. When approved by the building official, these foundation systems may be strengthened in accordance with the recommendations included with the evaluation in lieu of being replaced.

Exception: In lieu of testing existing foundations to determine material strengths, and when approved by the building official, a new nonperimeter foundation system designed for the forces noted in Section A301.3 may be used to resist all exterior wall lateral forces.

A304.2.3 Details for new perimeter foundations. All new perimeter foundations shall be continuous and constructed according to one of the details shown in Figure A3-1 or A3-2.

Exceptions:

1. When approved by the building official, the existing clearance between existing floor joists or girders and existing grade below the floor need not comply with the building code.
2. When approved by the building official, and when designed by an engineer or architect, partial perimeter foundations may be used in lieu of a continuous perimeter foundation.

A304.2.4 Required compressive strength. New concrete foundations shall have a minimum compressive strength of 2,500 pounds per square inch (17.24 MPa) at 28 days.

A304.2.5 New hollow-unit masonry foundations. New hollow-unit masonry foundations shall be solidly grouted. Mortar shall be Type M or S, and the grout and masonry units shall comply with the building code.

A304.2.6 Reinforcing steel. Reinforcing steel shall comply with the requirements of the building code.

A304.3 Foundation sill plate anchorage.

A304.3.1 Existing perimeter foundations. When the building has an existing continuous perimeter foundation, all perimeter wall sill plates shall be bolted to the foundation with chemical anchors or expansion bolts in accordance with Table A3-A.

Anchors or bolts shall be installed in accordance with Figure A3-3, with the plate washer installed between the nut and the sill plate. The nut shall be tightened to a snug-tight condition after curing is complete for chemical anchors and after expansion wedge engagement for expansion bolts. The installation of nuts on all bolts shall be subject to verification by the building official. Where existing conditions prevent anchor or bolt installation through the sill plate, this connection may be made in accordance with Figure A3-4A, A3-4B or A3-4C. The spacing of these alternate connections shall comply with the maximum spacing requirements of Table A3-A. Expansion bolts shall not be used when the installation causes surface cracking of the foundation wall at the location of the bolt.

A304.3.2 Placement of chemical anchors and expansion bolts. Chemical anchors or expansion bolts shall be placed within 12 inches (305 mm), but not less than 9 inches (229 mm), from the ends of sill plates and shall be placed in the center of the stud space closest to the required spacing. New sill plates may be installed in pieces when necessary because of existing conditions. For lengths of sill plate greater than 12 feet (3658 mm), anchors or bolts shall be spaced along the sill plate as noted in Table A3-A. For other lengths of sill plate, see Table A3-B. For lengths of sill plate less than 30 inches (762 mm), a minimum of one anchor or bolt shall be installed.

Exception: Where physical obstructions such as fireplaces, plumbing or heating ducts interfere with the placement of an anchor or bolt, the anchor or bolt shall be placed as close to the obstruction as possible, but not less than 9 inches (229 mm) from the end of the plate. Center-to-center spacing of the anchors or bolts shall be reduced as necessary to provide the minimum total number of anchors required based on the full length of the wall. Center-to-center spacing shall not be less than 12 inches (305 mm).

A304.3.3 New perimeter foundations. Sill plates for new perimeter foundations shall be bolted as required by Table A3-A and as shown in Figure A3-1 or A3-2.

A304.4 Cripple wall bracing.

A304.4.1 General. Exterior cripple walls not exceeding 4 feet (1219 mm) in height shall use the prescriptive bracing method listed below. Cripple walls over 4 feet (1219 mm) in height require analysis by an engineer or architect in accordance with Section A301.3.

A304.4.1.1 Sheathing installation requirements. Wood structural panel sheathing shall not be less than $^{15}/_{32}$-inch (12 mm) thick and shall be installed in accordance with Figure A3-5 or A3-6. All individual pieces of wood structural panels shall be nailed with 8d common nails spaced 4 inches (102 mm) on center at all edges and 12 inches (305 mm) on center at each intermediate support with not less than two nails for each stud. Nails shall be driven so that their heads are flush with the surface of the sheathing and shall penetrate the supporting member a minimum of $1^1/_2$ inches (38 mm). When a nail fractures the surface, it shall be left in place and not counted as part of the required nailing. A new 8d nail shall be located within 2 inches (51 mm) of the discounted nail and be hand-driven flush with the sheathing surface. All horizontal joints must occur over nominal 2-inch by 4-inch (51 mm by 102 mm) blocking installed with the nominal 4-inch (102 mm) dimension against the face of the plywood.

Vertical joints at adjoining pieces of wood structural panels shall be centered on existing studs such that there is a minimum $^1/_8$ inch (3.2 mm) between the panels, and such that the nails are placed a minimum of $^1/_2$ inch (12.7 mm) from the edges of the existing stud. Where such edge distances cannot be maintained because of the width of the existing stud, a new stud shall be added adjacent to the existing studs and connected in accordance with Figure A3-7.

A304.4.2 Distribution and amount of bracing. See Table A3-A and Figure A3-10 for the distribution and amount of bracing required for each wall line. Each braced panel length must be at least two times the height of the cripple stud. Where the minimum amount of bracing prescribed in Table A3-A cannot be installed along any walls, the bracing must be designed in accordance with Section A301.3.

Exception: Where physical obstructions such as fireplaces, plumbing or heating ducts interfere with the placement of cripple wall bracing, the bracing shall then be placed as close to the obstruction as possible. The total amount of bracing required shall not be reduced because of obstructions.

A304.4.3 Stud space ventilation. When bracing materials are installed on the interior face of studs forming an enclosed space between the new bracing and the existing exterior finish, each braced stud space must be ventilated. Adequate ventilation and access for future inspection shall be provided by drilling one 2-inch to 3-inch–diameter (51 mm to 76 mm) round hole through the sheathing, nearly centered between each stud at the top and bottom of the cripple wall. Such holes should be spaced a minimum of 1 inch (25 mm) clear from the sill or top plates. In stud spaces containing sill bolts, the hole shall be located on the center line of the sill bolt but not closer than 1 inch (25 mm) clear from the nailing edge of the sheathing. When existing blocking occurs within the stud space, additional ventilation holes shall be placed above and below the blocking, or the existing block shall be removed and a new nominal 2-inch by 4-inch (51 mm by 102 mm) block shall be installed with the nominal 4-inch (102 mm) dimension against the face of the plywood. For stud heights less than 18 inches (457 mm), only one ventilation hole need be provided.

A304.4.4 Existing underfloor ventilation. Existing underfloor ventilation shall not be reduced without providing equivalent new ventilation as close to the existing ventilation as possible. Braced panels may include underfloor ventilation openings when the height of the opening, measured from the top of the foundation wall to the top of the opening, does not exceed 25 percent of the height of the cripple stud wall; however, the length of the panel shall be increased a distance equal to the length of the opening or one stud space minimum. Where an opening exceeds 25 percent of the cripple wall height, braced panels shall not be located where the opening occurs. See Figure A3-7.

Exception: For homes with a post and pier foundation system where a new continuous perimeter foundation system is being installed, new ventilation shall be provided in accordance with the building code.

A304.5 Quality control. All work shall be subject to inspection by the building official including, but not limited to:

1. Placement and installation of new chemical anchors or expansion bolts installed in existing foundations. Special inspection is not required for chemical anchors installed in existing foundations regulated by the prescriptive provisions of this chapter.
2. Installation and nailing of new cripple wall bracing.
3. Any work may be subject to special inspection when required by the building official in accordance with the building code.

A304.6 Phasing of the strengthening work. When approved by the building official, the strengthening work contained in this chapter may be completed in phases. The strengthening work in any phase shall be performed on two parallel sides of the structure at the same time.

TABLE A3-A—SILL PLATE ANCHORAGE AND CRIPPLE WALL BRACING

NUMBER OF STORIES ABOVE CRIPPLE WALLS	MINIMUM SILL PLATE CONNECTION AND MAXIMUM SPACING[a, b]	AMOUNT OF BRACING FOR EACH WALL LINE[c,d,e]	
		A Combination of Exterior Walls Finished with Portland Cement Plaster and Roofing Using Clay Tile or Concrete Tile Weighing More than 6 psf (287 N/m²)	All Other Conditions
One story	$^{1}/_{2}$ inch (12.7 mm) spaced 6 feet, 0 inch (1829 mm) center-to-center with washer plate	Each end and not less than 50 percent of the wall length	Each end and not less than 40 percent of the wall length
Two stories	$^{1}/_{2}$ inch (12.7 mm) spaced 4 feet, 0 inch (1219 mm) center-to-center with washer plate; or $^{5}/_{8}$ inch (15.9 mm) spaced 6 feet, 0 inch (1829 mm) center-to-center with washer plate	Each end and not less than 70 percent of the wall length	Each end and not less than 50 percent of the wall length
Three stories	$^{5}/_{8}$ inch (15.9 mm) spaced 4 feet, 0 inch (1219 mm) center-to-center with washer plate	100 percent of the wall length[f]	Each end and not less than 80 percent of the wall length[f]

a. Sill plate anchors shall be chemical anchors or expansion bolts in accordance with Section A304.3.1.
b. All washer plates shall be 2 inches by 2 inches by $^{3}/_{16}$ inch (51 mm by 51 mm by 4.8 mm) minimum.
c. See Figure A3–10 for braced panel layout.
d. Braced panels at ends of walls shall be located as near to the end as possible.
e. All panels along a wall shall be nearly equal in length and shall be nearly equal in spacing along the length of the wall.
f. The minimum required underfloor ventilation openings are permitted in accordance with Section A304.4.4.

TABLE A3-B—SILL PLATE ANCHORAGE FOR VARIOUS LENGTHS OF SILL PLATE[a,b]

NUMBER OF STORIES	LENGTHS OF SILL PLATE		
	Less than 12 feet (3658 mm) to 6 feet (1829 mm)	Less than 6 feet (1829 mm) to 30 inches (762 mm)	Less than 30 inches footnote (762 mm)[c]
One story	Three connections	Two connections	One connection
Two stories	Four connections for $^{1}/_{2}$-inch (12.7 mm) anchors or bolts or Three connections for $^{5}/_{8}$-inch (15.9 mm) anchors or bolts	Two connections	One connection
Three stories	Four connections	Two connections	One connection

a. Connections shall be either chemical anchors or expansion bolts.
b. See Section A304.3.2 for minimum end distances.
c. Connections shall be placed as near to the center of the length of plate as possible.

TABLE A3–C—NOT USED

NUMBER OF STORIES	MINIMUM FOUNDATION DIMENSIONS					MINIMUM FOUNDATION REINFORCING	
	W	*F*	*D*[a, b, c]	T	*H*	VERTICAL REINFORCING	
1	12 inches (305 mm)	6 inches (152 mm)	12 inches (305 mm)	6 inches (152 mm)	≤ 24 inches (610 mm)	Single-pour wall and footing	Footing poured separate from wall
2	15 inches (381 mm)	7 inches (178 mm)	18 inches (457 mm)	8 inches (203 mm)	≥ 36 inches (914 mm)	#4 @ 48 inches (1219 mm) on center	#4 @ 32 inches (813 mm) on center
3	18 inches (457 mm)	8 inches (203 mm)	24 inches (610 mm)	10 inches (254 mm)	≥ 36 inches (914 mm)	#4 @ 48 inches (1219 mm) on center	#4 @ 18 inches (457 mm) on center

a. Where frost conditions occur, the minimum depth shall extend below the frost line.

b. The ground surface along the interior side of the foundation may be excavated to the elevation of the top of the footing.

c. When expansive soil is encountered, the foundation depth and reinforcement shall be as directed by the building official.

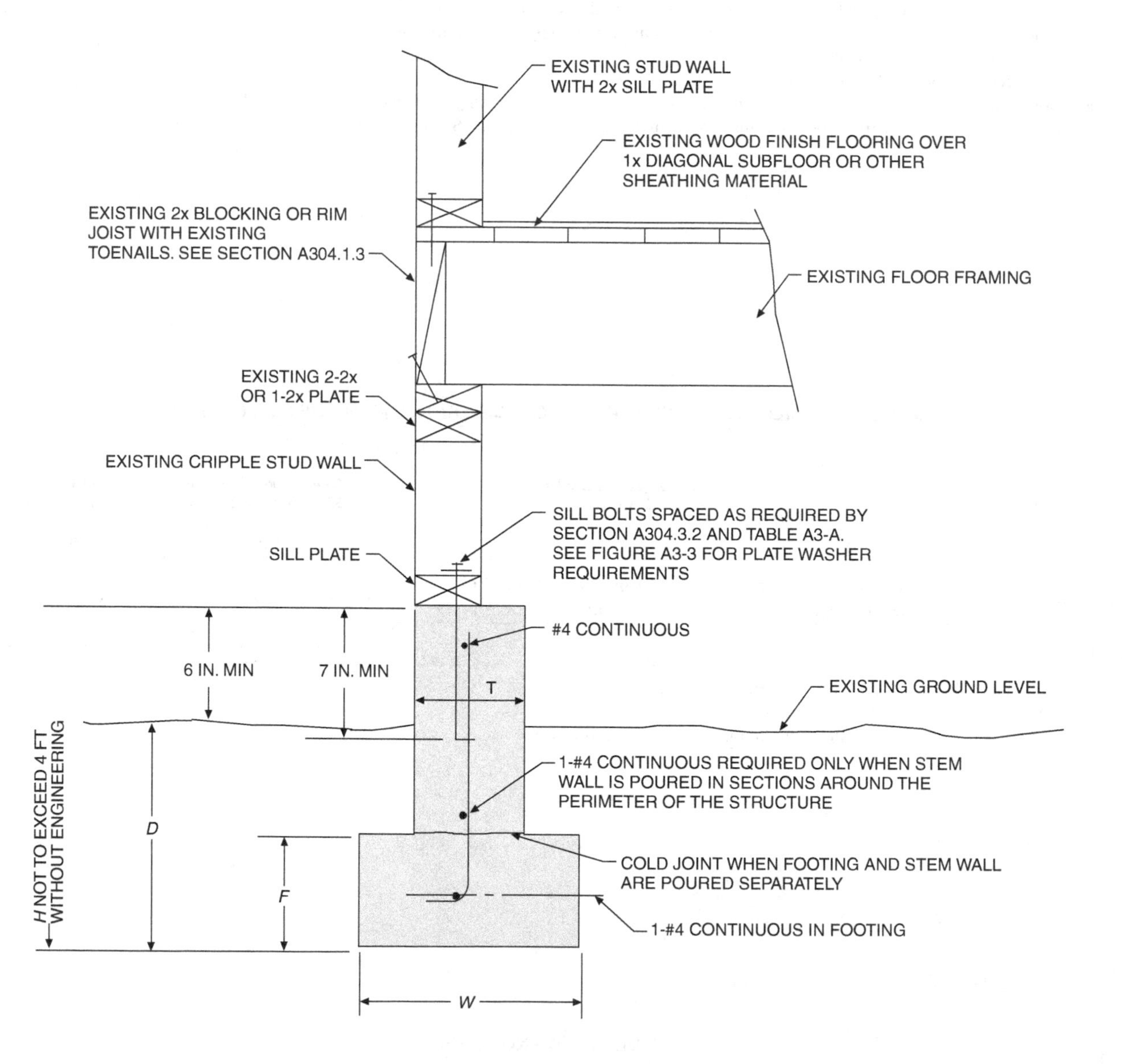

For SI: 1 inch = 25.4 mm, 1 foot = 304.8 mm.

NOTE: See Figure A3-5 or A3-6 for cripple wall bracing.

FIGURE A3-1—NEW REINFORCED CONCRETE FOUNDATION SYSTEM

	MINIMUM FOUNDATION DIMENSIONS					MINIMUM FOUNDATION REINFORCING	
NUMBER OF STORIES	*W*	*F*	*D*[a, b, c]	T	*H*	VERTICAL REINFORCING	HORIZONTAL REINFORCING
1	12 inches (305 mm)	6 inches (152 mm)	12 inches (305 mm)	6 inches (152 mm)	≤ 24 inches (610 mm)	#4 @ 24 inches (610 mm) on center	#4 continuous at top of stem wall
2	15 inches (381 mm)	7 inches (178 mm)	18 inches (457 mm)	8 inches (203 mm)	≥ 24 inches (610 mm)	#4 @ 24 inches (610 mm) on center	#4 @ 16 inches (406 mm) on center
3	18 inches (457 mm)	8 inches (203 mm)	24 inches (610 mm)	10 inches (254 mm)	≥ 36 inches (914 mm)	#4 @ 24 inches (610 mm) on center	#4 @ 16 inches (406 mm) on center

a. Where frost conditions occur, the minimum depth shall extend below the frost line.
b. The ground surface along the interior side of the foundation may be excavated to the elevation of the top of the footing.
c. When expansive soil is encountered, the foundation depth and reinforcement shall be as directed by the building official.

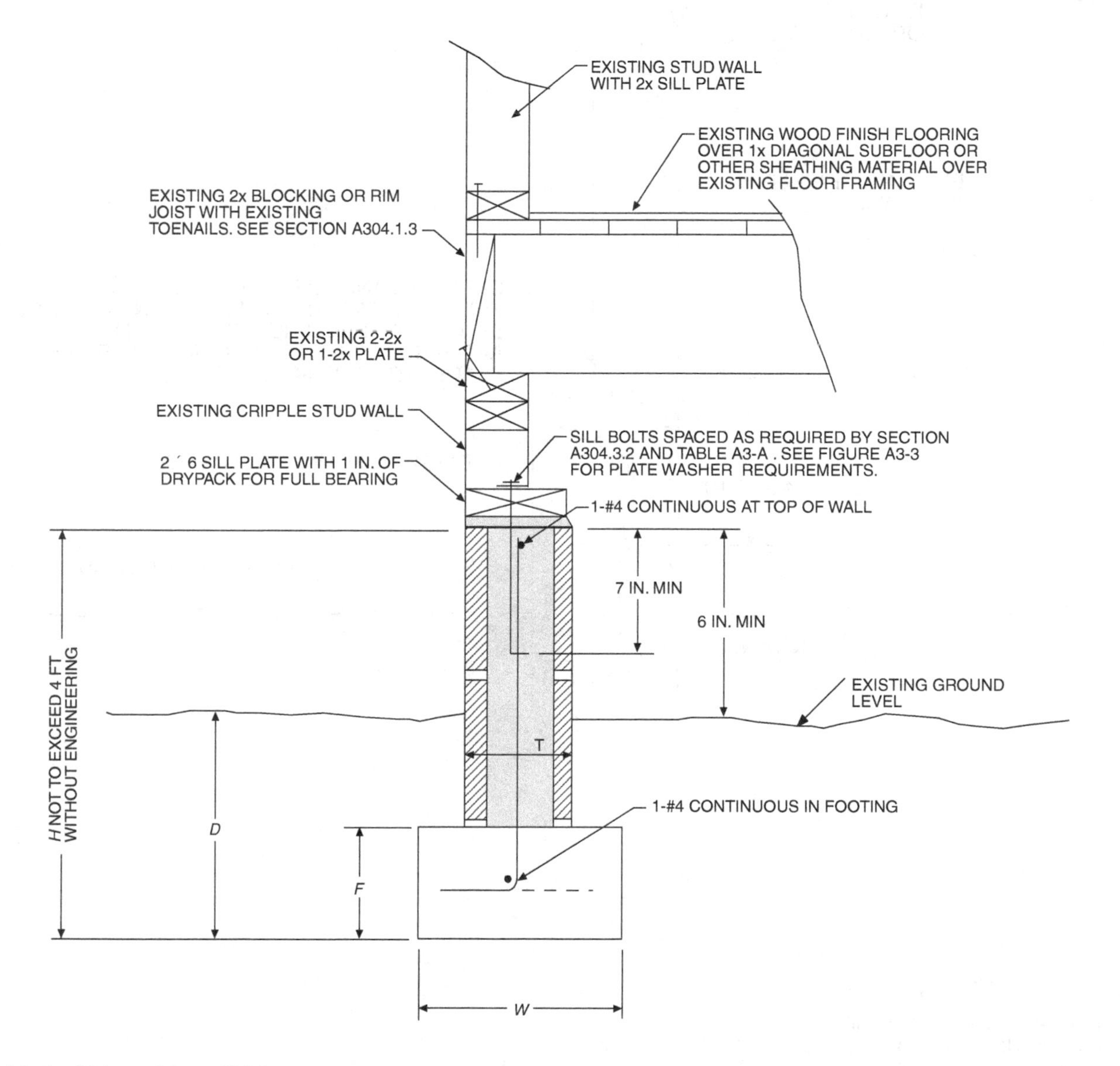

For SI: 1 inch = 25.4 mm, 1 foot = 304.8 mm.
NOTE: See Figure A3-5 or A3-6 for cripple wall bracing.

FIGURE A3-2—NEW HOLLOW-MASONRY UNIT FOUNDATION WALL

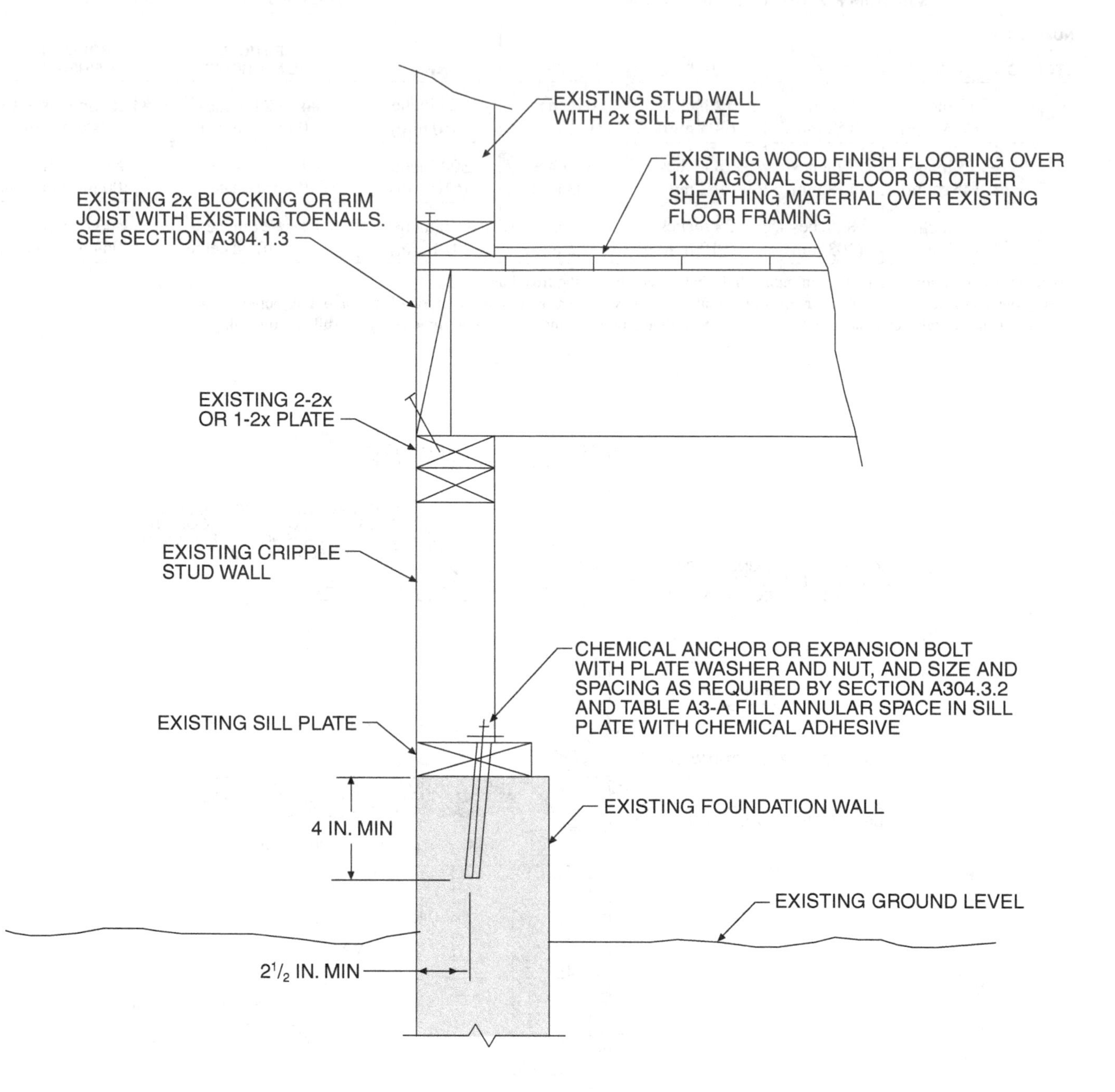

For SI: 1 inch = 25.4 mm.

NOTES:

1. Plate washers shall comply with the following:
   $^1/_2$ in. anchor or bolt—2 in. x 2 in. x $^3/_{16}$ in.
   $^5/_8$ in. anchor or bolt—2 in. x 2 in. x $^3/_{16}$ in.
2. See Figure A3-5 or A3-6 for cripple wall bracing.

FIGURE A3-3—SILL PLATE BOLTING TO EXISTING FOUNDATION

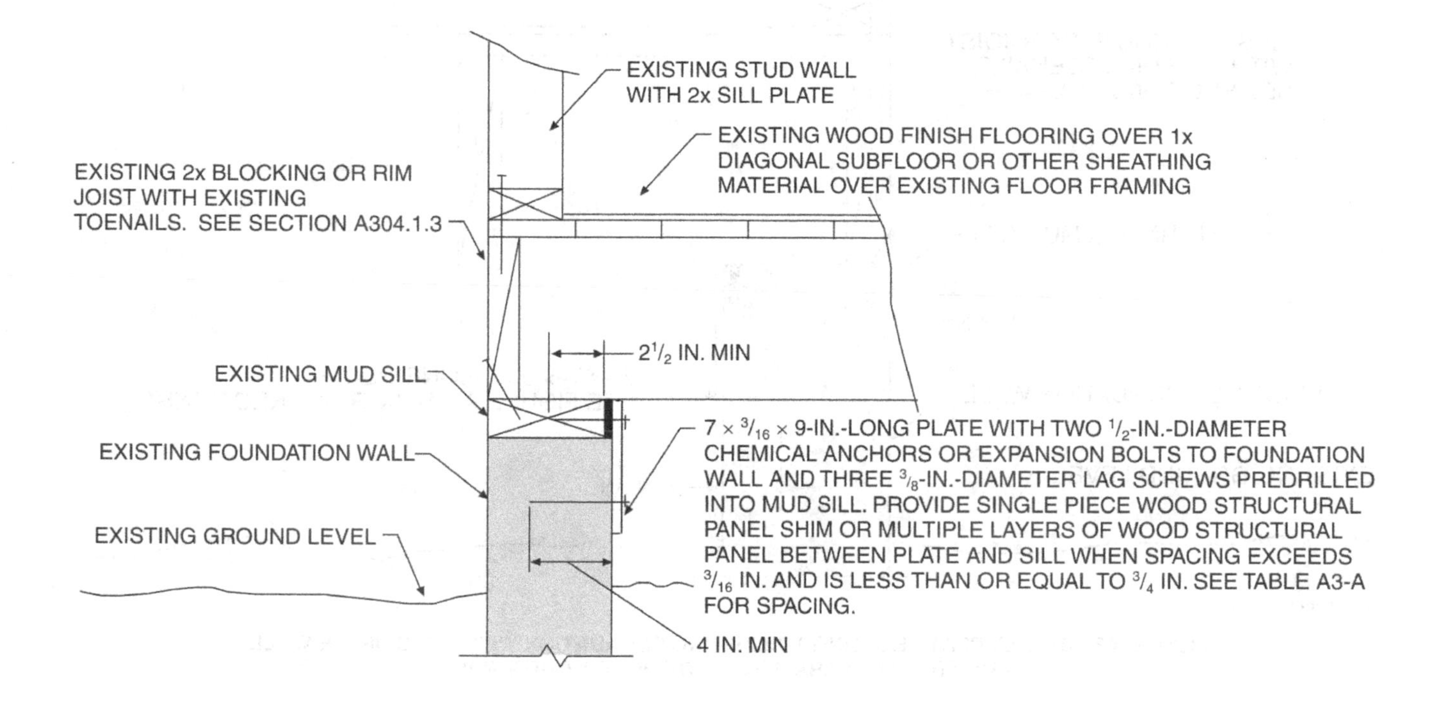

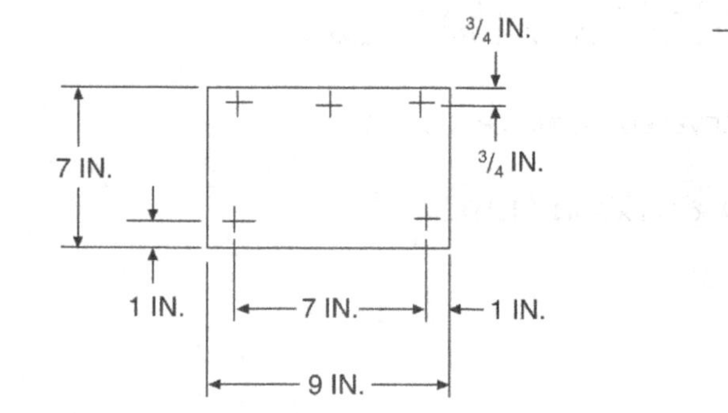

HOLE DIAMETER SHALL NOT EXCEED CONNECTOR DIAMETER BY MORE THAN $^1/_{16}$ IN.

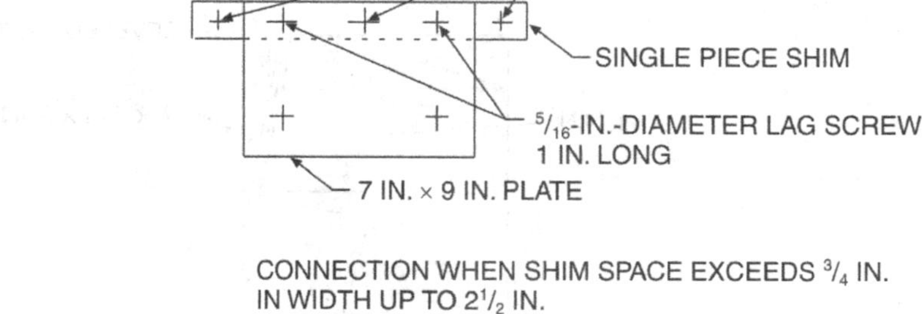

CONNECTION WHEN SHIM SPACE EXCEEDS $^3/_4$ IN. IN WIDTH UP TO $2^1/_2$ IN.

For SI: 1 inch = 25.4 mm, 1 foot = 304.8 mm.
NOTE: If shim space exceeds $2^1/_2$ in., alternate details will be required.

FIGURE A3-4A—SILL PLATE BOLTING IN EXISTING FOUNDATION—ALTERNATE

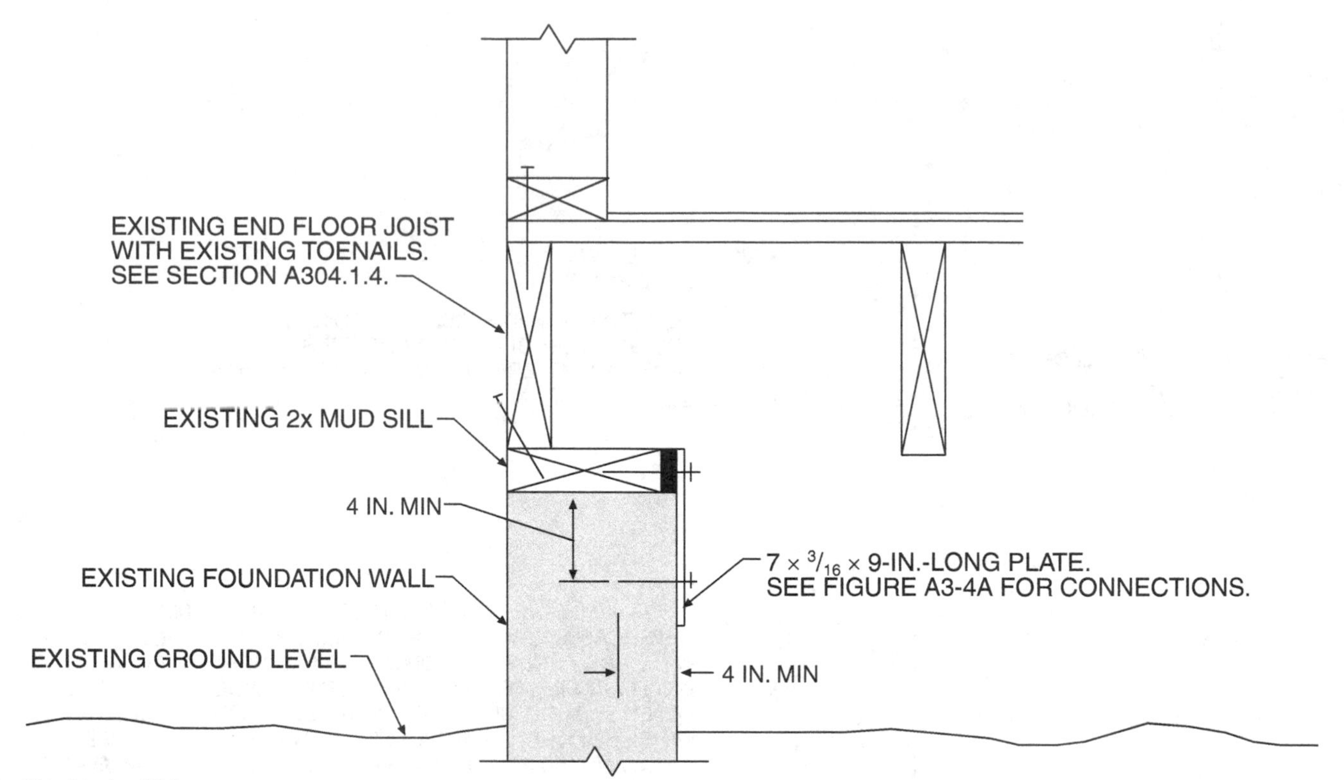

For SI: 1 inch = 25.4 mm.

FIGURE A3-4B—SILL PLATE BOLTING TO EXISTING FOUNDATION WITHOUT CRIPPLE WALL AND FRAMING PARALLEL TO THE FOUNDATION WALL

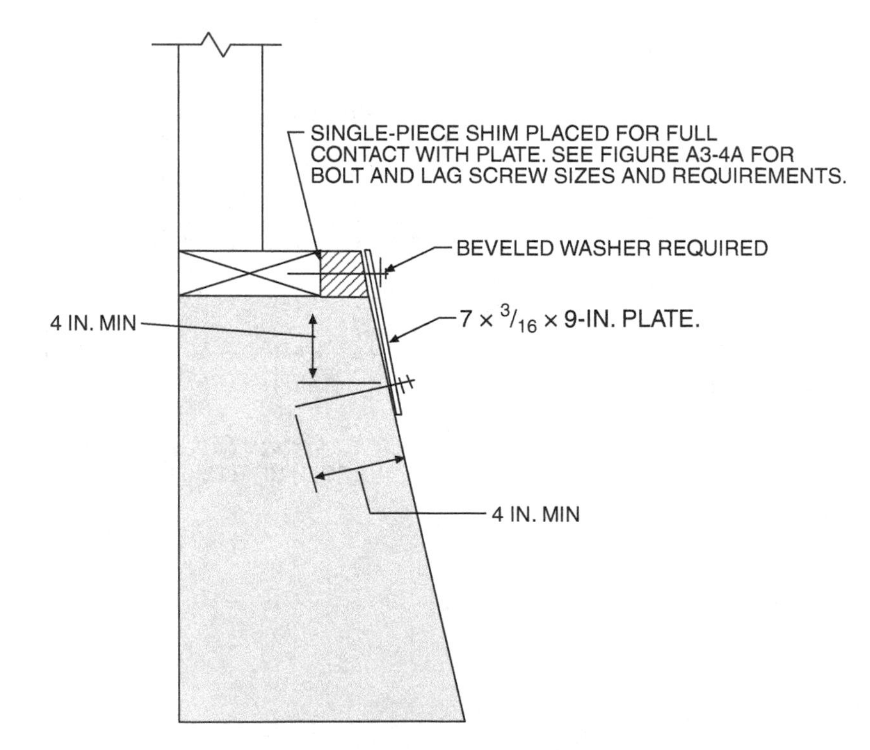

ALTERNATE CONNECTION FOR BATTERED FOOTING

For SI: 1 inch = 25.4 mm.

FIGURE A3-4C—SILL PLATE BOLTING IN EXISTING FOUNDATION—ALTERNATE

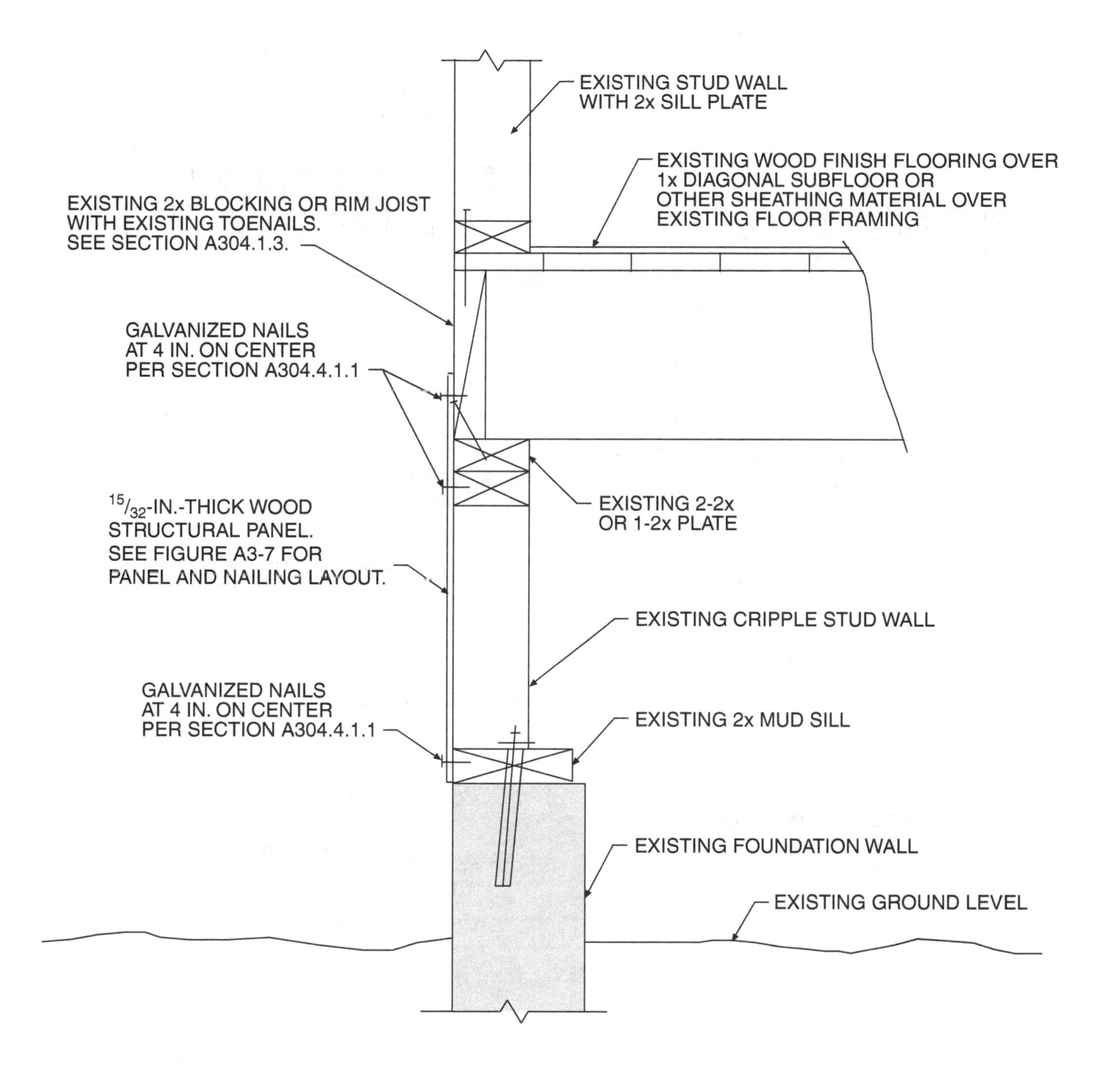

For SI: 1 inch = 25.4 mm.

NOTE: See Figure A3-3 for sill plate bolting.

FIGURE A3-5—CRIPPLE WALL BRACING WITH WOOD STRUCTURAL PANEL ON EXTERIOR FACE OF CRIPPLE STUDS

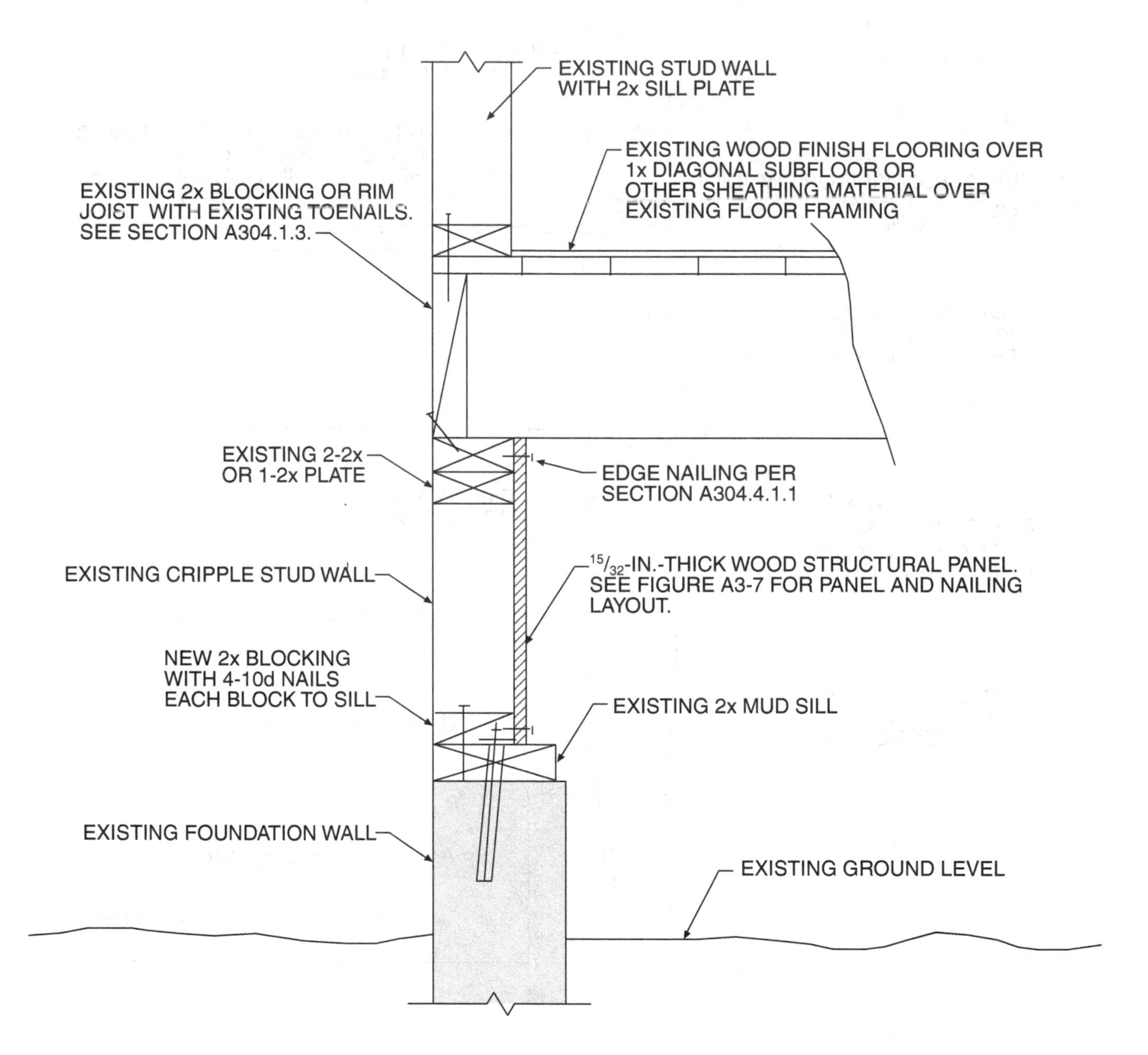

For SI: 1 inch = 25.4 mm.
NOTE: See Figure A3-3 for sill plate bolting.

FIGURE A3-6—CRIPPLE WALL BRACING WITH WOOD STRUCTURAL PANEL ON INTERIOR FACE OF CRIPPLE STUDS

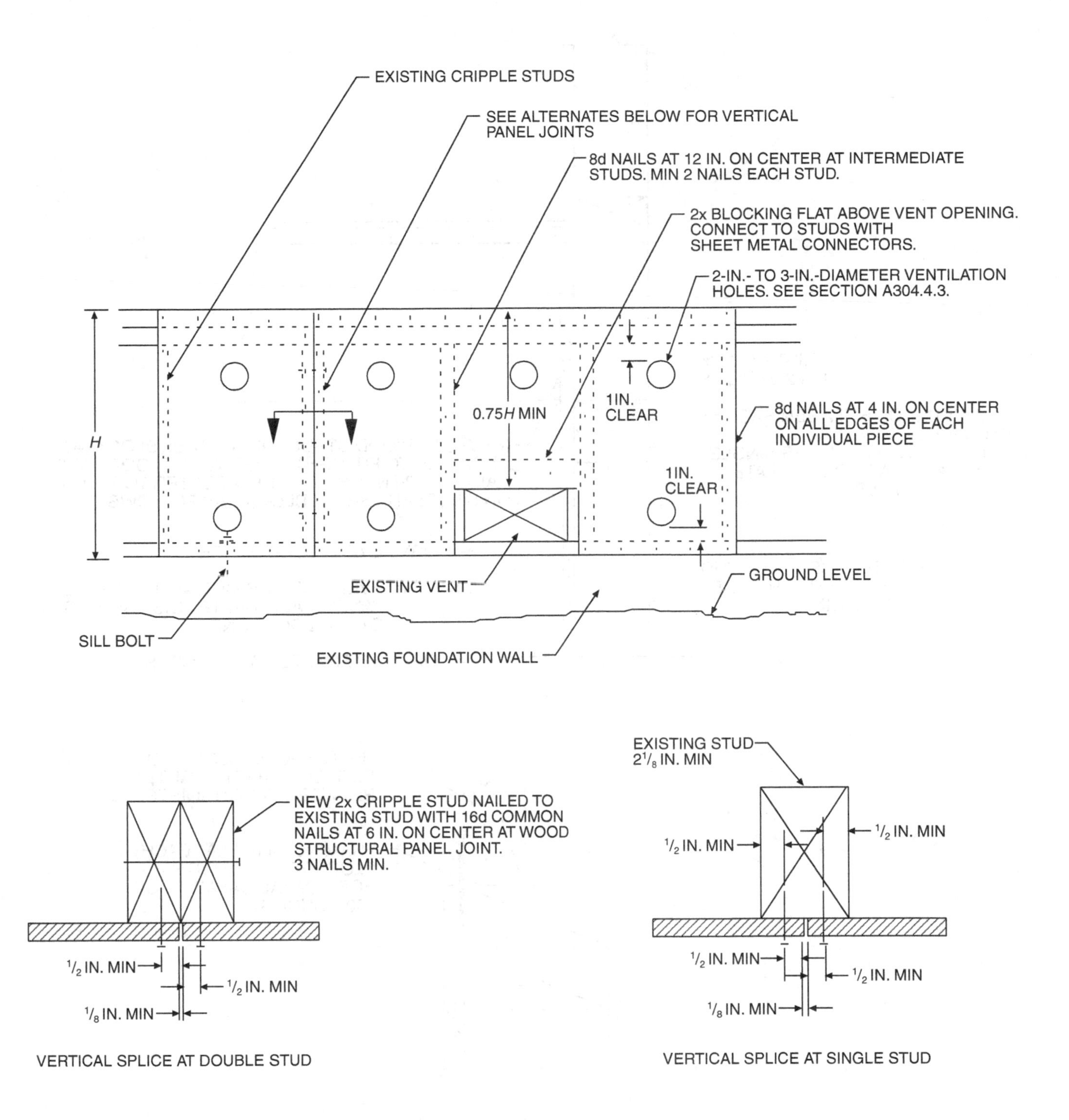

For SI: 1 inch = 25.4 mm.

FIGURE A3-7—PARTIAL CRIPPLE STUD WALL ELEVATION

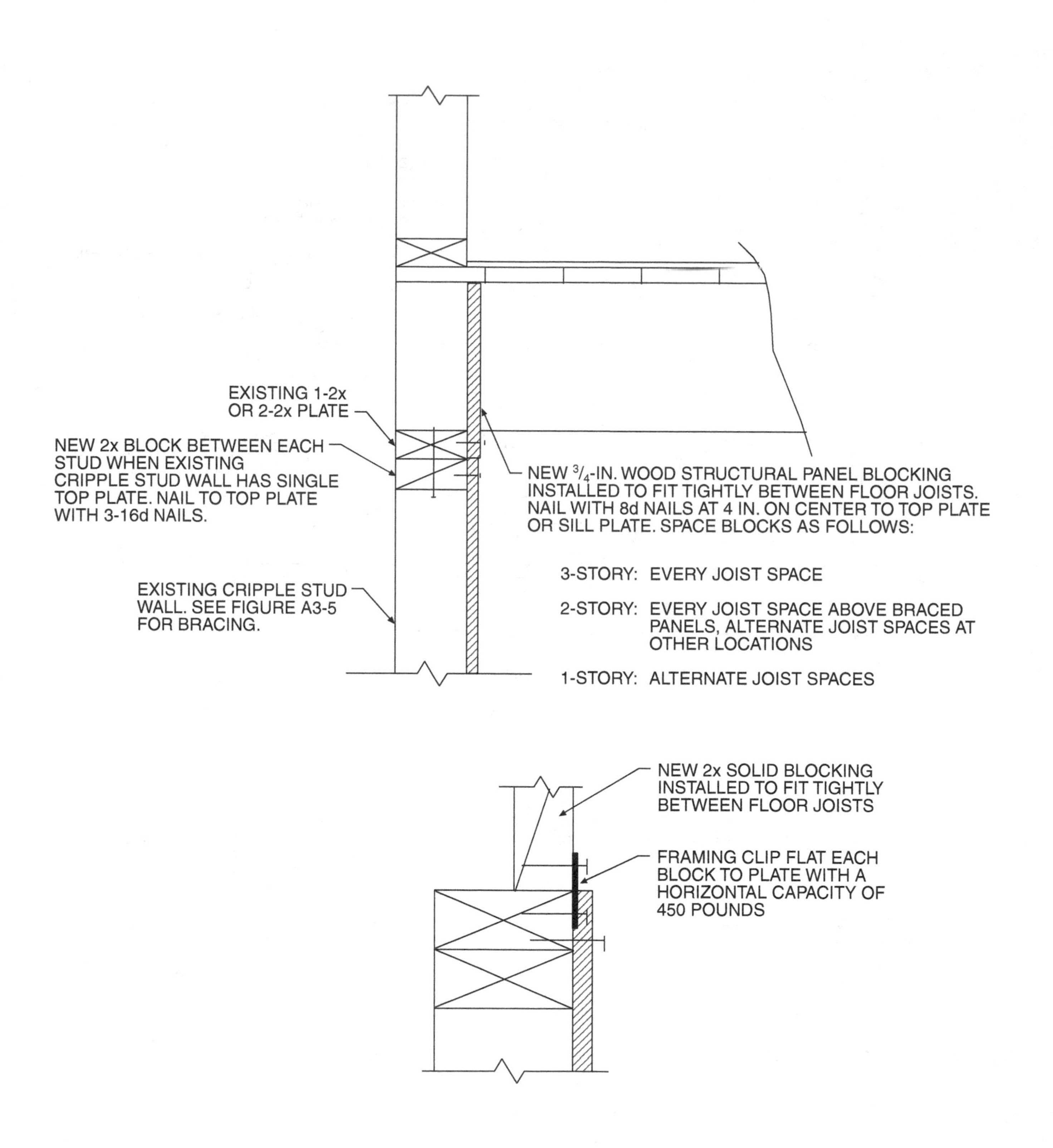

For SI: 1 inch = 25.4 mm, 1 pound = 4.4 N.

FIGURE A3-8—ALTERNATE BLOCKING WHERE RIM JOIST OR BLOCKING HAS BEEN OMITTED

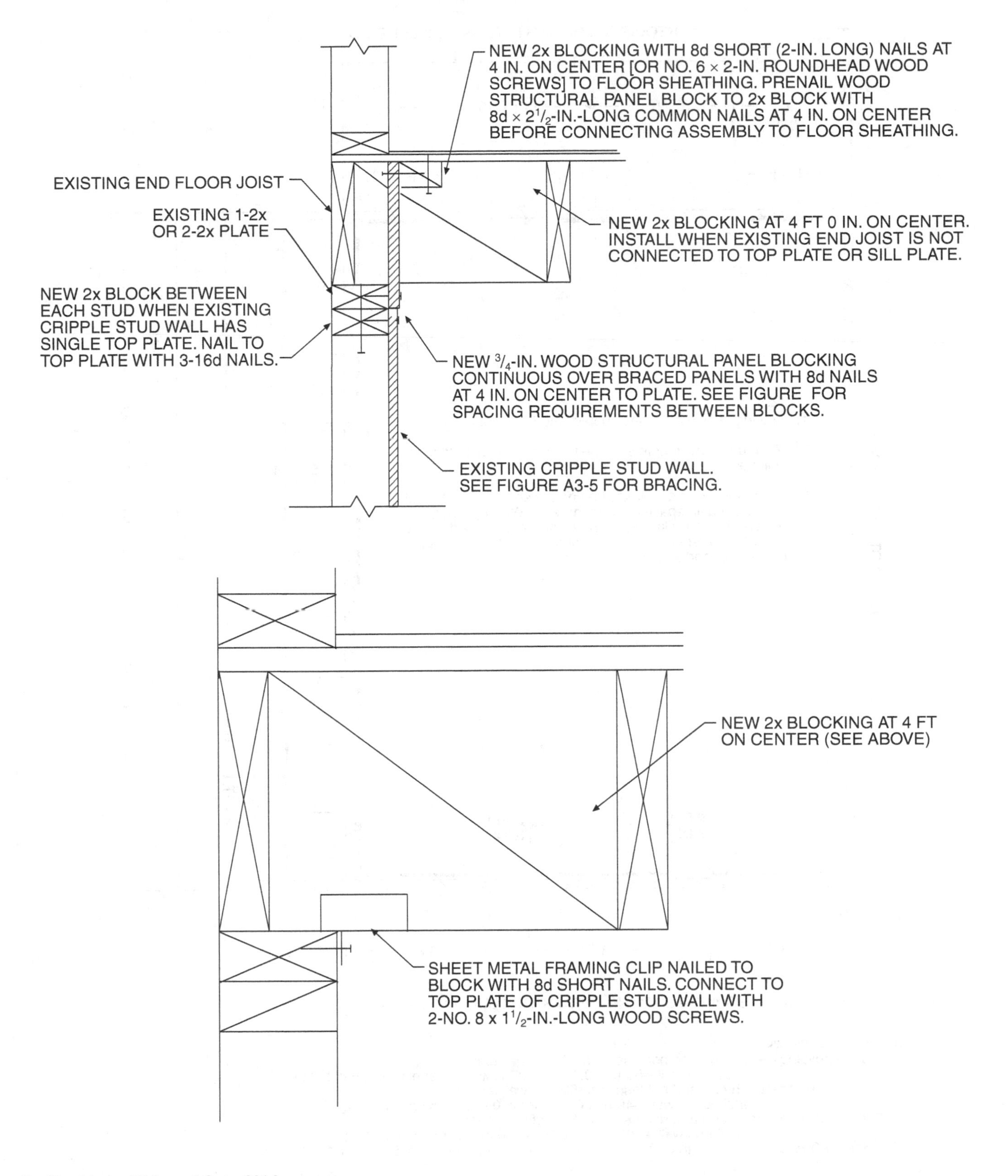

For SI: 1 inch = 25.4 mm, 1 foot = 304.8 mm.

FIGURE A3-9—CONNECTION OF CRIPPLE WALL TO FLOOR SHEATHING WHEN FLOOR FRAMING IS PARALLEL TO WALL

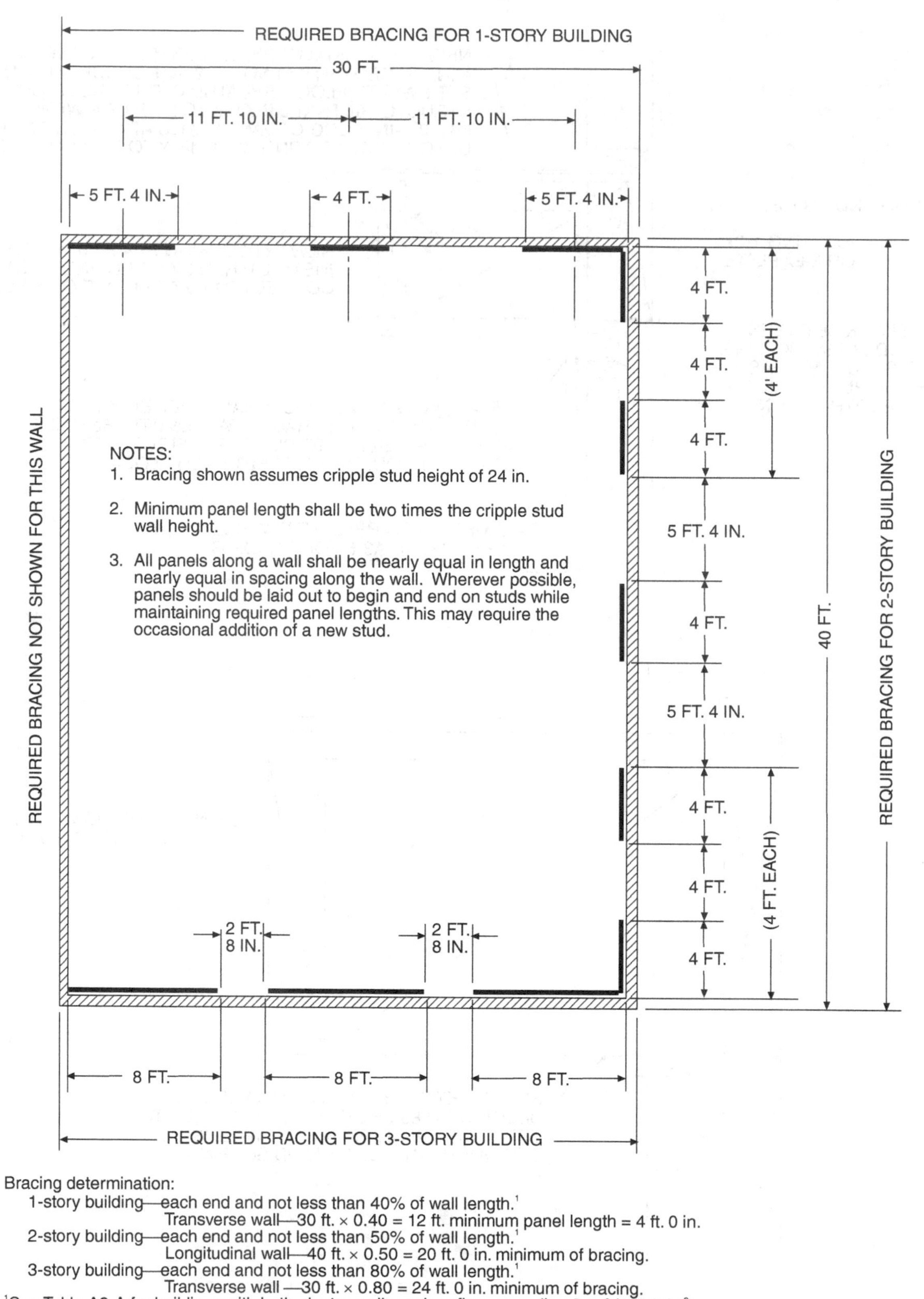

Bracing determination:

1-story building—each end and not less than 40% of wall length.[1]
Transverse wall—30 ft. × 0.40 = 12 ft. minimum panel length = 4 ft. 0 in.

2-story building—each end and not less than 50% of wall length.[1]
Longitudinal wall—40 ft. × 0.50 = 20 ft. 0 in. minimum of bracing.

3-story building—each end and not less than 80% of wall length.[1]
Transverse wall —30 ft. × 0.80 = 24 ft. 0 in. minimum of bracing.

[1]See Table A3-A for buildings with both plaster walls and roofing exceeding 6 psf (287 N/m^2).

For SI: 1 inch = 25.4 mm, 1 foot = 304.8 mm.

FIGURE A3-10—FLOOR PLAN—CRIPPLE WALL BRACING LAYOUT

CHAPTER A4

EARTHQUAKE HAZARD REDUCTION IN EXISTING WOOD-FRAME RESIDENTIAL BUILDINGS WITH SOFT, WEAK OR OPEN-FRONT WALLS

SECTION A401 GENERAL

A401.1 Purpose. The purpose of this chapter is to promote public welfare and safety by reducing the risk of death or injury that may result from the effects of earthquakes on existing wood-frame, multiunit residential buildings. The ground motions of past earthquakes have caused the loss of human life, personal injury and property damage in these types of buildings. This chapter creates minimum standards to strengthen the more vulnerable portions of these structures. When fully followed, these minimum standards will improve the performance of these buildings but will not necessarily prevent all earthquake-related damage.

A401.2 Scope. The provisions of this chapter shall apply to all existing Occupancy Group R-1 and R-2 buildings of wood construction or portions thereof where:

1. The ground floor portion of the wood-frame structure contains parking or other similar open floor space, which causes soft, weak or open-front wall lines as defined in this chapter, and there exists one or more stories above, or
2. The walls of any story or basement of wood construction are laterally braced with nonconforming structural materials as defined in this chapter, a soft or weak wall line exists as defined in this chapter and there exist two or more stories above.
3. The structure is assigned to Seismic Design Category C, D or E.

SECTION A402 DEFINITIONS

Notwithstanding the applicable definitions, symbols and notations in the building code, the following definitions shall apply for the purposes of this chapter:

APARTMENT HOUSE. Any building or portion thereof that contains three or more dwelling units. For the purposes of this chapter, "apartment house" includes residential condominiums.

ASPECT RATIO. The span-width ratio for horizontal diaphragms and the height-length ratio for vertical diaphragms.

CONGREGATE RESIDENCE. A congregate residence is any building or portion thereof for occupancy by other than a family that contains facilities for living, sleeping and sanitation as required by the building code and that may include facilities for eating and cooking. A congregate residence may be a shelter, convent, monastery, dormitory, fraternity or sorority house, but does not include jails, hospitals, nursing homes, hotels or lodging houses.

CRIPPLE WALL. A wood-frame stud wall extending from the top of the foundation wall to the underside of the lowest floor framing.

DWELLING UNIT. Any building or portion thereof for not more than one family that contains living facilities, including provisions for sleeping, eating, cooking and sanitation as required by the building code or congregate residence for 10 or fewer persons.

EXPANSION ANCHOR. An approved mechanical fastener placed in hardened concrete that is designed to expand in a self-drilled or pre-drilled hole of a specified size and engage the sides of the hole in one or more locations to develop shear and/or tension resistance to applied loads without grout, adhesive or drypack.

GROUND FLOOR. Any floor whose elevation is immediately accessible from an adjacent grade by vehicles or pedestrians. The ground floor portion of the structure does not include any floor that is completely below adjacent grades.

GUESTROOM. Any room or rooms used or intended to be used by a guest for sleeping purposes. Every 100 square feet (9.3 m^2) of superficial floor area in a congregate residence shall be considered a guestroom.

HOTEL. Any building containing six or more guestrooms intended or designed to be used, rented, hired out to be occupied, or that are occupied, for sleeping purposes by guests.

LIFE SAFETY PERFORMANCE LEVEL. The building performance level that includes significant damage to both structural and nonstructural components during a design earthquake, though at least some margin against either partial or total structural collapse remains. Injuries may occur, but the level of risk for life-threatening injury and entrapment is low.

LODGING HOUSE. Any building or portion thereof containing at least one but not more than five guest rooms where rent is paid in money, goods, labor or otherwise.

MOTEL. Motel shall mean a hotel as defined in this chapter.

MULTIUNIT RESIDENTIAL BUILDINGS. Hotels, lodging houses, congregate residences and apartment houses.

NONCONFORMING STRUCTURAL MATERIALS. Wall bracing materials other than wood structural panels or diagonal sheathing.

OPEN-FRONT WALL LINE. An exterior wall line, without vertical elements of the lateral-force-resisting system, that requires tributary seismic forces to be resisted by diaphragm rotation or excessive cantilever beyond parallel lines of shear walls. Diaphragms that cantilever more than 25 percent of the distance between lines of lateral-force-resisting elements from which the diaphragm cantilevers shall be considered excessive.

Exterior exit balconies of 6 feet (1829 mm) or less in width shall not be considered excessive cantilevers.

RETROFIT. An improvement of the lateral-force-resisting system by *alteration* of existing structural elements or *addition* of new structural elements.

SOFT WALL LINE. A wall line whose lateral stiffness is less than that required by story drift limitations or deformation compatibility requirements of this chapter. In lieu of analysis, a soft wall line may be defined as a wall line in a story where the story stiffness is less than 70 percent of the story above for the direction under consideration.

STORY. A story as defined by the building code, including any basement or underfloor space of a building with cripple walls exceeding 4 feet (1219 mm) in height.

STORY STRENGTH. The total strength of all seismic-resisting elements sharing the same story shear in the direction under consideration.

WALL LINE. Any length of wall along a principal axis of the building used to provide resistance to lateral loads. Parallel wall lines separated by less than 4 feet (1219 mm) shall be considered one wall line for the distribution of loads.

WEAK WALL LINE. A wall line in a story where the story strength is less than 80 percent of the story above in the direction under consideration.

SECTION A403 ANALYSIS AND DESIGN

A403.1 General. Buildings within the scope of this chapter shall be analyzed, designed and constructed in conformance with the building code, except as modified in this chapter.

> **Exception:** Buildings for which the prescriptive measures provided in Section A405 apply and are used.

No *alteration* of the existing lateral-force-resisting or vertical-load-carrying system shall reduce the strength or stiffness of the existing structure. When any portion of a building within the scope of this chapter is constructed on or into a slope steeper than one unit vertical in three units horizontal, the lateral-force-resisting system at and below the base level diaphragm shall be analyzed for the effects of concentrated lateral forces at the base caused by this hillside condition.

A403.2 Scope of analysis. This chapter requires the *alteration*, *repair*, replacement or *addition* of structural elements and their connections to meet the strength and stiffness requirements herein. The lateral-load-path analysis shall include the resisting elements and connections from the wood diaphragm immediately above any soft, weak or open-front wall lines to the foundation soil interface or to the uppermost floor or roof of a Type I structure below. Stories above the uppermost story with a soft, weak or open-front wall line need not be modified. The lateral-load-path analysis for added structural elements shall also include evaluation of the allowable soil-bearing and lateral pressures in accordance with the building code.

> **Exception:** When an open-front, weak or soft wall line exists because of parking at the ground floor of a two-story building and the parking area is less than 20 percent of the ground floor area, then only the wall lines in the open, weak or soft directions of the enclosed parking area need comply with the provisions of this chapter.

A403.3 Design base shear. The design base shear in a given direction shall be 75 percent of the value required for similar new construction in accordance with the building code.

A403.4 Vertical distribution of forces. The total seismic force shall be distributed over the height of the structure as for new construction in accordance with the building code. Distribution of force by story weight shall be permitted for two-story buildings. The value of R used in the design of any story shall be less than or equal to the value of R used in the given direction for the story above.

A403.5 Weak story limitation. Every weak story shall be strengthened to the lesser of:

1. Ω_0 times the story shear prescribed by Sections A403.3 and A403.4.
2. In two-story buildings up to 30 feet (9144 mm) in height, 65 percent of the strength of the story above. In all other buildings, 80 percent of the strength of the story above.

A403.6 Story drift limitation. The calculated story drift for each retrofitted story shall not exceed the allowable deformation compatible with all vertical-load-resisting elements and 0.025 times the story height. The calculated story drift shall not be reduced by the effects of horizontal diaphragm stiffness but shall be increased when these effects produce rotation. Drift calculations shall be in accordance with the building code.

The effects of rotation and soil stiffness shall be included in the calculated story drift when lateral loads are resisted by vertical elements whose required depth of embedment is determined by pole formulas. The coefficient of subgrade reaction used in the deflection calculations shall be provided from an approved geotechnical engineering report or other approved methods.

A403.7 P Δ effects. The requirements of the building code shall apply, except as modified herein. All structural framing elements and their connections not required by design to be part of the lateral-force-resisting system shall be designed and/or detailed to be adequate to maintain support of design dead plus live loads when subjected to the expected deformations caused by seismic forces. The stress analysis of cantilever columns shall use a buckling factor of 2.1 for the direction normal to the axis of the beam.

A403.8 Ties and continuity. All parts of the structure included in the scope of Section A403.2 shall be interconnected as required by the building code.

A403.8.1 Cripple walls. Cripple walls braced with nonconforming structural materials shall be braced in accordance with this chapter. When a single top plate exists in the cripple wall, all end joints in the top plate shall be tied. Ties shall be connected to each end of the discontinuous top plate and shall be equal to one of the following:

1. Three-inch by 6-inch (76 mm by 152 mm), 18-gage galvanized steel, nailed with six 8d common nails at each end.

2. One and one-fourth-inch by 12-inch (32 mm by 305 mm), 18-gage galvanized steel, nailed with six 16d common nails at each end.
3. Two-inch by 4-inch by 12-inch (51 mm by 102 mm by 305 mm) wood blocking, nailed with six 16d common nails at each end.

A403.9 Collector elements. Collector elements shall be provided that can transfer the seismic forces originating in other portions of the building to the elements within the scope of Section A403.2 that provide resistance to those forces.

A403.10 Horizontal diaphragms. The strength of an existing horizontal diaphragm sheathed with wood structural panels or diagonal sheathing need not be investigated unless the diaphragm is required to transfer lateral forces from vertical elements of the seismic-force-resisting system above the diaphragm to elements below the diaphragm because of an offset in placement of the elements.

Wood diaphragms with stories above shall not be allowed to transmit lateral forces by rotation or cantilever except as allowed by the building code; however, rotational effects shall be accounted for when unsymmetric wall stiffness increases shear demands.

Exception: Diaphragms that cantilever 25 percent or less of the distance between lines of lateral-load-resisting elements from which the diaphragm cantilevers may transmit their shears by cantilever, provided that rotational effects on shear walls parallel and perpendicular to the load are taken into account.

A403.11 Wood-framed shear walls. Wood-framed shear walls shall have strength and stiffness sufficient to resist the seismic loads and shall conform to the requirements of this section.

A403.11.1 Gypsum or cement plaster products. Gypsum or cement plaster products shall not be used to provide lateral resistance in a soft or weak story or in a story with an open-front wall line, whether or not new elements are added to mitigate the soft, weak or open-front condition.

A403.11.2 Wood structural panels.

A403.11.2.1 Drift limit. Wood structural panel shear walls shall meet the story drift limitation of Section A403.6. Conformance to the story drift limitation shall be determined by approved testing or calculation, not by the use of an aspect ratio. Calculated deflection shall be determined according to *International Building Code* Equation 23-1 and shall be increased by 25 percent. Contribution to the shear wall deflection from the anchor or tie-down slippage shall also be included. The slippage contribution shall include the vertical elongation of the connector metal components, the vertical slippage of the connectors to framing members, localized crushing of wood due to bearing loads and shrinkage of the wood elements because of changes in moisture content as a result of aging. The total vertical slippage shall be multiplied by the shear panel aspect ratio and added to the total horizontal deflection. Individual shear panels shall be permitted to exceed the maximum aspect ratio, provided the allowable story drift and allowable shear capacities are not exceeded.

A403.11.2.2 Openings. Shear walls are permitted to be designed for continuity around openings in accordance with the building code. Blocking and steel strapping shall be provided at corners of the openings to transfer forces from discontinuous boundary elements into adjoining panel elements. Alternatively, perforated shear wall provisions of the building code are permitted to be used.

A403.11.2.3 Wood species of framing members. Allowable shear values for wood structural panels shall consider the species of the framing members. When the allowable shear values are based on Douglas fir-larch framing members, and framing members are constructed of other species of lumber, the allowable shear values shall be multiplied by the following factors: 0.82 for species with specific gravities greater than or equal to 0.42 but less than 0.49, and 0.65 for species with specific gravities less than 0.42. Redwood shall use 0.65 and hem fir shall use 0.82, unless otherwise approved.

A403.11.3 Substitution for 3-inch (76 mm) nominal width framing members. Two 2-inch (51 mm) nominal width framing members shall be permitted in lieu of any required 3-inch (76 mm) nominal width framing member when the existing and new framing members are of equal dimensions, when they are connected as required to transfer the in-plane shear between them, and when the sheathing fasteners are equally divided between them.

A403.11.4 Hold-down connectors.

A403.11.4.1 Expansion anchors in tension. Expansion anchors that provide tension strength by friction resistance shall not be used to connect hold-down devices to existing concrete or masonry elements. Expansion anchors that provide tension strength by bearing (commonly referenced as "undercut" anchors) shall be permitted.

A403.11.4.2 Required depth of embedment. The required depth of embedment or edge distance for the anchor used in the hold-down connector shall be provided in the concrete or masonry below any plain concrete slab unless satisfactory evidence is submitted to the building official that shows that the concrete slab and footings are of monolithic construction.

A403.11.4.3 Required preload of bolted hold-down connectors. Bolted hold-down connectors shall be preloaded to reduce slippage of the connector. Preloading shall consist of tightening the nut on the tension anchor after the placement but before the tightening of the shear bolts in the panel boundary flange member. The tension anchor shall be tightened until the shear bolts are in firm contact with the edge of the hole nearest the direction of the tension anchor. Hold-down connectors with self-jigging bolt standoffs shall be installed in a manner to permit preloading.

SECTION A404
PHASED CONSTRUCTION

The work specified in this chapter shall be permitted to be done in the following phases. Work shall start with Phase 1 unless otherwise approved by the building official. When the building does not contain the conditions associated with the given phase, the work shall proceed to the next phase.

Phase 1 Work. The first phase shall include all work in the lowest story with a soft, weak or open-front wall line and all foundation work.

Phase 2 Work. The second phase shall include wood-framed walls in any story with two or more stories above that are laterally braced with nonconforming structural materials.

Phase 3 Work. The third and final phase shall include all required work not performed in Phase 1 or Phase 2.

SECTION A405
PRESCRIPTIVE MEASURES FOR WEAK STORY

A405.1 Limitation. These prescriptive measures shall apply only to two-story buildings and only when deemed appropriate by the *code official*. These prescriptive measures rely on rotation of the second floor diaphragm to distribute the seismic load between the side and rear walls of the ground floor open area. In the absence of an existing floor diaphragm of wood structural panel or diagonal sheathing, a new wood structural panel diaphragm of minimum thickness of $^3/_4$ inch (19 mm) and with 10d common nails at 6 inches (152 mm) on center shall be applied.

A405.1.1 Additional conditions. To qualify for these prescriptive measures, the following additional conditions need to be satisfied by the retrofitted structure:

1. Diaphragm aspect ratio *L/W* is less than 0.67, where *W* is the diaphragm dimension parallel to the soft, weak or open-front wall line and *L* is the distance in the orthogoal direction between that wall line and the rear wall of the ground floor open area.
2. Minimum length of side shear walls = 20 feet (6096 mm).
3. Minimum length of rear shear wall = three-fourth of rear wall.
4. No plan or vertical irregularities other than a soft, weak or open-front wall line.
5. Roofing weight less than or equal to 5 pounds per square foot (240 N/m^2).
6. Aspect ratio of the full second floor diaphragm meets the requirements of the building code for new construction.

A405.2 Minimum required retrofit.

A405.2.1 Anchor bolt size and spacing. The anchor bolt size and spacing shall be a minimum of $^3/_4$ inch (19 mm) in diameter at 32 inches (813 mm) on center. Where existing bolts are inadequate, new steel plates bolted to the side of the foundation and nailed to the sill may be used, such as an approved connector.

A405.2.2 Connection to floor above. Shear wall top plates shall be connected to blocking or rim joist at upper floor with a minimum of 18-gage galvanized steel angle clips 4$^1/_2$ inches (114 mm) long with 12-8d nails spaced no farther than 16 inches (406 mm) on center, or by equivalent shear transfer methods.

A405.2.3 Shear wall sheathing. The shear wall sheathing shall be a minimum of $^{15}/_{32}$ inch (11.9 mm) 5-Ply Structural I with 10d nails at 4 inches (102 mm) on center at edges and 12 inches (305 mm) on center at field; blocked all edges with 3 by 4 or larger. Where existing sill plates are less than 3-by thick, place flat 2-by on top of sill between studs, with flat 18-gage galvanized steel clips 4$^1/_2$ inches (114 mm) long with 12-8d nails or $^3/_8$-inch-diameter (9.5 mm) lags through blocking for shear transfer to sill plate. Stagger nailing from wall sheathing between existing sill and new blocking. Anchor new blocking to foundation as specified above.

A405.2.4 Shear wall hold-downs. Shear walls shall be provided with hold-down anchors at each end. Two hold-down anchors are required at intersecting corners. Hold-downs shall be approved connectors with a minimum $^5/_8$-inch-diameter (15.9 mm) threaded rod or other approved anchor with a minimum allowable load of 4,000 pounds (17.8 kN). Anchor embedment in concrete shall not be less than 5 inches (127 mm). Tie-rod systems shall not be less than $^5/_8$ inch (15.9 mm) in diameter unless using high strength cable. Threaded rod or high strength cable elongation shall not exceed $^5/_8$ inch (15.9 mm) using design forces.

SECTION A406
MATERIALS OF CONSTRUCTION

A406.1 New materials. All materials approved by the building code, including their appropriate allowable stresses and limiting aspect ratios, shall be permitted to meet the requirements of this chapter.

A406.2 Allowable foundation and lateral pressures. The use of default values from the building code for continuous and isolated concrete spread footings shall be permitted. For soil that supports embedded vertical elements, SectionA403.6 shall apply.

A406.3 Existing materials. All existing materials shall be in sound condition and constructed in general conformance to the building code before they are permitted to be used to resist the lateral loads prescribed in this chapter. The verification of existing materials conditions and their conformance to these requirements shall be made by physical observation reports, material testing or record drawings as determined by the structural designer and as approved by the building official.

A406.3.1 Horizontal wood diaphragms. Allowable shear values for existing horizontal wood diaphragms that require analysis under Section A403.10 are permitted to be taken from TableA4-A. The values in Table A4-A shall be used for allowable stress design. Design forces based on strength design shall be reduced to allowable stress levels before comparison with the limiting values in the table.

A406.3.2 Wood-structural-panel shear walls.

A406.3.2.1 Allowable nail slip values. The use of box nails and unseasoned lumber are permitted to be assumed. When the required drift calculations of Section A403.11.2.1 rely on the slip values for common nails or surfaced dry lumber, their use in construction shall be verified by exposure. The design value of the box nails shall be assumed to be similar to that of common nails having the same diameter. Verification of surfaced dry lumber shall be by identification conforming to the building code.

A406.3.2.2 Plywood panel construction. When verification of the existing plywood materials is by use of record drawings alone, the panel construction for plywood shall be assumed to be of three plies. The plywood modulus "G" shall be assumed equal to 50,000 pounds per square inch (345 MPa).

A406.3.3 Existing wood framing. Wood framing is permitted to use the design stresses specified in the building code under which the building was constructed or other stress criteria approved by the building official.

A406.3.4 Structural steel. All existing structural steel shall be permitted to use the allowable stresses for Grade A36. Existing pipe or tube columns shall be assumed to be of minimum wall thickness unless verified by testing or exposure.

A406.3.5 Strength of concrete. All existing concrete footings shall be permitted to be assumed to be plain concrete with a compressive strength of 2,000 pounds per square inch (13.8 MPa). Existing concrete compressive strength taken greater than 2,000 pounds per square inch (13.8 MPa) shall be verified by testing, record drawings or department records.

A406.3.6 Existing sill plate anchorage. Existing cast-in-place anchor bolts shall be permitted to use the allowable service loads for bolts with proper embedment when used for shear resistance to lateral loads.

SECTION A407
INFORMATION REQUIRED TO BE ON THE PLANS

A407.1 General. The plans shall show all information necessary for plan review and for construction and shall accurately reflect the results of the engineering investigation and design. The plans shall contain a note that states that this retrofit was designed in compliance with the criteria of this chapter.

A407.2 Existing construction. The plans shall show existing diaphragm and shear wall sheathing and framing materials; fastener type and spacing; diaphragm and shear wall connections; continuity ties; and collector elements. The plans shall also show the portion of the existing materials that needs verification during construction.

A407.3 New construction.

A407.3.1 Foundation plan elements. The foundation plan shall include the size, type, location and spacing of all anchor bolts with the required depth of embedment, edge and end distance; the location and size of all shear walls and all columns for braced frames or moment frames; referenced details for the connection of shear walls, braced frames or moment-resisting frames to their footing; and referenced sections for any grade beams and footings.

A407.3.2 Framing plan elements. The framing plan shall include the length, location and material of shear walls; the location and material of frames; references on details for the column-to-beam connectors, beam-to-wall connections and shear transfers at floor and roof diaphragms; and the required nailing and length for wall top plate splices.

A407.3.3 Shear wall schedule, notes and details. Shear walls shall have a referenced schedule on the plans that includes the correct shear wall capacity in pounds per foot (N/m); the required fastener type, length, gauge and head size; and a complete specification for the sheathing material and its thickness. The schedule shall also show the required location of 3-inch (76 mm) nominal or two 2-inch (51 mm) nominal edge members; the spacing of shear transfer elements such as framing anchors or added sill plate nails; the required hold-down with its bolt, screw or nail sizes; and the dimensions, lumber grade and species of the attached framing member.

Notes shall show required edge distance for fasteners on structural wood panels and framing members; required flush nailing at the plywood surface; limits of mechanical penetrations; and the sill plate material assumed in the design. The limits of mechanical penetrations shall also be detailed showing the maximum notching and drilled hole sizes.

A407.3.4 General notes. General notes shall show the requirements for material testing, special inspection and structural observation.

SECTION A408
QUALITY CONTROL

A408.1 Structural observation, testing and inspection. Structural observation, in accordance with Section 1709 of the *International Building Code*, shall be required for all structures in which seismic retrofit is being performed in accordance with this chapter. Structural observation shall include visual observation of work for conformance with the approved construction documents and confirmation of existing conditions assumed during design.

Structural testing and inspection for new construction materials shall be in accordance with the building code, except as modified by this chapter.

TABLE A4-A—ALLOWABLE VALUES FOR EXISTING MATERIALS

EXISTING MATERIALS OR CONFIGURATIONS OF MATERIALS[a]	ALLOWABLE VALUES
	× 14.594 for N/m
1. Horizontal diaphragms[b]	
1.1. Roofs with straight sheathing and roofing applied directly to the sheathing	-100 lbs. per ft. for seismic shear
1.2. Roofs with diagonal sheathing and roofing applied directly to the sheathing	250 lbs. per ft. for seismic shear
1.3. Floors with straight tongue-and-groove sheathing	100 lbs. per ft. for seismic shear
1.4. Floors with straight sheathing and finished wood flooring with board edges offset or perpendicular	500 lbs. per ft. for seismic shear
1.5. Floors with diagonal sheathing and finished wood flooring	600 lbs. per ft. for seismic shear
2. Crosswalls[b, c]	Per side:
2.1. Plaster on wood or metal lath	200 lbs. per ft. for seismic shear
2.2. Plaster on gypsum lath	175 lbs. per ft. for seismic shear
2.3. Gypsum wallboard, unblocked edges	75 lbs. per ft. for seismic shear
2.4. Gypsum wallboard, blocked edges	125 lbs. per ft. for seismic shear
3. Existing footings, wood framing, structural steel and reinforced steel	
3.1. Plain concrete footings	f'_c = 1,500 psi (10.3 MPa) unless otherwise shown by tests[d]
3.2. Douglas fir wood	Allowable stress same as D.F. No. 1[d]
3.3. Reinforcing steel	f_s = 18,000 psi (124 MPa) maximum[d]
3.4. Structural steel	f_s = 20,000 psi (138 MPa) maximum[d]

For SI: 1 foot = 304.8 mm.

a. Material must be sound and in good condition.

b. A one-third increase in allowable stress is not allowed.

c. Shear values of these materials may be combined, except the total combined value shall not exceed 300 pounds per foot.

d. Stresses given may be increased for combination of loads as specified in the building code.

CHAPTER A5

EARTHQUAKE HAZARD REDUCTION IN EXISTING CONCRETE BUILDINGS

SECTION A501 PURPOSE

The purpose of this chapter is to promote public safety and welfare by reducing the risk of death or injury that may result from the effects of earthquakes on concrete buildings and concrete frame buildings.

The provisions of this chapter are intended as minimum standards for structural seismic resistance, and are established primarily to reduce the risk of life loss or injury. Compliance with the provisions in this chapter will not necessarily prevent loss of life or injury or prevent earthquake damage to the rehabilitated buildings.

SECTION A502 SCOPE

The provisions of this chapter shall apply to all buildings having concrete floors or roofs supported by reinforced concrete walls or by concrete frames and columns. This chapter shall not apply to buildings with roof diaphragms that are defined as flexible diaphragms by the building code, and shall not apply to concrete frame buildings with masonry infilled walls.

Buildings that were designed and constructed in accordance with the seismic provisions of the 1993 *BOCA National Building Code*, the 1994 *Standard Building Code*, the 1976 *Uniform Building Code*, the 2000 *International Building Code* or later editions of these codes shall be deemed to comply with these provisions, unless the seismicity of the region has increased since the design of the building.

Exception: This chapter shall not apply to concrete buildings where Seismic Design Category A is permitted.

SECTION A503 GENERAL REQUIREMENTS

A503.1 General. This chapter provides a three-tiered procedure to evaluate the need for seismic rehabilitation of existing concrete buildings. The evaluation shall show that the *existing building* is in compliance with the appropriate part of the evaluation procedure as described in Sections A507, A508 and A509, or shall be modified to conform to the respective acceptance criteria. This chapter does not preclude a building from being evaluated or modified to conform to the acceptance criteria using other well-established procedures, based on rational methods of analysis in accordance with principles of mechanics and approved by the authority having jurisdiction.

A503.2 Properties of cast-in-place materials. Except where specifically permitted herein, the stress-strain relationship of concrete and reinforcement shall be determined from published data or by testing. All available information, including building plans, original calculations and design criteria, site observations, testing and records of typical materials and construction practices prevalent at the time of construction, shall be considered when determining material properties.

For Tier 3 analyses, expected material properties shall be used in lieu of nominal properties in the calculation of strength, stiffness and deformabiltity of building components.

The procedure for testing and determination of material properties shall be from Section 6.2 of ASCE 41-06.

A503.3 Structural observation, testing and inspection. Structural observation, in accordance with Section 1709 of the *International Building Code* shall be required for all structures in which seismic retrofit is being performed in accordance with this chapter. Structural observation shall include visual observation of work for conformance with the approved construction documents and confirmation of existing conditions assumed during design.

Structural testing and inspection for new construction materials shall be in accordance with the building code, except as modified by this chapter.

SECTION A504 SITE GROUND MOTION

A504.1 Site ground motion for Tier 1 analysis. The earthquake loading used for the determination of demand on elements of the structure shall correspond to that required by ASCE 31 Tier 1.

A504.2 Site ground motion for Tier 2 analysis. The earthquake loading used for the determination of demand on elements and the structure shall conform to 75 percent of that required by the building code.

A504.3 Site ground motion for Tier 3 analysis. The site ground motion shall be an elastic design response spectrum prepared in conformance with the building code but having spectral acceleration values equal to 75 percent of the code design response spectrum. The spectral acceleration values shall be increased by the occupancy importance factor when required by the building code.

SECTION A505 TIER 1 ANALYSIS PROCEDURE

A505.1 General. Structures conforming to the requirements of the ASCE 31 Tier 1, Screening Phase, are permitted to be shown to be in conformance with this chapter by submission of a report to the building official as described in this section.

A505.2 Evaluation report. The registered design professional shall prepare a report summarizing the analysis conducted in compliance with this section. As a minimum, the report shall include the following items:

1. Building description.

2. Site inspection summary.
3. Summary of reviewed record documents.
4. Earthquake design data used for the evaluation of the building.
5. Completed checklists.
6. Quick-check analysis calculations.
7. Summary of deficiencies.

SECTION A506 TIER 2 ANALYSIS PROCEDURE

A506.1 General. A Tier 2 analysis includes an analysis using the following linear methods: Static or equivalent lateral force procedures. A linear dynamic analysis may be used to determine the distribution of the base shear over the height of the structure. The analysis, as a minimum, shall address all potential deficiencies identified in Tier 1, using procedures specified in this section.

If a Tier 2 analysis identifies a nonconforming condition, such condition shall be modified to conform to the acceptance criteria. Alternatively, the design professional may choose to perform a Tier 3 analysis to verify the adequacy of the structure.

A506.2 Limitations. A Tier 2 analysis procedure may be used if:

1. There is no in-plane offset in the lateral-force-resisting system.
2. There is no out-of-plane offset in the lateral-force-resisting system.
3. There is no torsional irregularity present in any story. A torsional irregularity may be deemed to exist in a story when the maximum story drift, computed including accidental torsion, at one end of the structure transverse to an axis is more than 1.2 times the average of the story drifts at the two ends of the structure.
4. There is no weak story irregularity at any floor level on any axis of the building. A weak story is one in which the story strength is less than 80 percent of that in the story above. The story strength is the total strength of all seismic-resisting elements sharing the story shear for the direction under consideration.

   Exception: Static or equivalent lateral force procedures shall not be used if:

   1. The building is more than 100 feet (30 480 mm) in height.
   2. The building has a vertical mass or stiffness irregularity (soft story). Mass irregularity shall be considered to exist where the effective mass of any story is more than 150 percent of the effective mass of any adjacent story. A soft story is one in which the lateral stiffness is less than 70 percent of that in the story above or less than 80 percent of the average stiffness of the three stories above.
   3. The building has a vertical geometric irregularity. Vertical geometric irregularity shall be considered to exist where the horizontal dimension of the lateral-force-resisting system in any story is more than 130 percent of that in an adjacent story.
   4. The building has a nonorthogonal lateral-force-resisting system.

A506.3 Analysis procedure. A structural analysis shall be performed for all structures in accordance with the requirements of the building code, except as modified in Section A506. The response modification factor, *R*, shall be selected based on the type of seismic-force resisting system employed and shall comply with the requirements of Section 101.5.4.1.

A506.3.1 Mathematical model. The three-dimensional mathematical model of the physical structure shall represent the spatial distribution of mass and stiffness of the structure to an extent that is adequate for the calculation of the significant features of its distribution of lateral forces. All concrete and masonry elements shall be included in the model of the physical structure.

Exception: Concrete or masonry partitions that are isolated from the concrete frame members and the floor above.

Cast-in-place reinforced concrete floors with span-to-depth ratios less than three-to-one may be assumed to be rigid diaphragms. Other floors, including floors constructed of precast elements with or without a reinforced concrete topping, shall be analyzed in conformance with the building code to determine if they must be considered semi-rigid diaphragms. The effective in-plane stiffness of the diaphragm, including effects of cracking and discontinuity between precast elements, shall be considered. Parking structures that have ramps rather than a single floor level shall be modeled as having mass appropriately distributed on each ramp. The lateral stiffness of the ramp may be calculated as having properties based on the uncracked cross section of the slab exclusive of beams and girders.

A506.3.2 Component stiffness. Component stiffness shall be calculated based on the approximate values shown in Table 6-5 of ASCE 41.

A506.4 Design, detailing requirements and structural component load effects. The design and detailing of new components of the seismic-force-resisting system shall comply with the requirements of the *International Building Code*, unless specifically modified herein.

A506.5 Acceptance criteria. The calculated strength of a member shall not be less than the load effects on that member.

A506.5.1 Load combinations. For load and resistance factor design (strength design), structures and all portions thereof shall resist the most critical effects from the combinations of factored loads prescribed in the building code.

Exception: For concrete beams and columns, the shear effect shall be determined based on the most critical load combinations prescribed in the building code. The shear load effect, because of seismic forces, shall be multiplied by a factor of *Cd,* but combined shear load effect need not be greater than *Ve*, as calculated in accordance with Equation (A5-4). M_{pr1} and M_{pr2} are the end moments, assumed to be in the same direction (clockwise or counter clockwise), based on steel tensile stress being equal to 1.25 f_y, where f_y is the specified yield strength.

$$V_e = \frac{M_{pr1} + M_{pr2}}{L} \pm \frac{W_g}{2} \qquad \textbf{(Equation A5-4)}$$

where:

W_g = Total gravity loads on the beam.

A506.5.2 Determination of the strength of members. The strength of a member shall be determined by multiplying the nominal strength of the member by a strength reduction factor, ϕ. The nominal strength of the member shall be determined in accordance with the building code.

SECTION A507 TIER 3 ANALYSIS PROCEDURE

A507.1 General. A Tier 3 evaluation shall be performed using the nonlinear procedures of Section 6.3.1.2.2. of ASCE 41. The general assumptions and requirements of Section 6.0, excluding concrete frames with infills shall be used in the evaluation. Site-ground motions in accordance with Section A504.3 are permitted for this evaluation.

APPENDIX A
REFERENCED STANDARDS

ASCE

American Society of Civil Engineers
1801 Alexander Bell Drive
Reston, VA 20191-4400

Standard reference number	Title	Referenced in code section number
7—05	Minimum Design Loads for Buildings and Other Structures with Supplement No. 1	A104
31—03	Seismic Evaluation of Existing Buildings	A506.1, A507.1

ASTM

ASTM International
100 Barr Harbor Drive
West Conshohocken, PA 19428-2959

Standard reference number	Title	Referenced in code section number
C90—2003	Standard Specification for Load-bearing Concrete Masonry Units	A505.2.3
C496—96	Standard Test Method for Splitting Tensile Strength of Cylindrical Concrete Specimens	A104, A106.3.3.2
E519—00e1	Standard Test Method for Diagonal Tension (Shear) in Masonry Assemblages	A104, A106.3.3.2

DOC

U.S. Department of Commerce
National Institute of Standards and Technology
100 Bureau Drive Stop 3460
Gaithersburg, MD 20899

Standard reference number	Title	Referenced in code section number
PS-1—95	Construction and Industrial Plywood	A302
PS-2—92	Performance Standard for Wood-based Structural-use Panels	A302

ICC

International Code Council
500 New Jersey Avenue, NW, 6th Floor
Washington, DC 20001

Standard reference number	Title	Referenced in code section number
BNBC—93	BOCA National Building Code®	A502
BNBC—96	BOCA National Building Code®	A502
BNBC—99	BOCA National Building Code®	A202
IBC—00	International Building Code®	A202, A502
IBC—03	International Building Code®	A502
IBC—06	International Building Code®	A102.2, A103, A108.2, A203, A206.2, A301.2, A403.11.2.1, A408.1, A505.2.3, A505.3, A508.4, A205.4
SBC—94	Standard Building Code®	A502
SBC—97	Standard Building Code®	A502
SBC—99	Standard Building Code®	A202
UBC—76	Uniform Building Code®	A502
UBC—97	Uniform Building Code®	A102.2, A103, A104, A108.2, A202, A203, A206.2, A301.2, A401.2, A403.1, A403.2, A403.3, A403.4, A405.2.3, A403.7, A403.11.2.2, A406.2, A406.3.2.1, A408.1, A502
UBC—Standard 21-4	Hollow and Solid Load-bearing Concrete Masonry Units	A106.2
UBC—Standard 21-6	In-place Masonry Shear Tests	A104
UBC—Standard 21-7	Tests of Anchors in Unreinforced Masonry	A105.3, A107.3, A107.4, Table A1-E
UBC—Standard 21-8	Pointing of Unreinforced Masonry Walls	A103, A106.3.3.9
UBC—Standard 23-2	Construction and Industrial Plywood	A403.11.2.1

APPENDIX B

SUPPLEMENTARY ACCESSIBILITY REQUIREMENTS FOR EXISTING BUILDINGS AND FACILITIES

The provisions contained in this appendix are not mandatory unless specifically referenced in the adopting ordinance.

SECTION B101 QUALIFIED HISTORICAL BUILDINGS AND FACILITIES

B101.1 General. Qualified historic buildings and facilities shall comply with Sections B101.2 through B101.5.

B101.2 Qualified historic buildings and facilities. These procedures shall apply to buildings and facilities designated as historic structures that undergo alterations or a *change of occupancy*.

B101.3 Qualified historic buildings and facilities subject to Section 106 of the National Historic Preservation Act. Where an *alteration* or *change of occupancy* is undertaken to a qualified *historic building* or facility that is subject to Section 106 of the National Historic Preservation Act, the federal agency with jurisdiction over the undertaking shall follow the Section 106 process. Where the state historic preservation officer or Advisory Council on Historic Preservation determines that compliance with the requirements for accessible routes, ramps, entrances, or toilet facilities would threaten or destroy the historic significance of the building or facility, the alternative requirements of Section 310.9 for that element are permitted.

B101.4 Qualified historic buildings and facilities not subject to Section 106 of the National Historic Preservation Act. Where an *alteration* or *change of occupancy* is undertaken to a qualified *historic building* or facility that is not subject to Section 106 of the National Historic Preservation Act, and the entity undertaking the alterations believes that compliance with the requirements for accessible routes, ramps, entrances, or toilet facilities would threaten or destroy the historic significance of the building or facility, the entity shall consult with the state historic preservation officer. Where the state historic preservation officer determines that compliance with the accessibility requirements for accessible routes, ramps, entrances, or toilet facilities would threaten or destroy the historical significance of the building or facility, the alternative requirements of Section 310.9 for that element are permitted.

B101.4.1 Consultation with interested persons. Interested persons shall be invited to participate in the consultation process, including state or local accessibility officials, individuals with disabilities, and organizations representing individuals with disabilities.

B101.4.2 Certified local government historic preservation programs. Where the state historic preservation officer has delegated the consultation responsibility for purposes of this section to a local government historic preservation program that has been certified in accordance with Section 101 of the National Historic Preservation Act of 1966 [(16 U.S.C. 470a(c)] and implementing regulations (36 CFR 61.5), the responsibility shall be permitted to be carried out by the appropriate local government body or official.

B101.5 Displays. In qualified historic buildings and facilities where alternative requirements of Section 1005 are permitted, displays and written information shall be located where they can be seen by a seated person. Exhibits and signs displayed horizontally shall be 44 inches (1120 mm) maximum above the floor.

SECTION B102 FIXED TRANSPORTATION FACILITIES AND STATIONS

B102.1 General. Existing fixed transportation facilities and stations shall comply with Section B102.2.

B102.2 Existing facilities—key stations. Rapid rail, light rail, commuter rail, intercity rail, high-speed rail and other fixed guide-way systems, altered stations, and intercity rail and key stations, as defined under criteria established by the Department of Transportation in Subpart C of 49 CFR Part 37, shall comply with Sections B102.2.1 through B102.2.3.

B102.2.1 Accessible route. At least one accessible route from an accessible entrance to those areas necessary for use of the transportation system shall be provided. The accessible route shall include the features specified in Appendix E109.2 of the *International Building Code*, except that escalators shall comply with *International Building Code* Section 3005.2.2. Where technical unfeasibility in existing stations requires the accessible route to lead from the public way to a paid area of the transit system, an accessible fare collection machine complying with *International Building Code* Appendix E109.2.3 shall be provided along such accessible route.

B102.2.2 Platform and vehicle floor coordination. Station platforms shall be positioned to coordinate with vehicles in accordance with applicable provisions of 36 CFR Part 1192. Low-level platforms shall be 8 inches (250 mm) minimum above top of rail.

Exception: Where vehicles are boarded from sidewalks or street-level, low-level platforms shall be permitted to be less than 8 inches (250 mm).

B102.2.3 Direct connections. New direct connections to commercial, retail, or residential facilities shall, to the maximum extent feasible, have an accessible route complying with Section 605.2 from the point of connection to boarding platforms and transportation system elements used by the public. Any elements provided to facilitate future direct connections shall be on an accessible route connecting boarding platforms and transportation system elements used by the public.

SECTION B103 DWELLING UNITS AND SLEEPING UNITS

B103.1 Communication features. Where dwelling units and sleeping units are altered or added, the requirements of Section E104.3 of the *International Building Code* shall apply only to the units being altered or added until the number of units with accessible communication features complies with the minimum number required for new construction.

SECTION B104 REFERENCED STANDARDS

Y3.H626 2P National Historic Preservation J101.2, 43/933 Act of 1966, as amended J101.3, 3rd Edition, Washington, DC: J101.3.2 US Government Printing Office, 1993.

2009 *International Building Code*. Washington, DC: International Code Council, 2002.

49 CFR Part 37.43 (c), Alteration of Transportation Facilities by Public Entities, Department of Transportation, 400 7th Street SW, Room 8102, Washington, DC 20590-0001.

RESOURCE A

GUIDELINES ON FIRE RATINGS OF ARCHAIC MATERIALS AND ASSEMBLIES

Introduction

The *International Existing Building Code* (IEBC) is a comprehensive code with the goal of addressing all aspects of work taking place in existing buildings and providing user friendly methods and tools for regulation and improvement of such buildings. This resource document is included within the cover of the IEBC with that goal in mind and as a step towards accomplishing that goal.

In the process of *repair* and *alteration* of existing buildings, based on the nature and the extent of the work, the IEBC might require certain upgrades in the fire resistance rating of building elements, at which time it becomes critical for the designers and the code officials to be able to determine the fire resistance rating of the *existing building* elements as part of the overall evaluation for the assessment of the need for improvements. This resource document provides a guideline for such an evaluation for fire resistance rating of archaic materials that is not typically found in the modern model building codes.

Resource A is only a guideline and is not intended to be a document for specific adoption as it is not written in the format or language of ICC's *International Codes* and is not subject to the code development process.

PURPOSE

The *Guideline on Fire Ratings of Archaic Materials and Assemblies* focuses upon the fire-related performance of archaic construction. "Archaic" encompasses construction typical of an earlier time, generally prior to 1950. "Fire-related performance" includes fire resistance, flame spread, smoke production, and degree of combustibility.

The purpose of this guideline is to update the information which was available at the time of original construction, for use by architects, engineers, and code officials when evaluating the fire safety of a rehabilitation project. In addition, information relevant to the evaluation of general classes of materials and types of construction is presented for those cases when documentation of the fire performance of a particular archaic material or assembly cannot be found.

It has been assumed that the building materials and their fastening, joining, and incorporation into the building structure are sound mechanically. Therefore, some determination must be made that the original manufacture, the original construction practice, and the rigors of aging and use have not weakened the building. This assessment can often be difficult because process and quality control was not good in many industries, and variations among locally available raw materials and manufacturing techniques often resulted in a product which varied widely in its strength and durability. The properties of iron and steel, for example, varied widely, depending on the mill and the process used.

There is nothing inherently inferior about archaic materials or construction techniques. The pressures that promote fundamental change are most often economic or technological—matters not necessarily related to concerns for safety. The high cost of labor made wood lath and plaster uneconomical. The high cost of land and the congestion of the cities provided the impetus for high-rise construction. Improved technology made it possible. The difficulty with archaic materials is not a question of suitability, but familiarity.

Code requirements for the fire performance of key building elements (e.g., walls, floor/ceiling assemblies, doors, shaft enclosures) are stated in performance terms: hours of fire resistance. It matters not whether these elements were built in 1908 or 1980, only that they provide the required degree of fire resistance. The level of performance will be defined by the local community, primarily through the enactment of a building or rehabilitation code. This guideline is only a tool to help evaluate the various building elements, regardless of what the level of performance is required to be.

The problem with archaic materials is simply that documentation of their fire performance is not readily available. The application of engineering judgment is more difficult because building officials may not be familiar with the materials or construction method involved. As a result, either a full-scale fire test is required or the archaic construction in question removed and replaced. Both alternatives are time consuming and wasteful.

This guideline and the accompanying Appendix are designed to help fill this information void. By providing the necessary documentation, there will be a firm basis for the continued acceptance of archaic materials and assemblies.

1
FIRE-RELATED PERFORMANCE OF ARCHAIC MATERIALS AND ASSEMBLIES

1.1
FIRE PERFORMANCE MEASURES

This guideline does not specify the level of performance required for the various building components. These requirements are controlled by the building occupancy and use and are set forth in the local building or rehabilitation code.

The fire resistance of a given building element is established by subjecting a sample of the assembly to a "standard" fire test which follows a "standard" time-temperature curve. This test method has changed little since the 1920s. The test results tabulated in the Appendix have been adjusted to reflect current test methods.

The current model building codes cite other fire-related properties not always tested for in earlier years: flame spread, smoke production, and degree of combustibility. However, they can generally be assumed to fall within well defined values because the principal combustible component of archaic materials is cellulose. Smoke production is more important today because of the increased use of plastics. However, the early flame spread tests, developed in the early 1940s, also included a test for smoke production.

"Plastics," one of the most important classes of contemporary materials, were not found in the review of archaic materials. If plastics are to be used in a rehabilitated building, they should be evaluated by contemporary standards. Information and documentation of their fire-related properties and performance is widely available.

Flame spread, smoke production and degree of combustibility are discussed in detail below. Test results for eight common species of lumber, published in an Underwriter's Laboratories' report (104), are noted in the following table:

TUNNEL TEST RESULTS FOR EIGHT SPECIES OF LUMBER

SPECIES OF LUMBER	FLAME SPREAD	FUEL CONTRIBUTED	SMOKE DEVELOPED
Western White Pine	75	50-60	50
Northern White Pine	120-215	120-140	60-65
Ponderosa Pine	80-215	120-135	100-110
Yellow Pine	180-190	130-145	275-305
Red Gum	140-155	125-175	40-60
Yellow Birch	105-110	100-105	45-65
Douglas Fir	65-100	50-80	10-100

Flame Spread

The flame spread of interior finishes is most often measured by the ASTM E 84 "tunnel test." This test measures how far and how fast the flames spread across the surface of the test sample. The resulting flame spread rating (FSR) is expressed as a number on a continuous scale where cement-asbestos board is 0 and red oak is 100. (Materials with a flame spread greater than red oak have an FSR greater than 100.) The scale is divided into distinct groups or classes. The most commonly used flame spread classifications are: Class I or A*, with a 0-25 FSR; Class II or B, with a 26-75 FSR; and Class III or C, with a 76-200 FSR. The *NFPA Life Safety Code* also has a Class D (201-500 FSR) and Class E (over 500 FSR) interior finish.

These classifications are typically used in modern building codes to restrict the rate of fire spread. Only the first three classifications are normally permitted, though not all classes of materials can be used in all places throughout a building. For example, the interior finish of building materials used in exits or in corridors leading to exits is more strictly regulated than materials used within private dwelling units.

In general, inorganic archaic materials (e.g., bricks or tile) can be expected to be in Class I. Materials of whole wood are mostly Class II. Whole wood is defined as wood used in the same form as sawn from the tree. This is in contrast to the contemporary reconstituted wood products such as plywood, fiberboard, hardboard, or particle board. If the organic archaic material is not whole wood, the flame spread classification could be well over 200 and thus would be particularly unsuited for use in exits and other critical locations in a building. Some plywoods and various wood fiberboards have flame spreads over 200. Although they can be treated with fire retardants to reduce their flame spread, it would be advisable to assume that all such products have a flame spread over 200 unless there is information to the contrary.

Smoke Production

The evaluation of smoke density is part of the ASTM E 84 tunnel test. For the eight species of lumber shown in the table above, the highest levels are 275-305 for Yellow Pine, but most of the others are less smoky than red oak which has an index of 100. The advent of plastics caused substantial increases in the smoke density values measured by the tunnel test. The ensuing limitation of the smoke production for wall and ceiling materials by the model building codes has been a reaction to the introduction of plastic materials. In general, cellulosic materials fall in the 50-300 range of smoke density which is below the general limitation of 450 adopted by many codes.

Degree of Combustibility

The model building codes tend to define "noncombustibility" on the basis of having passed ASTM E 136 or if the material is totally inorganic. The acceptance of gypsum wallboard as noncombustible is based on limiting paper thickness to not over $^1/_8$ inch and a 0-50 flame spread rating by ASTM E 84. At times there were provisions to define a Class I or A material (0-25 FSR) as noncombustible, but this is not currently recognized by most model building codes.

If there is any doubt whether or not an archaic material is noncombustible, it would be appropriate to send out samples for evaluation. If an archaic material is determined to be noncombustible according to ASTM E 136, it can be expected that it will not contribute fuel to the fire.

* Some codes are Roman numerals, others use letters

1.2
COMBUSTIBLE CONSTRUCTION TYPES

One of the earliest forms of timber construction used exterior load-bearing masonry walls with columns and/or wooden walls supporting wooden beams and floors in the interior of the building. This form of construction, often called "mill" or "heavy timber" construction, has approximately 1 hour fire resistance. The exterior walls will generally contain the fire within the building.

With the development of dimensional lumber, there was a switch from heavy timber to "balloon frame" construction. The balloon frame uses load-bearing exterior wooden walls which have long timbers often extending from foundation to roof. When longer lumber became scarce, another form of construction, "platform" framing, replaced the balloon framing. The difference between the two systems is significant because platform framing is automatically fire-blocked at every floor while balloon framing commonly has concealed spaces that extend unblocked from basement to attic. The architect, engineer, and *code official* must be alert to the details of construction and the ease with which fire can spread in concealed spaces.

2
BUILDING EVALUATION

A given rehabilitation project will most likely go through several stages. The preliminary evaluation process involves the designer in surveying the prospective building. The fire resistance of *existing building* materials and construction systems is identified; potential problems are noted for closer study. The final evaluation phase includes: developing design solutions to upgrade the fire resistance of building elements, if necessary; preparing working drawings and specifications; and the securing of the necessary code approvals.

2.1
PRELIMINARY EVALUATION

A preliminary evaluation should begin with a building survey to determine the existing materials, the general arrangement of the structure and the use of the occupied spaces, and the details of construction. The designer needs to know "what is there" before a decision can be reached about what to keep and what to remove during the rehabilitation process. This preliminary evaluation should be as detailed as necessary to make initial plans. The fire-related properties need to be determined from the applicable building or rehabilitation code, and the materials and assemblies existing in the building then need to be evaluated for these properties. Two work sheets are shown below to facilitate the preliminary evaluation.

Two possible sources of information helpful in the preliminary evaluation are the original building plans and the building code in effect at the time of original construction. Plans may be on file with the local building department or in the offices of the original designers (e.g., architect, engineer) or their successors. If plans are available, the investigator should verify that the building was actually constructed as called for in the plans, as well as incorporate any later alterations or changes to the building. Earlier editions of the local building code should be on file with the building official. The code in effect at the time of construction will contain fire performance criteria. While this is no guarantee that the required performance was actually provided, it does give the investigator some guidance as to the level of performance which may be expected. Under some code administration and enforcement systems, the code in effect at the time of construction also defines the level of performance that must be provided at the time of rehabilitation.

Figure 1 illustrates one method for organizing preliminary field notes. Space is provided for the materials, dimensions, and condition of the principal building elements. Each floor of the structure should be visited and the appropriate information obtained. In practice, there will often be identical materials and construction on every floor, but the exception may be of vital importance. A schematic diagram should be prepared of each floor showing the layout of exits and hallways and indicating where each element described in the field notes fits into the structure as a whole. The exact arrangement of interior walls within apartments is of secondary importance from a fire safety point of view and need not be shown on the drawings unless these walls are required by code to have a fire resistance rating.

The location of stairways and elevators should be clearly marked on the drawings. All exterior means of escape (e.g., fire escapes) should be identified.*

The following notes explain the entries in Figure 1.

Exterior Bearing Walls: Many old buildings utilize heavily constructed walls to support the floor/ceiling assemblies at the exterior of the building. There may be columns and/or interior bearing walls within the structure, but the exterior walls are an important factor in assessing the fire safety of a building.

The field investigator should note how the floor/ceiling assemblies are supported at the exterior of the building. If columns are incorporated in the exterior walls, the walls may be considered non-bearing.

Interior Bearing Walls: It may be difficult to determine whether or not an interior wall is load bearing, but the field investigator should attempt to make this determination. At a later stage of the rehabilitation process, this question will need to be determined exactly. Therefore, the field notes should be as accurate as possible.

Exterior Nonbearing Walls: The fire resistance of the exterior walls is important for two reasons. These walls (both bearing and non-bearing) are depended upon to: a) contain a fire within the building of origin; or b) keep an exterior fire *outside* the building. It is therefore important to indicate on the drawings where any openings are located as well as the materials and construction of all doors or shutters. The drawings should indicate the presence of wired glass, its thickness and framing, and identify the materials used for windows and door frames. The protection of openings adjacent to exterior means of escape (e.g., exterior stairs, fire escapes) is particularly important. The ground floor drawing should locate the building on the property and indicate the precise distances to adjacent buildings.

* Problems providing adequate exiting are discussed at length in the *Egress Guideline for Residential Rehabilitation.*

FIGURE 1 **PRELIMINARY EVALUATION FIELD NOTES**

Building Element		Materials	Thickness	Condition	Notes
Exterior Bearing Walls					
Interior Bearing Walls					
Exterior Nonbearing Walls					
Interior Nonbearing Walls or Partitions:	A				
	B				
Structural Frame: Columns					
Beams					
Other					
Floor/Ceiling Structural System Spanning					
Roofs					
Doors (including frame and hardware): a) Enclosed vertical exitway					
b) Enclosed horizontal exitway					
c) Other					

Interior Nonbearing Walls (Partitions): A partition is a "wall that extends from floor to ceiling and subdivides space within any story of a building." (48) Figure 1 has two categories (A & B) for Interior Nonbearing Walls (Partitions) which can be used for different walls, such as hallway walls as compared to inter-apartment walls. Under some circumstances there may be only one type of wall construction; in others, three or more types of wall construction may occur.

The field investigator should be alert for differences in function as well as in materials and construction details. In general, the details within apartments are not as important as the major exit paths and stairwells. The preliminary field investigation should attempt to determine the thickness of all walls. A term introduced below called "thickness design" will depend on an accurate (± $^1/_4$ inch) determination. Even though this initial field survey is called "preliminary," the data generated should be as accurate and complete as possible.

The field investigator should note the exact location from which observations are recorded. For instance, if a hole is found through a stairwell wall which allows a cataloguing of the construction details, the field investigation notes should reflect the location of the "find." At the preliminary stage it is not necessary to core every wall; the interior details of construction can usually be determined at some location.

Structural Frame: There may or may not be a complete skeletal frame, but usually there are columns, beams, trusses, or other like elements. The dimensions and spacing of the structural elements should be measured and indicated on the drawings. For instance, if there are ten inch square columns located on a thirty foot square grid throughout the building, this should be noted. The structural material and cover or protective materials should be identified wherever possible. The thickness of the cover materials should be determined to an accuracy of ± $^1/_4$ inch. As discussed above, the preliminary field survey usually relies on accidental openings in the cover materials rather than a systematic coring technique.

Floor/Ceiling Structural Systems: The span between supports should be measured. If possible, a sketch of the cross-section of the system should be made. If there is no location where accidental damage has opened the floor/ceiling construction to visual inspection, it is necessary to make such an opening. An evaluation of the fire resistance of a floor/ceiling assembly requires detailed knowledge of the materials and their arrangement. Special attention should be paid to the cover on structural steel elements and the condition of suspended ceilings and similar membranes.

Roofs: The preliminary field survey of the roof system is initially concerned with water-tightness. However, once it is apparent that the roof is sound for ordinary use and can be retained in the rehabilitated building, it becomes necessary to evaluate the fire performance. The field investigator must measure the thickness and identify the types of materials which have been used. Be aware that there may be several layers of roof materials.

Doors: Doors to stairways and hallways represent some of the most important fire elements to be considered within a building. The uses of the spaces separated largely controls the level of fire performance necessary. Walls and doors enclosing stairs or elevator shafts would normally require a higher level of performance than between a the bedroom and bath. The various uses are differentiated in Figure 1.

Careful measurements of the thickness of door panels must be made, and the type of core material within each door must be determined. It should be noted whether doors have self-closing devices; the general operation of the doors should be checked. The latch should engage and the door should fit tightly in the frame. The hinges should be in good condition. If glass is used in the doors, it should be identified as either plain glass or wired glass mounted in either a wood or steel frame.

Materials: The field investigator should be able to identify ordinary building materials. In situations where an unfamiliar material is found, a sample should be obtained. This sample should measure at least 10 cubic inches so that an ASTM E 136 fire test can be conducted to determine if it is combustible.

Thickness: The thickness of all materials should be measured accurately since, under certain circumstances, the level of fire resistance is very sensitive to the material thickness.

Condition: The method of attaching the various layers and facings to one another or to the supporting structural element should be noted under the appropriate building element. The "secureness" of the attachment and the general condition of the layers and facings should be noted here.

Notes: The "Notes" column can be used for many purposes, but it might be a good idea to make specific references to other field notes or drawings.

After the building survey is completed, the data collected must be analyzed. A suggested work sheet for organizing this information is given below as Figure 2.

The required fire resistance and flame spread for each building element are normally established by the local building or rehabilitation code. The fire performance of the existing materials and assemblies should then be estimated, using one of the techniques described below. If the fire performance of the *existing building* element(s) is equal to or greater than that required, the materials and assemblies may remain. If the fire performance is less than required, then corrective measures must be taken.

The most common methods of upgrading the level of protection are to either remove and replace the *existing building* element(s) or to *repair* and upgrade the existing materials and assemblies. Other fire protection measures, such as automatic sprinklers or detection and alarm systems, also could be considered, though they are beyond the scope of this guideline. If the upgraded protection is still less than that required or deemed to be acceptable, additional corrective measures must be taken. This process must continue until an acceptable level of performance is obtained.

FIGURE 2 PRELIMINARY EVALUATION WORKSHEET

Building Element		Required Fire Resistance	Required Flame Spread	Estimated Fire Resistance	Estimated Flame Spread	Method of Upgrading	Estimated Upgraded Protection	Notes
Exterior Bearing Walls								
Interior Bearing Walls								
Exterior Nonbearing Walls								
Interior Nonbearing Walls or Partitions:	A							
	B							
Structural Frame: Columns								
Beams								
Other								
Floor/Ceiling Structural System Spanning								
Roofs								
Doors (including frame and hardware): a) Enclosed vertical exitway								
b) Enclosed horizontal exitway								
c) Others								

2.2 FIRE RESISTANCE OF EXISTING BUILDING ELEMENTS

The fire resistance of the *existing building* elements can be estimated from the tables and histograms contained in the Appendix. The Appendix is organized first by type of building element: walls, columns, floor/ceiling assemblies, beams, and doors. Within each building element, the tables are organized by type of construction (e.g., masonry, metal, wood frame), and then further divided by minimum dimensions or thickness of the building element.

A histogram precedes every table that has 10 or more entries. The X-axis measures fire resistance in hours; the Y-axis shows the number of entries in that table having a given level of fire resistance. The histograms also contain the location of each entry within that table for easy cross-referencing.

The histograms, because they are keyed to the tables, can speed the preliminary investigation. For example, Table 1.3.2, *Wood Frame Walls 4" to Less Than 6" Thick*, contains 96 entries. Rather than study each table entry, the histogram shows that every wall assembly listed in that table has a fire resistance of less than 2 hours. If the building code required the wall to have 2 hours fire resistance, the designer, with a minimum of effort, is made aware of a problem that requires closer study.

Suppose the code had only required a wall of 1 hour fire resistance. The histogram shows far fewer complying elements (19) than noncomplying ones (77). If the existing assembly is not one of the 19 complying entries, there is a strong possibility the existing assembly is deficient. The histograms can also be used in the converse situation. If the existing assembly is not one of the smaller number of entries with a lower than required fire resistance, there is a strong possibility the existing assembly will be acceptable.

At some point, the *existing building* component or assembly must be located within the tables. Otherwise, the fire resistance must be determined through one of the other techniques presented in the guideline. Locating the building component in the Appendix Tables not only guarantees the accuracy of the fire resistance rating, but also provides a source of documentation for the building official.

2.3 EFFECTS OF PENETRATIONS IN FIRE RESISTANT ASSEMBLIES

There are often many features in existing walls or floor/ceiling assemblies which were not included in the original certification or fire testing. The most common examples are pipes and utility wires passed through holes poked through an assembly. During the life of the building, many penetrations are added, and by the time a building is ready for rehabilitation it is not sufficient to just consider the fire resistance of the assembly as originally constructed. It is necessary to consider all penetrations and their relative impact upon fire performance. For instance, the fire resistance of the corridor wall may be less important than the effect of plain glass doors or transoms. In fact, doors are the most important single class of penetrations.

A fully developed fire generates substantial quantities of heat and excess gaseous fuel capable of penetrating any holes which might be present in the walls or ceiling of the fire compartment. In general, this leads to a severe degradation of the fire resistance of those building elements and to a greater potential for fire spread. This is particularly applicable to penetrations located high in a compartment where the positive pressure of the fire can force the unburned gases through the penetration.

Penetrations in a floor/ceiling assembly will generally completely negate the barrier qualities of the assembly and will lead to rapid spread of fire to the space above. It will not be a problem, however, if the penetrations are filled with noncombustible materials strongly fastened to the structure. The upper half of walls are similar to the floor/ceiling assembly in that a positive pressure can reasonably be expected in the top of the room, and this will push hot and/or burning gases through the penetration unless it is completely sealed.

Building codes require doors installed in fire resistive walls to resist the passage of fire for a specified period of time. If the door to a fully involved room is not closed, a large plume of fire will typically escape through the doorway, preventing anyone from using the space outside the door while allowing the fire to spread. This is why door closers are so important. Glass in doors and transoms can be expected to rapidly shatter unless constructed of listed or approved wire glass in a steel frame. As with other building elements, penetrations or non-rated portions of doors and transoms must be upgraded or otherwise protected.

Table 5.1 in Section V of the Appendix contains 41 entries of doors mounted in sound tightfitting frames. Part 3.4 below outlines one procedure for evaluating and possibly upgrading existing doors.

3 FINAL EVALUATION AND DESIGN SOLUTION

The final evaluation begins after the rehabilitation project has reached the final design stage and the choices made to keep certain archaic materials and assemblies in the rehabilitated building. The final evaluation process is essentially a more refined and detailed version of the preliminary evaluation. The specific fire resistance and flame spread requirements are determined for the project. This may involve local building and fire officials reviewing the preliminary evaluation as depicted in Figures 1 and 2 and the field drawings and notes. When necessary, provisions must be made to upgrade *existing building* elements to provide the required level of fire performance.

There are several approaches to design solutions that can make possible the continued use of archaic materials and assemblies in the rehabilitated structure. The simplest case occurs when the materials and assembly in question are found within the Appendix Tables and the fire performance properties satisfy code requirements. Other approaches must be used, though, if the assembly cannot be found within the Appendix or the fire performance needs to be upgraded. These approaches have been grouped into two classes: experimental and theoretical.

3.1 THE EXPERIMENTAL APPROACH

If a material or assembly found in a building is not listed in the Appendix Tables, there are several other ways to evaluate fire performance. One approach is to conduct the appropriate fire test(s) and thereby determine the fire-related properties directly. There are a number of laboratories in the United States which routinely conduct the various fire tests. A current list can be obtained by writing the Center for Fire Research, National Bureau of Standards, Washington, D.C. 20234.

The contract with any of these testing laboratories should require their observation of specimen preparation as well as the testing of the specimen. A complete description of where and how the specimen was obtained from the building, the transportation of the specimen, and its preparation for testing should be noted in detail so that the building official can be satisfied that the fire test is representative of the actual use.

The test report should describe the fire test procedure and the response of the material or assembly. The laboratory usually submits a cover letter with the report to describe the provisions of the fire test that were satisfied by the material or assembly under investigation. A building official will generally require this cover letter, but will also read the report to confirm that the material or assembly complies with the code requirements. Local code officials should be involved in all phases of the testing process.

The experimental approach can be costly and time consuming because specimens must be taken from the building and transported to the testing laboratory. When a load bearing assembly has continuous reinforcement, the test specimen must be removed from the building, transported, and tested in one piece. However, when the fire performance cannot be determined by other means, there may be no alternative to a full-scale test.

A "nonstandard" small-scale test can be used in special cases. Sample sizes need only be 10-25 square feet (0.93-2.3 m^2), while full-scale tests require test samples of either 100 or 180 square feet (9.3 or 17 m^2) in size. This small-scale test is best suited for testing nonload-bearing assemblies against thermal transmission only.

3.2
THE THEORETICAL APPROACH

There will be instances when materials and assemblies in a building undergoing rehabilitation cannot be found in the Appendix Tables. Even where test results are available for more or less similar construction, the proper classification may not be immediately apparent. Variations in dimensions, loading conditions, materials, or workmanship may markedly affect the performance of the individual building elements, and the extent of such a possible effect cannot be evaluated from the tables.

Theoretical methods being developed offer an alternative to the full-scale fire tests discussed above. For example, Section 4302(b) of the 1979 edition of the *Uniform Building Code* specifically allows an engineering design for fire resistance in lieu of conducting full-scale tests. These techniques draw upon computer simulation and mathematical modeling, thermodynamics, heat-flow analysis, and materials science to predict the fire performance of building materials and assemblies.

One theoretical method, known as the "Ten Rules of Fire Endurance Ratings," was published by T. Z. Harmathy in the May, 1965 edition of *Fire Technology.* (35) Harmathy's Rules provide a foundation for extending the data within the Appendix Tables to analyze or upgrade current as well as archaic building materials or assemblies.

HARMATHY'S TEN RULES

Rule 1: The "thermal" fire endurance of a construction consisting of a number of parallel layers is greater than the sum of the "thermal" fire endurances characteristic of the individual layers when exposed separately to fire.*

The minimum performance of an untested assembly can be estimated if the fire endurance of the individual components is known. Though the exact rating of the assembly cannot be stated, the endurance of the assembly is greater than the sum of the endurance of the components.

When a building assembly or component is found to be deficient, the fire endurance can be upgraded by providing a protective membrane. This membrane could be a new layer of brick, plaster, or drywall. The fire endurance of this membrane is called the "finish rating." Appendix Tables 1.5.1 and 1.5.2 contain the finish ratings for the most commonly employed materials. (See also the notes to Rule 2).

The test criteria for the finish rating is the same as for the thermal fire endurance of the total assembly: average temperature increases of 250°F (121°C) above ambient or 325°F (163°C) above ambient at any one place with the membrane being exposed to the fire. The temperature is measured at the interface of the assembly and the protective membrane.

Rule 2: The fire endurance of a construction does not decrease with the addition of further layers.

Harmathy notes that this rule is a consequence of the previous rule. Its validity follows from the fact that the additional layers increase both the resistance to heat flow and the heat capacity of the construction. This, in turn, reduces the rate of temperature rise at the unexposed surface.

This rule is not just restricted to "thermal" performance but affects the other fire test criteria: direct flame passage, cotton waste ignition, and load bearing performance. This means that certain restrictions must be imposed on the materials to be added and on the loading conditions. One restriction is that a new layer, if applied to the exposed surface, must not produce additional thermal stresses in the construction, i.e., its thermal expansion characteristics must be similar to those of the adjacent layer. Each new layer must also be capable of contributing enough additional strength to the assembly to sustain the added dead load. If this requirement is not fulfilled, the allowable live load must be reduced by an amount equal to the weight of the new layer. Because of these limitations, this rule should not be applied without careful consideration.

Particular care must be taken if the material added is a good thermal insulator. Properly located, the added insulation could improve the "thermal" performance of the assembly. Improperly located, the insulation could block necessary thermal transmission through the assembly, thereby subjecting the structural elements to greater temperatures for longer periods of time, and could cause premature structural failure of the supporting members.

Rule 3: The fire endurance of constructions containing continuous air gaps or cavities is greater than the fire endurance of similar constructions of the same weight, but containing no air gaps or cavities.

By providing for voids in a construction, additional resistances are produced in the path of heat flow. Numerical heat flow analyses indicate that a 10 to 15 percent increase in fire endurance can be achieved by creating an air gap at the midplane of a brick wall. Since the gross volume is also increased by the presence of voids, the air gaps and cavities have a beneficial effect on stability as well. However, constructions containing combustible materials within an air gap may be regarded as exceptions to this rule because of the possible development of burning in the gap.

There are numerous examples of this rule in the tables. For instance:

Table 1.1.4; Item W-8-M-82: Cored concrete masonry, nominal 8 inch thick wall with one unit in wall thickness and with 62 percent minimum of solid material in each unit, load bearing (80 PSI). Fire endurance: $2^1/_2$ hours.

Table 1.1.5; Item W-10-M-11: Cored concrete mansonry, nominal 10 inch thick wall with two units in wall thickness and a 2-inch (51 mm) air space, load bearing (80 PSI). The units are essentially the same as item W-8-M-82. Fire endurance: $3^1/_2$ hours.

These walls show 1 hour greater fire endurance by the addition of the 2-inch (51 mm) air space.

Rule 4: The farther an air gap or cavity is located from the exposed surface, the more beneficial is its effect on the fire endurance.

* The "thermal" fire endurance is the time at which the average temperature on the unexposed side of a construction exceeds its initial value by 250° when the other side is exposed to the "standard" fire specified by ASTM Test Method E-19.

Radiation dominates the heat transfer across an air gap or cavity, and it is markedly higher where the temperature is higher. The air gap or cavity is thus a poor insulator if it is located in a region which attains high temperatures during fire exposure.

Some of the clay tile designs take advantage of these factors. The double cell design, for instance, ensures that there is a cavity near the unexposed face. Some floor/ceiling assemblies have air gaps or cavities near the top surface and these enhance their thermal performance.

Rule 5: The fire endurance of a construction cannot be increased by increasing the thickness of a completely enclosed air layer.

Harmathy notes that there is evidence that if the thickness of the air layer is larger than about $^1/_2$ inch (12.7 mm), the heat transfer through the air layer depends only on the temperature of the bounding surfaces, and is practically independent of the distance between them. This rule is not applicable if the air layer is not completely enclosed, i.e., if there is a possibility of fresh air entering the gap at an appreciable rate.

Rule 6: Layers of materials of low thermal conductivity are better utilized on that side of the construction on which fire is more likely to happen.

As in Rule 4, the reason lies in the heat transfer process, though the conductivity of the solid is much less dependent on the ambient temperature of the materials. The low thermal conductor creates a substantial temperature differential to be established across its thickness under transient heat flow conditions. This rule may not be applicable to materials undergoing physico-chemical changes accompanied by significant heat absorption or heat evolution.

Rule 7: The fire endurance of asymmetrical constructions depends on the direction of heat flow.

This rule is a consequence of Rules 4 and 6 as well as other factors. This rule is useful in determining the relative protection of corridors and stairwells from the surrounding spaces. In addition, there are often situations where a fire is more likely, or potentially more severe, from one side or the other.

Rule 8: The presence of moisture, if it does not result in explosive spalling, increases the fire endurance.

The flow of heat into an assembly is greatly hindered by the release and evaporation of the moisture found within cementitious materials such as gypsum, portland cement, or magnesium oxychloride. Harmathy has shown that the gain in fire endurance may be as high as 8 percent for each percent (by volume) of moisture in the construction. It is the moisture chemically bound within the construction material at the time of manufacture or processing that leads to increased fire endurance. There is no direct relationship between the relative humidity of the air in the pores of the material and the increase in fire endurance.

Under certain conditions there may be explosive spalling of low permeability cementitious materials such as dense concrete. In general, one can assume that extremely old concrete has developed enough minor cracking that this factor should not be significant.

Rule 9: Load-supporting elements, such as beams, girders and joists, yield higher fire endurances when subjected to fire endurance tests as parts of floor, roof, or ceiling assemblies than they would when tested separately.

One of the fire endurance test criteria is the ability of a load-supporting element to carry its design load. The element will be deemed to have failed when the load can no longer be supported.

Failure usually results for two reasons. Some materials, particularly steel and other metals, lose much of their structural strength at elevated temperatures. Physical deflection of the supporting element, due to decreased strength or thermal expansion, causes a redistribution of the load forces and stresses throughout the element. Structural failure often results because the supporting element is not designed to carry the redistributed load.

Roof, floor, and ceiling assemblies have primary (e.g., beams) and secondary (e.g., floor joists) structural members. Since the primary load-supporting elements span the largest distances, their deflection becomes significant at a stage when the strength of the secondary members (including the roof or floor surface) is hardly affected by the heat. As the secondary members follow the deflection of the primary load-supporting element, an increasingly larger portion of the load is transferred to the secondary members.

When load-supporting elements are tested separately, the imposed load is constant and equal to the design load throughout the test. By definition, no distribution of the load is possible because the element is being tested by itself. Without any other structural members to which the load could be transferred, the individual elements cannot yield a higher fire endurance than they do when tested as parts of a floor, roof or ceiling assembly.

Rule 10: The load-supporting elements (beams, girders, joists, etc.) of a floor, roof, or ceiling assembly can be replaced by such other load-supporting elements which, when tested separately, yielded fire endurances not less than that of the assembly.

This rule depends on Rule 9 for its validity. A beam or girder, if capable of yielding a certain performance when tested separately, will yield an equally good or better performance when it forms a part of a floor, roof, or ceiling assembly. It must be emphasized that the supporting element of one assembly must not be replaced by the supporting element of another assembly if the performance of this latter element is not known from a separate (beam) test. Because of the load-reducing effect of the secondary elements that results from a test performed on an assembly, the performance of the supporting element alone cannot be evaluated by simple arithmetic. This rule also indicates the advantage of performing separate fire tests on primary load-supporting elements.

ILLUSTRATION OF HARMATHY'S RULES

Harmathy provided one schematic figure which illustrated his Rules.* It should be useful as a quick reference to assist in applying his Rules.

* Reproduced from the May 1065 *Fire Technology* (Vol. 1, No. 2). Copyright National Fire Protection Association, Boston. Reproduced by permission.

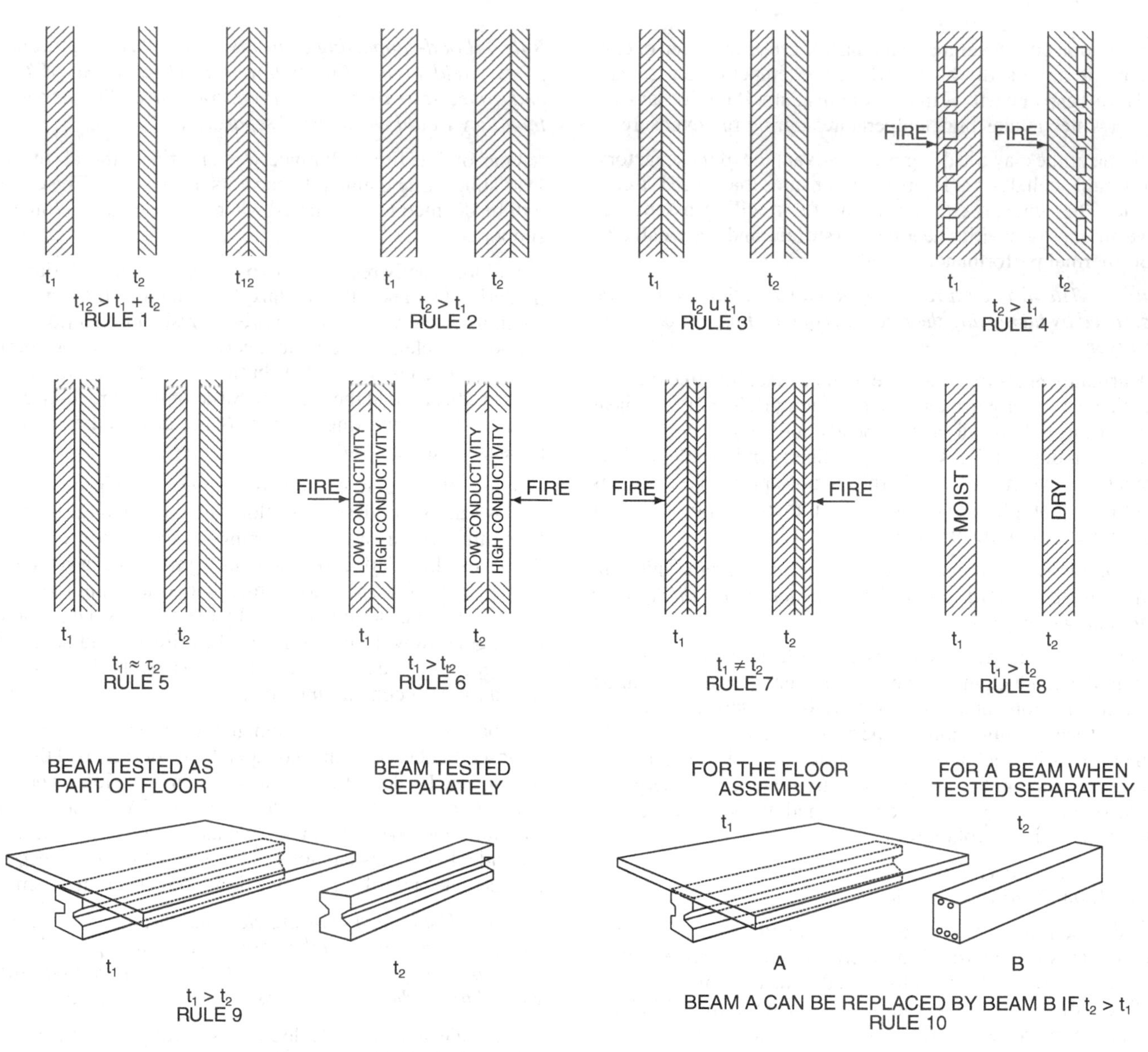

Diagrammatic illustration of ten rules.
t = fire endurance

EXAMPLE APPLICATION OF HARMATHY'S RULES

The following examples, based in whole or in part upon those presented in Harmathy's paper (35), show how the Rules can be applied to practical cases.

Example 1

Problem

A contractor would like to keep a partition which consists of a $3^3/_4$-inch (95 mm) thick layer of red clay brick, a $1^1/_4$-inch (32 mm) thick layer of plywood, and a $^3/_8$ inch (9.5 mm) thick layer of gypsum wallboard, at a location where 2-hour fire endurance is required. Is this assembly capable of providing a 2-hour protection?

Solution

(1) This partition does not appear in the Appendix Tables.

(2) Bricks of this thickness yield fire endurances of approximately 75 minutes (Table 1.1.2, Item W-4-M-2).

(3) The $1^1/_4$-inch (32 mm) thick plywood has a finish rating of 30 minutes.

(4) The $^3/_8$-inch (9.5 mm) gypsum wallboard has a finish rating of 10 minutes.

(5) Using the recommended values from the tables and applying Rule 1, the fire endurance (FI) of the assembly is larger than the sum of the individual layers, or

$$FI > 75 + 30 + 10 = 115 \text{ minutes}$$

Discussion

This example illustrates how the Appendix Tables can be utilized to determine the fire resistance of assemblies not explicitly listed.

Example 2

Problem

(1) A number of buildings to be rehabilitated have the same type of roof slab which is supported with different structural elements.

(2) The designer and contractor would like to determine whether or not this roof slab is capable of yielding a 2-hour fire endurance. According to a rigorous interpretation of ASTM E 119, however, only the roof assembly, including the roof slab as well as the cover and the supporting elements, can be subjected to a fire test. Therefore, a fire endurance classification cannot be issued for the slabs separately.

(3) The designer and contractor believe this slab will yield a 2-hour fire endurance even without the cover, and any beam of at least 2-hour fire endurance will provide satisfactory support. Is it possible to obtain a classification for the slab separately?

Solution

(1) The answer to the question is yes.

(2) According to Rule 10 it is not contrary to common sense to test and classify roofs and supporting elements separately. Furthermore, according to Rule 2, if the roof slabs actually yield a 2 hour fire endurance, the endurance of an assembly, including the slabs, cannot be less than 2 hours.

(3) The recommended procedure would be to review the tables to see if the slab appears as part of any tested roof or floor/ceiling assembly. The supporting system can be regarded as separate from the slab specimen, and the fire endurance of the assembly listed in the table is at least the fire endurance of the slab. There would have to be an adjustment for the weight of the roof cover in the allowable load if the test specimen did not contain a cover.

(4) The supporting structure or element would have to have at least a 2-hour fire endurance when tested separately.

Discussion

If the tables did not include tests on assemblies which contained the slab, one procedure would be to assemble the roof slabs on any convenient supporting system (not regarded as part of the specimen) and to subject them to a load which, besides the usually required superimposed load, includes some allowances for the weight of the cover.

Example 3

Problem

A steel-joisted floor and ceiling assembly is known to have yielded a fire endurance of 1 hour and 35 minutes. At a certain location, a 2-hour endurance is required. What is the most economical way of increasing the fire endurance by at least 25 minutes?

Solution

(1) The most effective technique would be to increase the ceiling plaster thickness. Existing coats of paint would have to be removed and the surface properly prepared before the new plaster could be applied. Other materials (e.g., gypsum wallboard) could also be considered.

(2) There may be other techniques based on other principles, but an examination of the drawings would be necessary.

Discussion

(1) The additional plaster has at least three effects:

a) The layer of plaster is increased and thus there is a gain of fire endurance (Rule 1).

b) There is a gain due to shifting the air gap farther from the exposed surface (Rule 4).

c) There is more moisture in the path of heat flow to the structural elements (Rules 7 and 8).

(2) The increase in fire endurance would be at least as large as that of the finish rating for the added thickness of plaster. The combined effects in (1) above would further increase this by a factor of 2 or more, depending upon the geometry of the assembly.

Example 4

Problem

The fire endurance of item W-l0-M-l in Table 1.1.5 is 4 hours. This wall consists of two $3^3/_4$-inch (95 mm) thick layers of structural tiles separated by a 2-inch (51 mm) air gap and $^3/_4$-inch (19 mm) portland cement plaster or stucco on both sides. If the actual wall in the building is identical to item W-10-M-1 except that it has a 4-inch (102 mm) air gap, can the fire endurance be estimated at 5 hours?

Solution

The answer to the question is no for the reasons contained in Rule 5.

Example 5

Problem

In order to increase the insulating value of its precast roof slabs, a company has decided to use two layers of different concretes. The lower layer of the slabs, where the strength of the concrete is immaterial (all the tensile load is carried by the steel reinforcement), would be made with a concrete of low strength but good insulating value. The upper layer, where the concrete is supposed to carry the compressive load, would remain the original high strength, high thermal conductivity concrete. How will the fire endurance of the slabs be affected by the change?

Solution

The effect on the thermal fire endurance is beneficial:

(1) The total resistance to heat flow of the new slabs has been increased due to the replacement of a layer of high thermal conductivity by one of low conductivity.

(2) The layer of low conductivity is on the side more likely to be exposed to fire, where it is more effectively utilized according to Rule 6. The layer of low thermal conductivity also provides better protection for the steel reinforcement, thereby extending the time before reaching the temperature at which the creep of steel becomes significant.

3.3 "THICKNESS DESIGN" STRATEGY

The "thickness design" strategy is based upon Harmathy's Rules 1 and 2. This design approach can be used when the construction materials have been identified and measured, but the specific assembly cannot be located within the tables. The tables should be surveyed again for thinner walls of like material and construction detail that have yielded the desired or greater fire endurance. If such an assembly can be found, then the thicker walls in the building have more than enough fire resistance. The thickness of the walls thus becomes the principal concern.

This approach can also be used for floor/ceiling assemblies, except that the thickness of the cover* and the slab become the central concern. The fire resistance of the untested assembly will be at least the fire resistance of an assembly listed in the table having a similar design but with less cover and/or thinner slabs. For other structural elements (e.g., beams and columns), the element listed in the table must also be of a similar design but with less cover thickness.

3.4 EVALUATION OF DOORS

A separate section on doors has been included because the process for evaluation presented below differs from those suggested previously for other building elements. The impact of unprotected openings or penetrations in fire resistant assemblies has been detailed in Part 2.3 above. It is sufficient to note here that openings left unprotected will likely lead to failure of the barrier under actual fire conditions.

For other types of building elements (e.g., beams, columns), the Appendix Tables can be used to establish a minimum level of fire performance. The benefit to rehabilitation is that the need for a full-scale fire test is then eliminated. For doors, however, this cannot be done. The data contained in Appendix Table 5.1, Resistance of Doors to Fire Exposure, can only provide guidance as to whether a successful fire test is even feasible.

For example, a door required to have 1 hour fire resistance is noted in the tables as providing only 5 minutes. The likelihood of achieving the required 1 hour, even if the door is upgraded, is remote. The ultimate need for replacement of the doors is reasonably clear, and the expense and time needed for testing can be saved. However, if the performance documented in the table is near or in excess of what is being required, then a fire test should be conducted. The test documentation can then be used as evidence of compliance with the required level of performance.

The table entries cannot be used as the sole proof of performance of the door in question because there are too many unknown variables which could measurably affect fire performance. The wood may have dried over the years; coats of flammable varnish could have been added. Minor deviations in the internal construction of a door can result in significant differences in performance. Methods of securing inserts in panel doors can vary. The major non-destructive method of analysis, an x-ray, often cannot provide the necessary detail. It is for these, and similar reasons, that a fire test is still felt to be necessary.

It is often possible to upgrade the fire performance of an existing door. Sometimes, "as is" and modified doors are evaluated in a single series of tests when failure of the unmodified door is expected. Because doors upgraded after an initial failure must be tested again, there is a potential savings of time and money.

The most common problems encountered are plain glass, panel inserts of insufficient thickness, and improper fit of a door in its frame. The latter problem can be significant because a fire can develop a substantial positive pressure, and the fire will work its way through otherwise innocent-looking gaps between door and frame.

One approach to solving these problems is as follows. The plain glass is replaced with approved or listed wire glass in a steel frame. The panel inserts can be upgraded by adding an additional layer of material. Gypsum wallboard is often used for this purpose. Intumescent paint applied to the edges of the door and frame will expand when exposed to fire, forming an effective seal around the edges. This seal, coupled with the generally even thermal expansion of a wood door in a wood frame, can prevent the passage of flames and other fire gases. Figure 3 below illustrates these solutions.

Because the interior construction of a door cannot be determined by a visual inspection, there is no absolute guarantee that the remaining doors are identical to the one(s) removed from the building and tested. But the same is true for doors constructed today, and reason and judgment must be applied. Doors that appear identical upon visual inspection can be weighed. If the weights are reasonably close, the doors can be assumed to be identical and therefore provide the same level of fire performance. Another approach is to fire test more than one door or to dismantle doors selected at random to see if they had been constructed in the same manner. Original building plans showing door details or other records showing that doors were purchased at one time or obtained from a single supplier can also be evidence of similar construction.

More often though, it is what is visible to the eye that is most significant. The investigator should carefully check the condition and fit of the door and frame, and for frames out of plumb or separating from the wall. Door closers, latches, and hinges must be examined to see that they function properly and are tightly secured. If these are in order and the door and frame have passed a full-scale test, there can be a reasonable basis for allowing the existing doors to remain.

4 SUMMARY

This section summarizes the various approaches and design solutions discussed in the preceding sections of the guideline. The term "structural system" includes: frames, beams, columns, and other structural elements. "Cover" is a protective layer(s) of materials or membrane which slows the flow of heat to the structural elements. It cannot be stressed too strongly that the fire endurance of actual building elements can be greatly reduced or totally negated by removing part of the cover to allow pipes, ducts, or conduits to pass through the element. This must be repaired in the rehabilitation process.

The following approaches shall be considered equivalent.

* Cover: the protective layer or membrane of material which slows the flow of heat to the structural elements.

FIGURE 3

MODIFICATION DETAILS

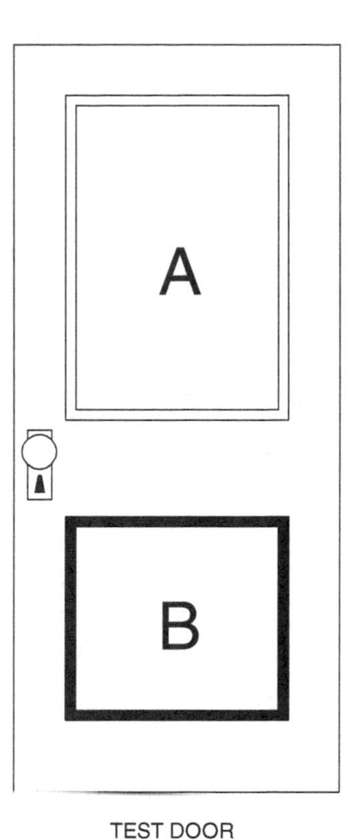

TEST DOOR

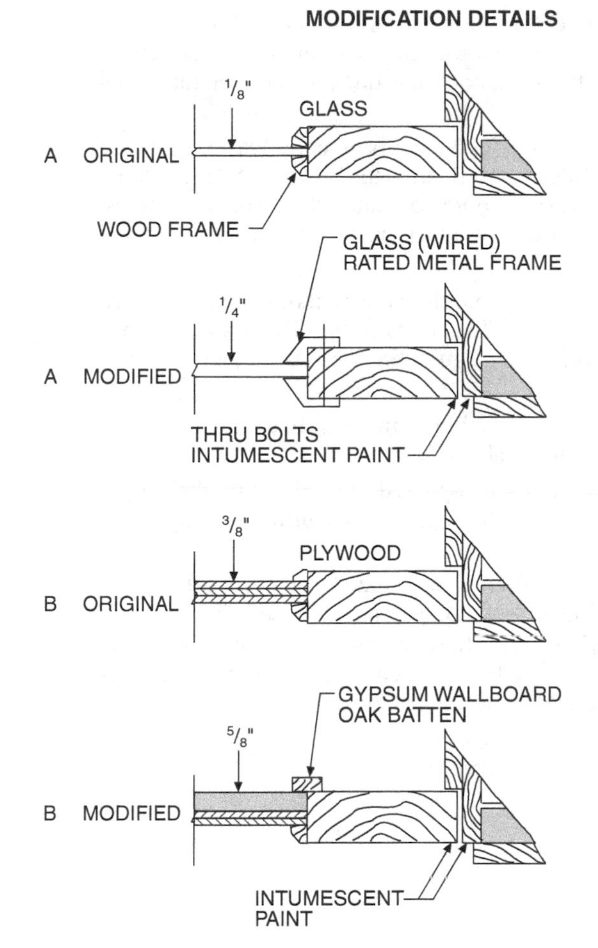

4.1 The fire resistance of a building element can be established from the Appendix Tables. This is subject to the following limitations:

The building element in the rehabilitated building shall be constructed of the same materials with the same nominal dimensions as stated in the tables.

All penetrations in the building element or its cover for services such as electricity, plumbing, and HVAC shall be packed with noncombustible cementitious materials and so fixed that the packing material will not fall out when it loses its water of hydration.

The effects of age and wear and tear shall be repaired so that the building element is sound and the original thickness of all components, particularly covers and floor slabs, is maintained.

This approach essentially follows the approach taken by model building codes. The assembly must appear in a table either published in or accepted by the code for a given fire resistance rating to be recognized and accepted.

4.2 The fire resistance of a building element which does not explicitly appear in the Appendix Tables can be established if one or more elements of same design but different dimensions have been listed in the tables. For walls, the existing element must be thicker than the one listed. For floor/ceiling assemblies, the assembly listed in the table must have the same or less cover and the same or thinner slab constructed of the same material as the actual floor/ceiling assembly. For other structural elements, the element listed in the table must be of a similar design but with less cover thickness. The fire resistancc in all instances shall be the fire resistance recommended in the table. This is subject to the following limitations:

The actual element in the rehabilitated building shall be constructed of the same materials as listed in the table. Only the following dimensions may vary from those specified: for walls, the overall thickness must exceed that specified in the table; for floor/ceiling assemblies, the thickness of the cover and the slab must be greater than, or equal to, that specified in the table; for other structural elements, the thickness of the cover must be greater than that specified in the table.

All penetrations in the building element or its cover for services such as electricity, plumbing, or HVAC shall be packed with noncombustible cementitious materials and so fixed that the packing material will not fall out when it loses its water of hydration.

The effects of age and wear and tear shall be repaired so that the building element is sound and the original thickness of all components, particularly covers and floor slabs, is maintained.

This approach is an application of the "thickness design" concept presented in Part 3.3 of the guideline. There should be many instances when a thicker building element was utilized than the one listed in the Appendix Tables. This guideline recognizes the inherent superiority of a thicker design. Note: "thickness design" for floor/ceiling assemblies and structural elements refers to cover and slab thickness rather than total thickness.

The "thickness design" concept is essentially a special case of Harmathy's Rules (specifically Rules 1 and 2). It should be recognized that the only source of data is the Appendix Tables. If other data are used, it must be in connection with the approach below.

4.3 The fire resistance of building elements can be established by applying Harmathy's Ten Rules of Fire Resistance Ratings as set forth in Part 3.2 of the guideline. This is subject to the following limitations:

The data from the tables can be utilized subject to the limitations in 4.2 above.

Test reports from recognized journals or published papers can be used to support data utilized in applying Harmathy's Rules.

Calculations utilizing recognized and well established computational techniques can be used in applying Harmathy's Rules. These include, but are not limited to, analysis of heat flow, mechanical properties, deflections, and load bearing capacity.

APPENDIX

Introduction

The fire resistance tables that follow are a part of Resource A and provide a tabular form of assigning fire resistance ratings to various archaic building elements and assemblies.

These tables for archaic materials and assemblies do for archaic materials what Tables 720.1(1), 720.1(2), and 720.1(3) of the *International Building Code* do for more modern building elements and assemblies. The fire resistance tables of Resource A should be used as described in the "Purpose and Procedure" that follows the table of contents for these tables.

RESOURCE A TABLE OF CONTENTS

PURPOSE AND PROCEDURE

The tables and histograms which follow are to be used only within the analytical framework detailed in the main body of this guideline.

Histograms precede any table with 10 or more entries. The use and interpretation of these histograms is explained in Part 2 of the guideline. The tables are in a format similar to that found in the model building codes. The following example, taken from an entry in Table 1.1.2, best explains the table format.

1. Item Code: The item code consists of a four place series in the general form w-x-y-z in which each member of the series denotes the following:

   w = Type of building element (e.g., W=Walls; F=Floors, etc.)

   x = The building element thickness rounded down to the nearest one inch increment (e.g., $4^5/_8$ inches is rounded off to 4 inches)

   y = The general type of material from which the building element is constructed (e.g., M=Masonry; W=Wood, etc.)

   z = The item number of the particular building element in a given table

   The item code shown in the example W-4-M-50 denotes the following:

   W = Wall, as the building element

   4 = Wall thickness in the range of 4 inches (102 mm) to less than 5 inches (127 mm)

   M = Masonry construction

   50 = The 50th entry in Table 1.1.2

2. The specific name or heading of this column identifies the dimensions which, if varied, has the greatest impact on fire resistance. The critical dimension for walls, the example here, is thickness. It is different for other building elements (e.g., depth for beams; membrane thickness for some floor/ceiling assemblies). The table entry is the named dimension of the building element measured at the time of actual testing to within ± $^1/_8$ inch (3.2 mm) tolerance. The thickness tabulated includes facings where facings are a part of the wall construction.

3. Construction Details: The construction details provide a brief description of the manner in which the building element was constructed.

4. Performance: This heading is subdivided into two columns. The column labeled "Load" will either list the load that the building element was subjected to during the fire test or it will contain a note number which will list the load and any other significant details. If the building element was not subjected to a load during the test, this column will contain "n/a," which means "not applicable."

   The second column under performance is labeled "Time" and denotes the actual fire endurance time observed in the fire test.

5. Reference Number: This heading is subdivided into three columns: Pre-BMS-92; BMS-92; and Post-BMS-92. The table entry under this column is the number in the Bibliography of the original source reference for the test data.

6. Notes: Notes are provided at the end of each table to allow a more detailed explanation of certain aspects of the test. In certain tables the notes given to this column have also been listed under the "Construction Details" and/or "Load" columns.

7. Rec Hours: This column lists the recommended fire endurance rating, in hours, of a building element. In some cases, the recommended fire endurance will be less than that listed under the "Time" column. In no case is the "Rec Hours" greater than given in the "Time" column.

ITEM CODE	THICKNESS	CONSTRUCTION DETAILS	PERFORMANCE		REFERENCE NUMBER			NOTES	REC. HOURS
			LOAD	TIME	PRE-BMS-92	BMS-92	POST-BMS-92		
W-4-M-50	$4^5/_8''$	Core: structural clay tile, See notes 12, 16, 21; Facings on unexposed side only, see note 18	n/a	25 min.		1		3, 4, 24	$^1/_3$

SECTION I - WALLS
FIGURE 1.1.1—WALLS—MASONRY 0″ TO LESS THAN 4″ THICK

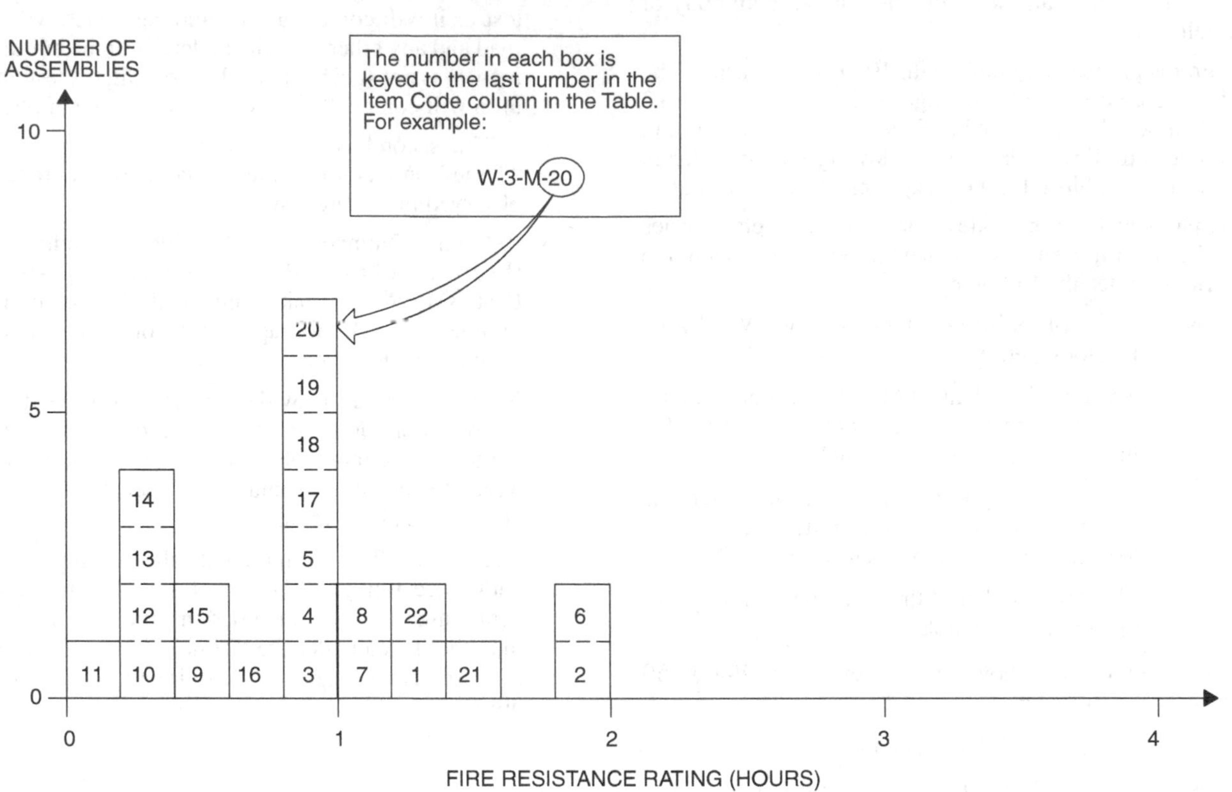

TABLE 1.1.1—MASONRY WALLS 0″ TO LESS THAN 4″ THICK

ITEM CODE	THICKNESS	CONSTRUCTION DETAILS	PERFORMANCE		REFERENCE NUMBER			NOTES	REC. HOURS
			LOAD	TIME	PRE-BMS-92	BMS-92	POST-BMS-92		
W-2-M-1	2 1/4″	Solid partition; 3/4″ gypsum plank- 10′ × 1′6″; 3/4″ plus gypsum plaster each side.	N/A	1 hr. 22 min.			7	1	1 1/4
W-3-M-2	3″	Concrete block (18″ × 9″ × 3″) of fuel ash, portland cement and plasticizer; cement/sand mortar.	N/A	2 hrs.			7	2, 3	2
W-2-M-3	2″	Solid gypsum block wall; No facings	N/A	1 hr.		1		4	1
W-3-M-4	3″	Solid gypsum blocks, laid in 1:3 sanded gypsum mortar.	N/A	1 hr.		1		4	1
W-3-M-5	3″	Magnesium oxysulfate wood fiber blocks; 2″ thick, laid in portland cement-lime mortar; Facings: 1/2″ of 1:3 sanded gypsum plaster on both sides.	N/A	1 hr.		1		4	1
W-3-M-6	3″	Magnesium oxysulfate bound wood fiber blocks; 3″ thick; laid in portland cement-lime mortar; Facings: 1/2″ of 1:3 sanded gypsum plaster on both sides.	N/A	2 hrs.		1		4	2

(Continued)

TABLE 1.1.1—MASONRY WALLS
0″ TO LESS THAN 4″ THICK—continued

ITEM CODE	THICKNESS	CONSTRUCTION DETAILS	PERFORMANCE		REFERENCE NUMBER			NOTES	REC. HOURS
			LOAD	TIME	PRE-BMS-92	BMS-92	POST-BMS-92		
W-3-M-7	3″	Clay tile; Ohio fire clay; single cell thick; Face plaster: $^{5}/_{8}$″ (both sides) 1:3 sanded gypsum; Design "E," Construction "A."	N/A	1 hr. 6 min.	0		2	5, 6, 7, 11, 12, 39	1
W-3-M-8	3″	Clay tile; Illinois surface clay; single cell thick; Face plaster: $^{5}/_{8}$" (both sides) 1:3 sanded gypsum; Design "A," Construction "E."	N/A	1 hr. 1 min			2	5, 8, 9, 11, 12, 39	1
W-3-M-9	3″	Clay tile; Illinois surface clay; single cell thick; No face plaster; Design "A," Construction "C."	N/A	25 min.			2	5, 10, 11, 12, 39	$^{1}/_{3}$
W-3-M-10	3$^{7}/_{8}$″	8″ × 4$^{7}/_{8}$″ glass blocks; weight 4 lbs. each; portland cement-lime mortar; horizontal mortar joints reinforced with metal lath.	N/A	15 min.		1		4	$^{1}/_{4}$
W-3-M-11	3″	Core: structural clay tile; see Notes 14, 18, 13; No facings.	N/A	10 min.		1		5, 11, 26	$^{1}/_{6}$
W-3-M-12	3″	Core: structural clay tile; see Notes 14, 19, 23; No facings.	N/A	20 min.		1		5, 11, 26	$^{1}/_{3}$
W-3-M-13	3$^{5}/_{8}$″	Core: structural clay tile; see Notes 14, 18, 23; Facings: unexposed side; see Note 20.	N/A	20 min.		1		5, 11, 26	$^{1}/_{3}$
W-3-M-14	3$^{5}/_{8}$″	Core: structural clay tile; see Notes 14, 19, 23; Facings: unexposed side only; see Note 20.	N/A	20 min.		1		5, 11, 26	$^{1}/_{3}$
W-3-M-15	3$^{5}/_{8}$″	Core: clay structural tile; see Notes 14, 18, 23; Facings: side exposed to fire; see Note 20.	N/A	30 min.		1		5, 11, 26	$^{1}/_{2}$
W-3-M-16	3$^{5}/_{8}$″	Core: clay structural tile; see Notes 14, 19, 23; Facings: side exposed to fire; see Note 20.	N/A	45 min.		1		5, 11, 26	$^{3}/_{4}$
W-2-M-17	2″	2″ thick solid gypsum blocks; see Note 27.	N/A	1 hr.		1		27	1
W-3-M-18	3″	Core: 3″ thick gypsum blocks 70% solid; see Note 2; No facings.	N/A	1 hr.		1		27	1
W-3-M-19	3″	Core: hollow concrete units; see Notes 29, 35, 36, 38; No facings.	N/A	1 hr.		1		27	1
W-3-M-20	3″	Core: hollow concrete units; see Notes 28, 35, 36, 37, 38; No facings.	N/A	1 hr.		1			1
W-3-M-21	3$^{1}/_{2}$″	Core: hollow concrete units; see Notes 28, 35, 36, 37, 38; Facings: one side; see Note 37.	N/A	1$^{1}/_{2}$ hrs.		1			1$^{1}/_{2}$

(Continued)

TABLE 1.1.1—MASONRY WALLS
0″ TO LESS THAN 4″ THICK—continued

ITEM CODE	THICKNESS	CONSTRUCTION DETAILS	PERFORMANCE		REFERENCE NUMBER			NOTES	REC. HOURS
			LOAD	TIME	PRE-BMS-92	BMS-92	POST-BMS-92		
W-3-M-22	$3^1/_2$″	Core: hollow concrete units; see Notes 29, 35, 36, 38; Facings: one side, see Note 37.	N/A	$1^1/_4$ hrs.		1			$1^1/_4$

For SI: 1 inch = 25.4 mm, 1 pound per square inch = 0.00689 MPa, °C = [(°F) - 32]/1.8.

Notes:

1. Failure mode - flame thru.
2. Passed 2-hour fire test (Grade "C" fire res. - British).
3. Passed hose stream test.
4. Tested at NBS under ASA Spec. No. A2-1934. As nonload bearing partitions.
5. Tested at NBS under ASA Spec. No. 42-1934 (ASTM C 19-33) except that hose stream testing where carried was run on test specimens exposed for full test duration, not for a reduced period as is contemporarily done.
6. Failure by thermal criteria - maximum temperature rise 325°F.
7. Hose stream failure.
8. Hose stream - pass.
9. Specimen removed prior to any failure occurring.
10. Failure mode - collapse.
11. For clay tile walls, unless the source or density of the clay can be positively identified or determined, it is suggested that the lowest hourly rating for the fire endurance of a clay tile partition of that thickness be followed. Identified sources of clay showing longer fire endurance can lead to longer time recommendations.
12. See appendix for construction and design details for clay tile walls.
13. Load: 80 psi for gross wall area.
14. One cell in wall thickness.
15. Two cells in wall thickness.
16. Double shells plus one cell in wall thickness.
17. One cell in wall thickness, cells filled with broken tile, crushed stone, slag cinders or sand mixed with mortar.
18. Dense hard-burned clay or shale tile.
19. Medium-burned clay tile.
20. Not less than $^5/_8$ inch thickness of 1:3 sanded gypsum plaster.
21. Units of not less than 30 percent solid material.
22. Units of not less than 40 percent solid material.
23. Units of not less than 50 percent solid material.
24. Units of not less than 45 percent solid material.
25. Units of not less than 60 percent solid material.
26. All tiles laid in portland cement-lime mortar.
27. Blocks laid in 1:3 sanded gypsum mortar voids in blocks not to exceed 30 percent.
28. Units of expanded slag or pumice aggregate.
29. Units of crushed limestone, blast furnace, slag, cinders and expanded clay or shale.
30. Units of calcareous sand and gravel. Coarse aggregate, 60 percent or more calcite and dolomite.
31. Units of siliceous sand and gravel. Ninety percent or more quartz, chert or flint.
32. Unit at least 49 percent solid.
33. Unit at least 62 percent solid.
34. Unit at least 65 percent solid.
35. Unit at least 73 percent solid.
36. Ratings based on one unit and one cell in wall thickness.
37. Minimum of $^1/_2$ inch - 1:3 sanded gypsum plaster.
38. Nonload bearing.
39. See Clay Tile Partition Design Construction drawings, below.

(Continued)

TABLE 1.1.1—MASONRY WALLS
0″ TO LESS THAN 4″ THICK—continued

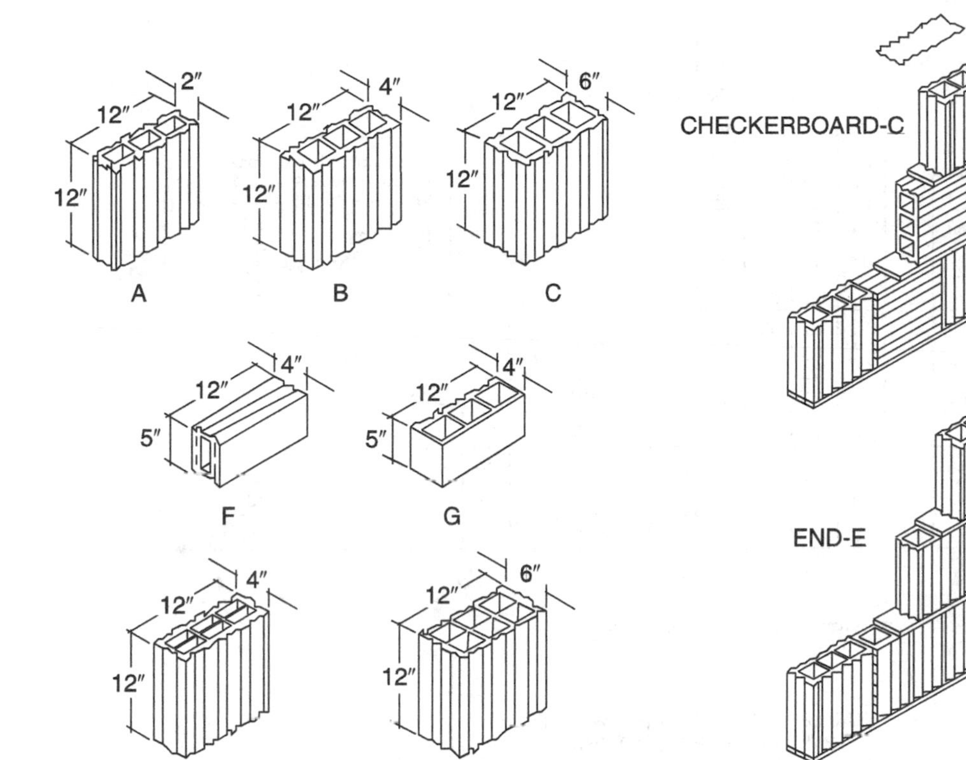

DESIGNS OF TILES USED IN FIRE-TEST PARTITIONS

CHECKERBOARD-C

END-E

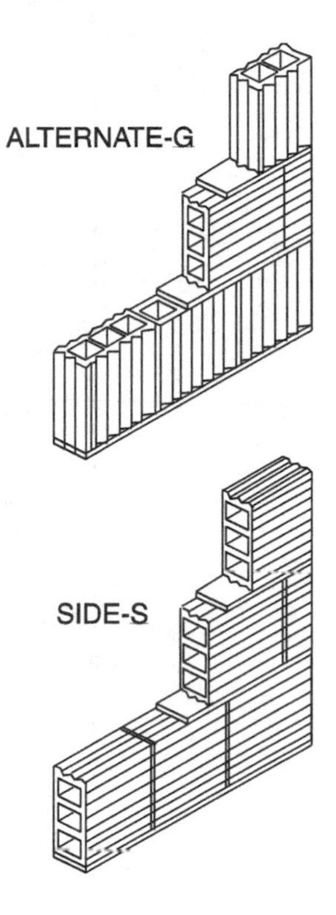

THE FOUR TYPES OF CONSTRUCTION USED IN FIRE-TEST PARTITIONS

FIGURE 1.1.2—MASONRY WALLS
4″ TO LESS THAN 6″ THICK

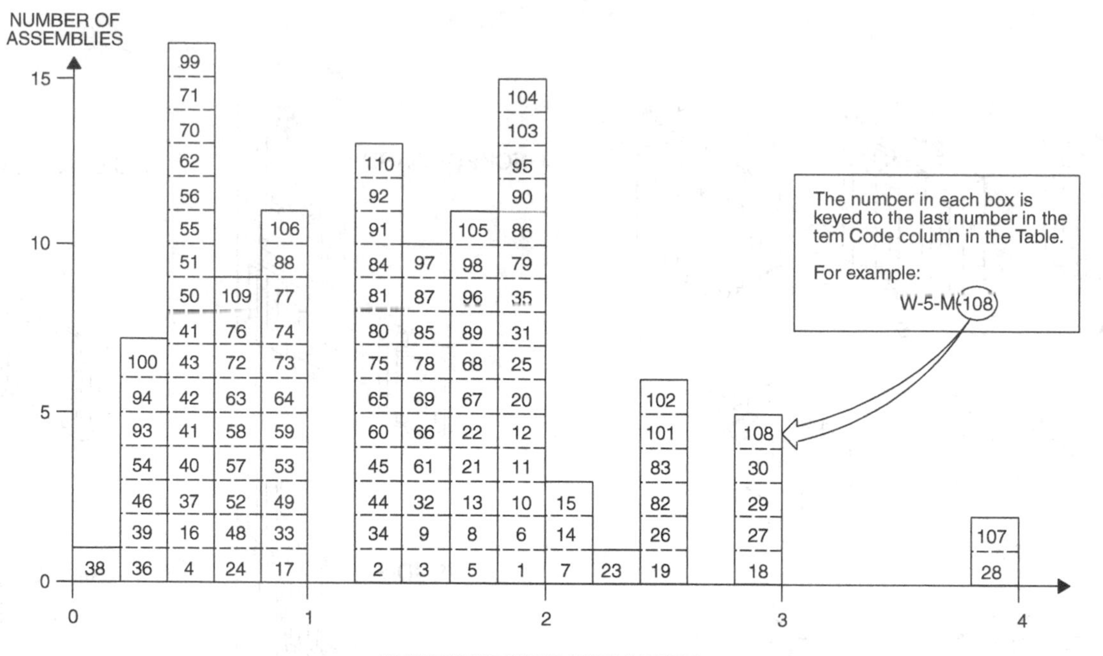

TABLE 1.1.2—MASONRY WALLS
4″ TO LESS THAN 6″ THICK

ITEM CODE	THICKNESS	CONSTRUCTION DETAILS	PERFORMANCE		REFERENCE NUMBER			NOTES	REC. HOURS
			LOAD	TIME	PRE-BMS-92	BMS-92	POST-BMS-92		
W-4-M-1	4″	Solid 3″ thick, gypsum blocks laid in 1:3 sanded gypsum mortar; Facings: $^1/_2$″ of 1:3 sanded gypsum plaster (both sides).	N/A	2 hrs.		1		1	2
W-4-M-2	4″	Solid clay or shale brick.	N/A	1 hr. 15 min		1		1, 2	$1^1/_4$
W-4-M-3	4″	Concrete; No facings.	N/A	1 hr. 30 min.		1		1	$1^1/_2$
W-4-M-4	4″	Clay tile; Illinois surface clay; single cell thick; No face plaster; Design "B," Construction "C."	N/A	25 min.			2	3-7, 36	$^1/_3$
W-4-M-5	4″	Solid sand-lime brick.	N/A	1 hr. 45 min.		1		1	$1^3/_4$
W-4-M-6	4″	Solid wall; 3″ thick block; $^1/_2$″ plaster each side; $17^3/_4$″ × $8^3/_4$″ × 4″ "Breeze Blocks"; portland cement/sand mortar.	N/A	1 hr. 52 min.			7	2	$1^3/_4$
W-4-M-7	4″	Concrete (4020 psi); Reinforcement: vertical $^3/_8$″; horizontal $^1/_4$″; 6″ × 6″ grid.	N/A	2 hrs. 10 min.			7	2	2
W-4-M-8	4″	Concrete wall (4340 psi crush); reinforcement $^1/_4$″ diameter rebar on 8″ centers (vertical and horizontal).	N/A	1 hr. 40 min.			7	2	$1^2/_3$

(Continued)

TABLE 1.1.2—MASONRY WALLS
4″ TO LESS THAN 6″ THICK—continued

ITEM CODE	THICKNESS	CONSTRUCTION DETAILS	PERFORMANCE		REFERENCE NUMBER			NOTES	REC. HOURS
			LOAD	TIME	PRE-BMS-92	BMS-92	POST-BMS-92		
W-4-M-9	$4^{3}/_{16}''$	$4^{3}/_{16}'' \times 2^{5}/_{8}''$ cellular fletton brick (1873 psi) with $^{1}/_{2}''$ sand mortar; bricks are U-shaped yielding hollow cover (approx. $2'' \times 4''$) in final cross-section configuration.	N/A	1 hr. 25 min.			7	2	$1^{1}/_{3}$
W-4-M-10	$4^{1}/_{4}''$	$4^{1}/_{4}'' \times 2^{1}/_{2}''$ fletton (1831 psi) brick in $^{1}/_{2}''$ sand mortar.	N/A	1 hr. 53 min			7	2	$1^{3}/_{4}$
W-4-M-11	$4^{1}/_{4}''$	$4^{1}/_{4}'' \times 2^{1}/_{2}''$ London stock (683 psi) brick; $^{1}/_{2}''$ grout.	N/A	1 hr. 52 min.			7	2	$1^{3}/_{4}$
W-4-M-12	$4^{1}/_{2}''$	$4^{1}/_{4}'' \times 2^{1}/_{2}''$ Leicester red, wire-cut brick (4465 psi) in $^{1}/_{2}''$ sand mortar.	N/A	1 hr. 56 min.			7	6	$1^{3}/_{4}$
W-4-M-13	$4^{1}/_{4}''$	$4^{1}/_{4}'' \times 2^{1}/_{2}''$ stairfoot brick (7527 psi) $^{1}/_{2}''$ sand mortar.	N/A	1 hr. 37 min.			7	2	$1^{1}/_{2}$
W-4-M-14	$4^{1}/_{4}''$	$4^{1}/_{4}'' \times 2^{1}/_{2}''$ sand-lime brick (2603 psi) $^{1}/_{2}''$ sand mortar.	N/A	2 hrs. 6 min.			7	2	2
W-4-M-15	$4^{1}/_{4}''$	$4^{1}/_{4}'' \times 2^{1}/_{2}''$ concrete brick (2527 psi) $^{1}/_{2}''$ sand mortar.	N/A	2 hrs. 10 min.			7	2	2
W-4-M-16	$4^{1}/_{2}''$	4″ thick clay tile; Ohio fire clay; single cell thick; No plaster exposed face; $^{1}/_{2}''$ 1:2 gypsum back face; Design "F," Construction "S."	N/A	31 min.			2	3-6, 36	$^{1}/_{2}$
W-4-M-17	$4^{1}/_{2}''$	4″ thick clay tile; Ohio fire clay; single cell thick; Plaster exposed face; $^{1}/_{2}''$ 1:2 sanded gypsum; Back Face: none; Construction "S," Design "F."	80 psi	50 min.			2	3-5, 8, 36	$^{3}/_{4}$
W-4-M-18	$4^{1}/_{2}''$	Core: solid sand-lime brick; $^{1}/_{2}''$ sanded gypsum plaster facings on both sides.	80 psi	3 hrs.		1		1, 11	3
W-4-M-19	$4^{1}/_{2}''$	Core: solid sand-lime brick; $^{1}/_{2}''$ sanded gypsum plaster facings on both sides.	80 psi	2 hrs. 30 min.		1		1, 11	$2^{1}/_{2}$
W-4-M-20	$4^{1}/_{2}''$	Core: concrete brick $^{1}/_{2}''$ of 1:3 sanded gypsum plaster facings on both sides.	80 psi	2 hrs.		1		1, 11	2
W-4-M-21	$4^{1}/_{2}''$	Core: solid clay or shale brick; $^{1}/_{2}''$ thick, 1:3 sanded gypsum plaster facings on fire sides.	80 psi	1 hr. 45 min.		1		1, 2, 11	$1^{3}/_{4}$
W-4-M-22	$4^{3}/_{4}''$	4″ thick clay tile; Ohio fire clay; single cell thick; cells filled with cement and broken tile concrete; Plaster on exposed face; none on unexposed face; $^{3}/_{4}''$ 1:3 sanded gypsum; Design "G," Construction "E."	N/A	1 hr. 48 min.			2	2, 3-5, 9, 36	$1^{3}/_{4}$
W-4-M-23	$4^{3}/_{4}''$	4″ thick clay tile; Ohio fire clay; single cell thick; cells filled with cement and broken tile concrete; No plaster exposed faced; $^{3}/_{4}''$ neat gypsum plaster on unexposed face; Design "G," Construction "E."	N/A	2 hrs. 14 min.			2	2, 3-5, 9, 36	2

(Continued)

TABLE 1.1.2—MASONRY WALLS
4″ TO LESS THAN 6″ THICK—continued

ITEM CODE	THICKNESS	CONSTRUCTION DETAILS	PERFORMANCE		REFERENCE NUMBER			NOTES	REC. HOURS
			LOAD	TIME	PRE-BMS-92	BMS-92	POST-BMS-92		
W-5-M-24	5″	3″ × 13″ air space; 1″ thick metal reinforced concrete facings on both sides; faces connected with wood splines.	2,250 lbs./ft.	45 min.		1		1	$^3/_4$
W-5-M-25	5″	Core: 3″ thick void filled with "nondulated" mineral wool weighing 10 lbs./ft.^{3}; 1″ thick metal reinforced concrete facings on both sides.	2,250 lbs./ft.	2 hrs.		1		1	2
W-5-M-26	5″	Core: solid clay or shale brick; $^1/_2$″ thick, 1:3 sanded gypsum plaster facings on both sides.	40 psi	2 hrs. 30 min.		1		1, 2, 11	$2^1/_2$
W-5-M-27	5″	Core: solid 4″ thick gypsum blocks, laid in 1:3 sanded gypsum mortar; $^1/_2$″ of 1:3 sanded gypsum plaster facings on both sides.	N/A	3 hrs.		1		1	3
W-5-M-28	5″	Core: 4″ thick hollow gypsum blocks with 30% voids; blocks laid in 1:3 sanded gypsum mortar; No facings.	N/A	4 hrs.		1		1	4
W-5-M-29	5″	Core: concrete brick; $^1/_2$″ of 1:3 sanded gypsum plaster facings on both sides.	160 psi	3 hrs.		1		1	3
W-5-M-30	$5^1/_4$″	4″ thick clay tile; Illinois surface clay; double cell thick; Plaster: $^5/_8$″ sanded gypsum 1:3 both faces; Design "D," Construction "S."	N/A	2 hrs. 53 min.			2	2-5, 9, 36	$2^3/_4$
W-5-M-31	$5^1/_4$″	4″ thick clay tile; New Jersey fire clay; double cell thick; Plaster: $^5/_8$″ sanded gypsum 1:3 both faces; Design "D," Construction "S."	N/A	1 hr. 52 min.			2	2-5, 9, 36	$1^3/_4$
W-5-M-32	$5^1/_4$″	4″ thick clay tile; New Jersey fire clay; single cell thick; Plaster: $^5/_8$″ sanded gypsm 1:3 both faces; Design "D," Construction "S."	N/A	1 hr. 34 min.	2		2	2-5, 9, 36	$1^1/_2$
W-5-M-33	$5^1/_4$″	4″ thick clay tile; New Jersey fire clay; single cell thick; Face plaster: $^5/_8$″ both sides; 1:3 sanded gypsum; Design "B," Construction "S."	N/A	50 min.			2	3-5, 8, 36	$^3/_4$
W-5-M-34	$5^1/_4$″	4″ thick clay tile; Ohio fire clay; single cell thick; Face plaster: $^5/_8$″ both sides; 1:3 sanded gypsum; Design "B," Construction "A."	N/A	1 hr. 19 min.			2	2-5, 9, 36	$1^1/_4$
W-5-M-35	$5^1/_4$″	4″ thick clay tile; Illinois surface clay; single cell thick; Face plaster: $^5/_8$″ both sides; 1:3 sanded gypsum; Design "B," Construction "S."	N/A	1 hr. 59 min.			2	2-5, 10 36	$1^3/_4$
W-5-M-36	4″	Core: structural clay tile; see Notes 12, 16, 21; No facings.	N/A	15 min.		1		3, 4, 24	$^1/_4$

(Continued)

TABLE 1.1.2—MASONRY WALLS
4″ TO LESS THAN 6″ THICK—continued

ITEM CODE	THICKNESS	CONSTRUCTION DETAILS	PERFORMANCE		REFERENCE NUMBER			NOTES	REC. HOURS
			LOAD	TIME	PRE-BMS-92	BMS-92	POST-BMS-92		
W-4-M-37	4″	Core: structural clay tile; see Notes 12, 17, 21; No facings.	N/A	25 min.		1		3, 4, 24	$^{1}/_{3}$
W-4-M-38	4″	Core: structural clay tile; see Notes 12, 16, 20; No facings.	N/A	10 min.		1		3, 4, 24	$^{1}/_{6}$
W-4-M-39	4″	Core: structural clay tile; see Notes 12, 17, 20; No facings.	N/A	20 min.		1		3, 4, 24	$^{1}/_{3}$
W-4-M-40	4″	Core: structural clay tile; see Notes 13, 16, 23; No facings.	N/A	30 min.		1		3, 4, 24	$^{1}/_{2}$
W-4-M-41	4″	Core: structural clay tile; see Notes 13, 17, 23; No facings.	N/A	35 min.		1		3, 4, 24	$^{1}/_{2}$
W-4-M-42	4″	Core: structural clay tile; see Notes 13, 16, 21; No facings.	N/A	25 min.		1		3, 4, 24	$^{1}/_{3}$
W-4-M-43	4″	Core: structural clay tile; see Notes 13, 17, 21; No facings.	N/A	30 min.		1		3, 4, 24	$^{1}/_{2}$
W-4-M-44	4″	Core: structural clay tile; see Notes 15, 16, 20; No facings	N/A	1 hr. 15 min.		1		3, 4, 24	$1^{1}/_{4}$
W-4-M-45	4″	Core: structural clay tile; see Notes 15, 17, 20; No facings.	N/A	1 hr. 15 min.		1		3, 4, 24	$1^{1}/_{4}$
W-4-M-46	4″	Core: structural clay tile; see Notes 14, 16, 22; No facings.	N/A	20 min.		1		3, 4, 24	$^{1}/_{3}$
W-4-M-47	4″	Core: structural clay tile; see Notes 14, 17, 22; No facings.	N/A	25 min.		1		3, 4, 24	$^{1}/_{3}$
W-4-M-48	$4^{1}/_{4}$″	Core: structural clay tile; see Notes 12, 16, 21; Facings: both sides; see Note 18.	N/A	45 min.		1		3, 4, 24	$^{3}/_{4}$
W-4-M-49	$4^{1}/_{4}$″	Core: structural clay tile; see Notes 12, 17, 21; Facings: both sides; see Note 18.	N/A	1 hr.		1		3, 4, 24	1
W-4-M-50	$4^{5}/_{8}$″	Core: structural clay tile; see Notes 12, 16, 21; Facings: unexposed side only; see Note 18.	N/A	25 min.		1		3, 4, 24	$^{1}/_{3}$
W-4-M-51	$4^{5}/_{8}$″	Core: structural clay tile; see Notes 12, 17, 21; Facings: unexposed side only; see Note 18.	N/A	30 min.		1		3, 4, 24	$^{1}/_{2}$
W-4-M-52	$4^{5}/_{8}$″	Core: structural clay tile; see Notes 12, 16, 21; Facings: unexposed side only; see Note 18.	N/A	45 min.		1		3, 4, 24	$^{3}/_{4}$
W-4-M-53	$4^{5}/_{8}$″	Core: structural clay tile; see Notes 12, 17, 21; Facings: fire side only; see Note 18.	N/A	1 hr.		1		3, 4, 24	1
W-4-M-54	$4^{5}/_{8}$″	Core: structural clay tile; see Notes 12, 16, 20; Facings: unexposed side; see Note 18.	N/A	20 min.		1		3, 4, 24	$^{1}/_{3}$
W-4-M-55	$4^{5}/_{8}$″	Core: structural clay tile; see Notes 12, 17, 20; Facings: exposed side; see Note 18.	N/A	25 min.		1		3, 4, 24	$^{1}/_{3}$
W-4-M-56	$4^{5}/_{8}$″	Core: structural clay tile; see Notes 12, 16, 20; Facings: fire side only; see Note 18.	N/A	30 min.		1		3, 4, 24	$^{1}/_{2}$
W-4-M-57	$4^{5}/_{8}$″	Core: structural clay tile; see Notes 12, 17, 20; Facings: fire side only; see Note 18.	N/A	45 min.		1		3, 4, 24	$^{3}/_{4}$

(Continued)

TABLE 1.1.2—MASONRY WALLS
4″ TO LESS THAN 6″ THICK—continued

ITEM CODE	THICKNESS	CONSTRUCTION DETAILS	PERFORMANCE		REFERENCE NUMBER			NOTES	REC. HOURS
			LOAD	TIME	PRE-BMS-92	BMS-92	POST-BMS-92		
W-4-M-58	$4^5/_8$″	Core: structural clay tile; see Notes 13, 16, 23; Facings: unexposed side only; see Note 18.	N/A	40 min.		1		3, 4, 24	$^2/_3$
W-4-M-59	$4^5/_8$″	Core: structural clay tile; see Notes 13, 17, 23; Facings: unexposed side only; see Note 18.	N/A	1 hr.		1		3, 4, 24	1
W-4-M-60	$4^5/_8$″	Core: structural clay tile; see Notes 13, 16, 23; Facings: fire side only; see Note 18.	N/A	1 hr. 15 min.		1		3, 4, 24	$1^1/_4$
W-4-M-61	$4^5/_8$″	Core: structural clay tile; see Notes 13, 17, 23; Facings: fire side only; see Note 18.	N/A	1 hr. 30 min.		1		3, 4, 24	$1^1/_2$
W-4-M-62	$4^5/_8$″	Core: structural clay tile; see Notes 13, 16, 21; Facings: unexposed side only; see Note 18.	N/A	35 min.		1		3, 4, 24	$^1/_2$
W-4-M-63	$4^5/_8$″	Core: structural clay tile; see Notes 13, 17, 21; Facings: unexposed face only; see Note 18.	N/A	45 min.		1		3, 4, 24	$^3/_4$
W-4-M-64	$4^5/_8$″	Core: structural clay tile; see Notes 13, 16, 23; Facings: exposed face only; see Note 18.	N/A	1 hr.		1		3, 4, 24	1
W-4-M-65	$4^5/_8$″	Core: structural clay tile; see Notes 13, 17, 21; Facings: exposed side only; see Note 18.	N/A	1 hr. 15 min.		1		3, 4, 24	$1^1/_4$
W-4-M-66	$4^5/_8$″	Core: structural clay tile; see Notes 15, 17, 20; Facings: unexposed side only; see Note 18	N/A	1 hr. 30 min.		1		3, 4, 24	$1^1/_2$
W-4-M-67	$4^5/_8$″	Core: structural clay tile; see Notes 15, 16, 20; Facings: exposed side only; see Note 18.	N/A	1 hr. 45 min.		1		3, 4, 24	$1^3/_4$
W-4-M-68	$4^5/_8$″	Core: structural clay tile; see Notes 15, 17, 20; Facings: exposed side only; see Note 18.	N/A	1 hr. 45 min.		1		3, 4, 24	$1^3/_4$
W-4-M-69	$4^5/_8$″	Core: structural clay tile; see Notes 15, 16, 20; Facings: unexposed side only; see Note 18.	N/A	1 hr. 30 min.		1		3, 4, 24	$1^3/_4$
W-4-M-70	$4^5/_8$″	Core: structural clay tile; see Notes 14, 16, 22; Facings: unexposed side only; see Note 18.	N/A	30 min.		1		3, 4, 24	$^1/_2$
W-4-M-71	$4^5/_8$″	Core: structural clay tile; see Notes 14, 17, 22; Facings: exposed side only; see Note 18.	N/A	35 min.		1		3, 4, 24	$^1/_2$
W-4-M-72	$4^5/_8$″	Core: structural clay tile; see Notes 14, 16, 22; Facings: fire side of wall only; see Note 18.	N/A	45 min.		1		3, 4, 24	$^3/_4$
W-4-M-73	$4^5/_8$″	Core: structural clay tile; see Notes 14, 17, 22; Facings: fire side of wall only; see Note 18.	N/A	1 hr.		1		3, 4, 24	1
W-4-M-74	$5^1/_4$″	Core: structural clay tile; see Notes 12, 16, 21; Facings: both sides; see Note 18.	N/A	1 hr.		1		3, 4, 24	1

(Continued)

TABLE 1.1.2—MASONRY WALLS
4″ TO LESS THAN 6″ THICK—continued

ITEM CODE	THICKNESS	CONSTRUCTION DETAILS	PERFORMANCE		REFERENCE NUMBER			NOTES	REC. HOURS
			LOAD	TIME	PRE-BMS-92	BMS-92	POST-BMS-92		
W-5-M-75	$5^1/_4$″	Core: structural clay tile; see Notes 12, 17, 21; Facings: both sides; see Note 18	N/A	1 hr. 15 min.		1		3, 4, 24	$1^1/_4$
W-5-M-76	$5^1/_4$″	Core: structural clay tile; see Notes 12, 16, 20; Facings: both sides; see Note 18.	N/A	45 min.		1		3, 4, 24	$^3/_4$
W-5-M-77	$5^1/_4$″	Core: structural clay tile; see Notes 12, 17, 20; Facings: both sides; see Note 18.	N/A	1 hr.		1		3, 4, 24	1
W-5-M-78	$5^1/_4$″	Core: structural clay tile; see Notes 13, 16, 23; Facings: both sides of wall; see Note 18.	N/A	1 hr. 30 min.		1		3, 4, 24	$1^1/_2$
W-5-M-79	$5^1/_4$″	Core: structural clay tile; see Notes 13, 17, 23; Facings: both sides of wall; see Note 18.	N/A	2 hrs.		1		3, 4, 24	2
W-5-M-80	$5^1/_4$″	Core: structural clay tile; see Notes 13, 16, 21; Facings: both sides of wall; see Note 18.	N/A	1 hr. 15 min.		1		3, 4, 24	$1^1/_4$
W-5-M-81	$5^1/_4$″	Core: structural clay tile; see Notes 13, 16, 21; Facings: both sides of wall; see Note 18.	N/A	1 hr. 30 min.		1		3, 4, 24	$1^1/_2$
W-5-M-82	$5^1/_4$″	Core: structural clay tile; see Notes 15, 16, 20; Facings: both sides; see Note 18.	N/A	2 hrs. 30 min.		1		3, 4, 24	$2^1/_2$
W-5-M-83	$5^1/_4$″	Core: structural clay tile; see Notes 15, 17, 20; Facings: both sides; see Note 18.	N/A	2 hrs. 30 min.		1		3, 4, 24	$2^1/_2$
W-5-M-84	$5^1/_4$″	Core: structural clay tile; see Notes 14, 16, 22; Facings: both sides of wall; see Note 18.	N/A	1 hr. 15 min.		1		3, 4, 24	$1^1/_4$
W-5-M-85	$5^1/_4$″	Core: structural clay tile; see Notes 14, 17, 22; Facings: both sides of wall; see Note 18.	N/A	1 hr. 30 min.		1		3, 4, 24	$1^1/_2$
W-4-M-86	4″	Core: 3" thick gypsum blocks 70% solid; see Note 26; Facings: both sides; see Note 25.	N/A	2 hrs.		1			2
W-4-M-87	4″	Core: hollow concrete units; see Notes 27, 34, 35; No facings.	N/A	1 hr. 30 min.		1			$1^1/_2$
W-4-M-88	4″	Core: hollow concrete units; see Notes 28, 33, 35; No facings.	N/A	1 hr.		1			1
W-4-M-89	4″	Core: hollow concrete units; see Notes 28, 34, 35; Facings: both sides; see Note 25.	N/A	1 hr. 45 min.		1			$1^3/_4$
W-4-M-90	4″	Core: hollow concrete units; see Notes 27, 34, 35; Facings: both sides; see Note 25.	N/A	2 hrs.		1			2
W-4-M-91	4″	Core: hollow concrete units; see Notes 27, 32, 35; No facings.	N/A	1 hr. 15 min.		1			$1^1/_4$
W-4-M-92	4″	Core: hollow concrete units; see Notes 28, 34, 35; No facings.	N/A	1 hr. 15 min.		1			$1^1/_4$
W-4-M-93	4″	Core: hollow concrete units; see Notes 29, 32, 35; No facings.	N/A	20 min.		1			$^1/_3$

(Continued)

TABLE 1.1.2—MASONRY WALLS
4″ TO LESS THAN 6″ THICK—continued

ITEM CODE	THICKNESS	CONSTRUCTION DETAILS	PERFORMANCE		REFERENCE NUMBER			NOTES	REC. HOURS
			LOAD	TIME	PRE-BMS-92	BMS-92	POST-BMS-92		
W-4-M-94	4″	Core: hollow concrete units; see Notes 30, 34, 35; No facings.	N/A	15 min.		1			$^1/_4$
W-4-M-95	$4^1/_2$″	Core: hollow concrete units; see Notes 27, 34, 35; Facings: one side only; see Note 25.	N/A	2 hrs.		1			2
W-4-M-96	$4^1/_2$″	Core: hollow concrete units; see Notes 27, 32, 35; Facings: one side only; see Note 25.	N/A	1 hr. 45 min.		1			$1^3/_4$
W-4-M-97	$4^1/_2$″	Core: hollow concrete units; see Notes 28, 33, 35; Facings: one side; see Note 25.	N/A	1 hr. 30 min.		1			$1^1/_2$
W-4-M-98	$4^1/_2$″	Core: hollow concrete units; see Notes 28, 34, 35; Facings: one side only; see Note 25.	N/A	1 hr. 45 min.		1			$1^3/_4$
W-4-M-99	$4^1/_2$″	Core: hollow concrete units; see Notes 29, 32, 35; Facings: one side; see Note 25.	N/A	30 min.		1			$^1/_2$
W-4-M-100	$4^1/_2$″	Core: hollow concrete units; see Notes 30, 34, 35; Facings: one side; see Note 25.	N/A	20 min.		1			$^1/_3$
W-5-M-101	5″	Core: hollow concrete units; see Notes 27, 34, 35; Facings: both sides; see Note 25.	N/A	2 hrs. 30 min.		1			$2^1/_2$
W-5-M-102	5″	Core: hollow concrete units; see Notes 27, 32, 35; Facings: both sides; see Note 25.	N/A	2 hrs. 30 min.		1			$2^1/_2$
W-5-M-103	5″	Core: hollow concrete units; see Notes 28, 33, 35; Facings: both sides; see Note 25.	N/A	2 hrs.		1			2
W-5-M-104	5″	Core: hollow concrete units; see Notes 28, 31, 35; Facings: both sides; see Note 25.	N/A	2 hrs.		1			2
W-5-M-105	5″	Core: hollow concrete units; see Notes 29, 32, 35; Facings: both sides; see Note 25.	N/A	1 hr. 45 min.		1			$1^3/_4$
W-5-M-106	5″	Core: hollow concrete units; see Notes 30, 34, 35; Facings: both sides; see Note 25.	N/A	1 hr.		1			1
W-5-M-107	5″	Core: 5″ thick solid gypsum blocks; see Note 26; No facings.	N/A	4 hrs.		1			4
W-5-M-108	5″	Core: 4″ thick hollow gypsum blocks; see Note 26; Facings: both sides; see Note 25.	N/A	3 hrs.		1			3
W-5-M-109	4″	Concrete with 4″ × 4″ No. 6 welded wire mesh at wall center.	100 psi	45 min.			43	2	$^3/_4$

(Continued)

TABLE 1.1.2—MASONRY WALLS
4″ TO LESS THAN 6″ THICK—continued

ITEM CODE	THICKNESS	CONSTRUCTION DETAILS	PERFORMANCE		REFERENCE NUMBER			NOTES	REC. HOURS
			LOAD	TIME	PRE-BMS-92	BMS-92	POST-BMS-92		
W-4-M-110	4″	Concrete with 4″ × 4″ No. 6 welded wire mesh at wall center.	N/A	1 hr. 15 min.			43	2	$1^1/_4$

For SI: 1 inch = 25.4 mm, 1 pound per square inch = 0.00689 MPa.

Notes:

1. Tested as NBS under ASA Spec. No. A 2-1934.
2. Failure mode - maximum temperature rise.
3. Treated at NBS under ASA Spec. No. 42-1934 (ASTM C 19-53) except that hose stream testing where carried out was run on test specimens exposed for full test duration, not for or reduced period as is contemporarily done.
4. For clay tile walls, unless the source the clay can be positively identified, it is suggested that the most pessimistic hour rating for the fire endurance of a clay tile partition of that thickness to be followed. Identified sources of clay showing longer fire endurance can lead to longer time recommendations.
5. See appendix for construction and design details for clay tile walls.
6. Failure mode - flame thru or crack formation showing flames.
7. Hole formed at 25 minutes; partition collapsed at 42 minutes or removal from furnace.
8. Failure mode - collapse.
9. Hose stream pass.
10. Hose stream hole formed in specimen.
11. Load: 80 psi for gross wall cross sectional area.
12. One cell in wall thickness.
13. Two cells in wall thickness.
14. Double cells plus one cell in wall thickness.
15. One cell in wall thickness, cells filled with broken tile, crushed stone, slag, cinders or sand mixed with mortar.
16. Dense hard-burned clay or shale tile.
17. Medium-burned clay tile.
18. Not less than $^5/_8$ inch thickness of 1:3 sanded gypsum plaster.
19. Units of not less than 30 percent solid material.
20. Units of not less than 40 percent solid material.
21. Units of not less than 50 percent solid material.
22. Units of not less than 45 percent solid material.
23. Units of not less than 60 percent solid material.
24. All tiles laid in portland cement-lime mortar.
25. Minimum $^1/_2$ inch - 1:3 sanded gypsum plaster.
26. Laid in 1:3 sanded gypsum mortar. Voids in hollow units not to exceed 30 percent.
27. Units of expanded slag or pumice aggregate.
28. Units of crushed limestone, blast furnace slag, cinders and expanded clay or shale.
29. Units of calcareous sand and gravel. Coarse aggregate, 60 percent or more calcite and dolomite.
30. Units of siliceous sand and gravel. Ninety percent or more quartz, chert or flint.
31. Unit at least 49 percent solid.
32. Unit at least 62 percent solid.
33. Unit at least 65 percent solid.
34. Unit at least 73 percent solid.
35. Ratings based on one unit and one cell in wall thickness.
36. See Clay Tile Partition Design Construction drawings, below.

(Continued)

TABLE 1.1.2—MASONRY WALLS
4″ TO LESS THAN 6″ THICK—continued

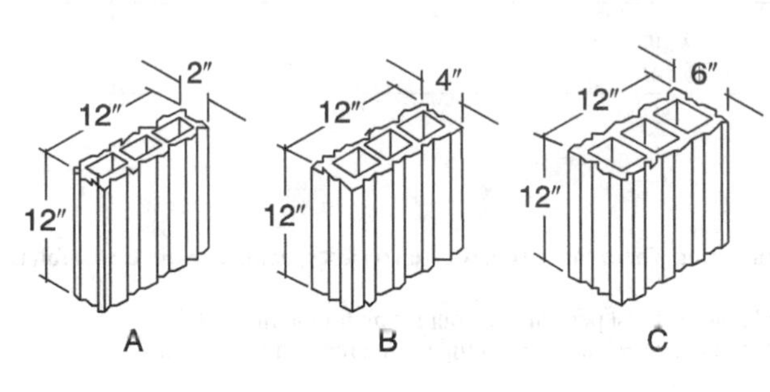

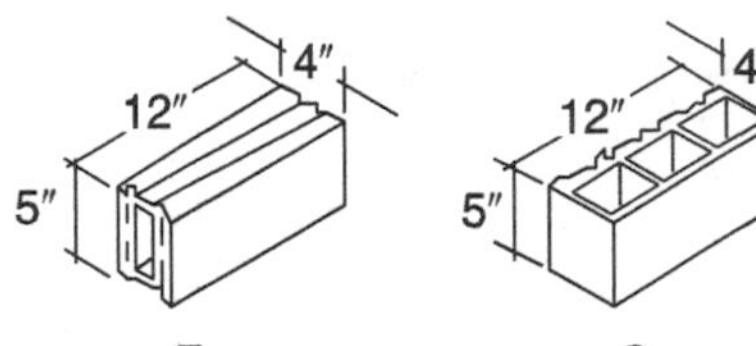

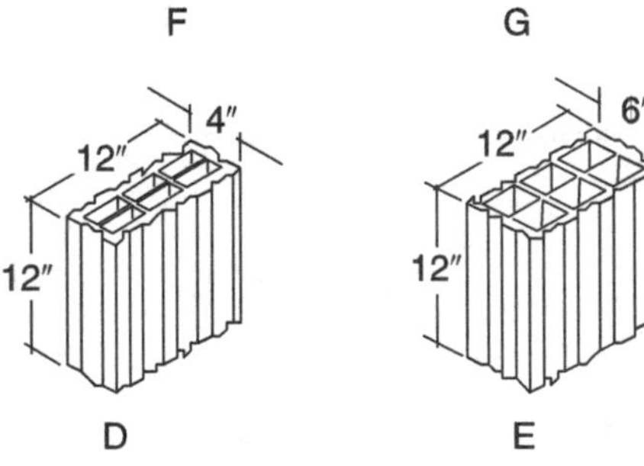

DESIGNS OF TILES USED IN FIRE-TEST PARTITIONS

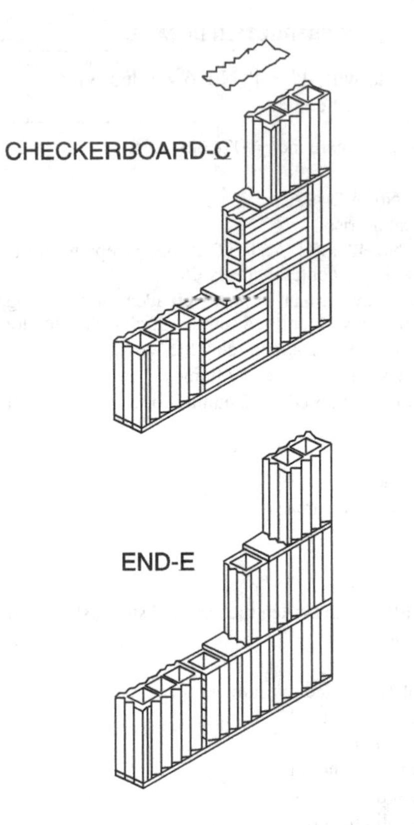

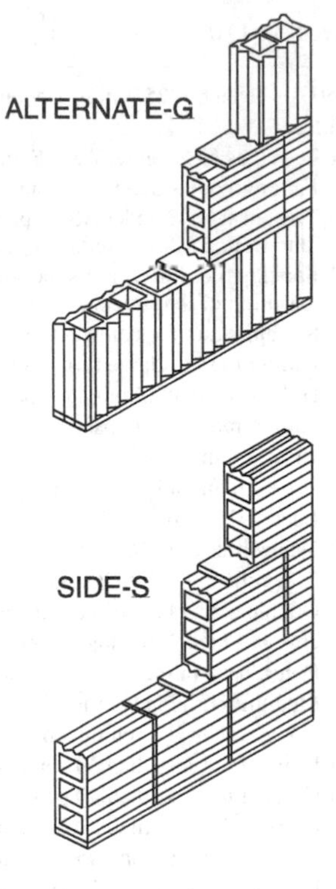

THE FOUR TYPES OF CONSTRUCTION USED IN FIRE-TEST PARTITIONS

FIGURE 1.1.3—MASONRY WALLS
6″ TO LESS THAN 8″ THICK

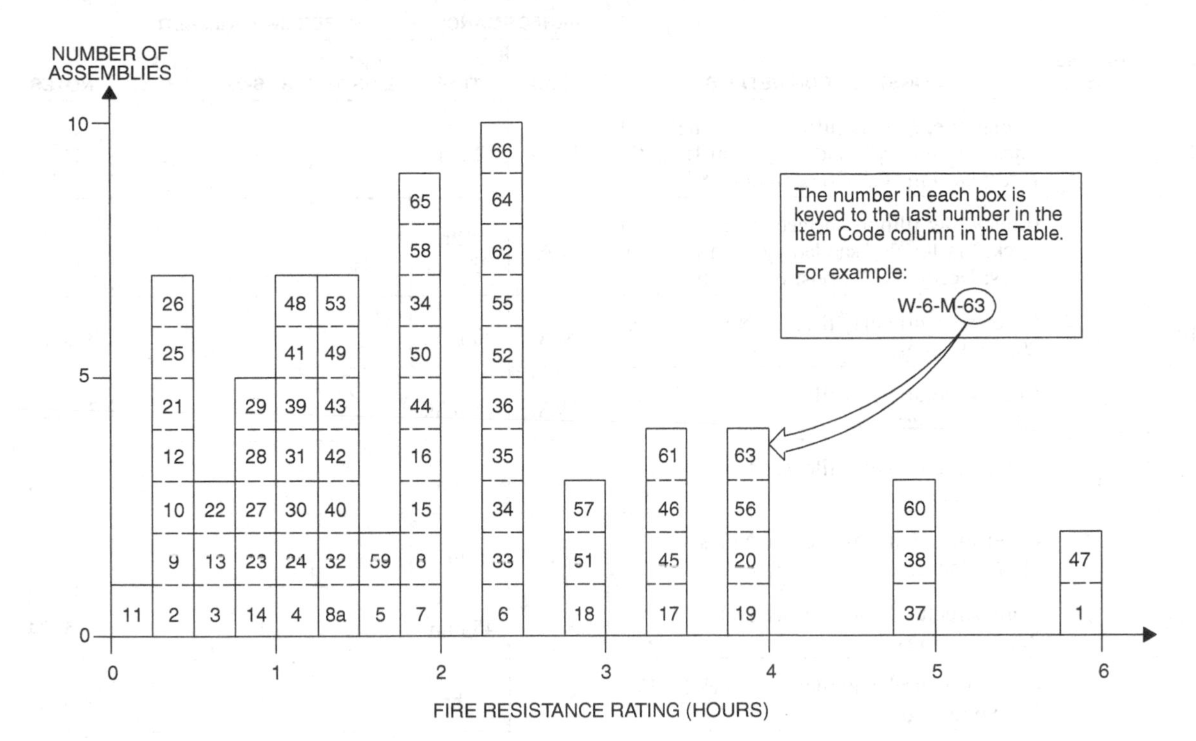

TABLE 1.1.3—MASONRY WALLS
6″ TO LESS THAN 8″ THICK

ITEM CODE	THICKNESS	CONSTRUCTION DETAILS	PERFORMANCE		REFERENCE NUMBER			NOTES	REC. HOURS
			LOAD	TIME	PRE-BMS-92	BMS-92	POST-BMS-92		
W-6-M-1	6″	Core: 5″ thick, solid gypsum blocks laid in 1:3 sanded gypsum mortar; $^{1}/_{2}$″ of 1:3 sanded gypsum plaster facings on both sides.	N/A	6 hrs.		1			6
W-6-M-2	6″	6″ clay tile; Ohio fire clay; single cell thick; No plaster; Design "C," Construction "A."	N/A	17 min.			2	1, 3, 4, 6, 55	$^{1}/_{4}$
W-6-M-3	6″	6″ clay tile; Illinois surface clay; double cell thick; No plaster; Design "E," Construction "C."	N/A	45 min.			2	1-4, 7, 55	$^{3}/_{4}$
W-6-M-4	6″	6″ clay tile; New Jersey fire clay; double cell thick; No plaster; Design "E," Construction "S."	N/A	1 hr. 1 min.			2	1-4, 8, 55	1
W-7-M-5	$7^{1}/_{4}$″	6″ clay tile; Illinois surface clay; double cell thick; Plaster: $^{5}/_{8}$″ - 1:3 sanded gypsum both faces; Design "E," Construction "A."	N/A	1 hr. 41 min.			2	1-4, 55	$1^{2}/_{3}$
W-7-M-6	$7^{1}/_{4}$″	6″ clay tile; New Jersey fire clay; double cell thick; Plaster: $^{5}/_{8}$″ - 1:3 sanded gypsum both faces; Design "E," Construction "S."	N/A	2 hrs. 23 min.			2	1-4, 9, 55	$2^{1}/_{3}$
W-7-M-7	$7^{1}/_{4}$″	6″ clay tile; Ohio fire clay; single cell thick; Plaster: $^{5}/_{8}$″ sanded gypsum; 1:3 both faces; Design "C," Construction "A."	N/A	1 hr. 54 min.			2	1-4, 9, 55	$2^{3}/_{4}$

(Continued)

TABLE 1.1.3—MASONRY WALLS
6″ TO LESS THAN 8″ THICK—continued

ITEM CODE	THICKNESS	CONSTRUCTION DETAILS	PERFORMANCE		REFERENCE NUMBER			NOTES	REC. HOURS
			LOAD	TIME	PRE-BMS-92	BMS-92	POST-BMS-92		
W-7-M-8	$7^1/_4$″	6″ clay tile; Illinois surface clay; single cell thick; Plaster: $^5/_8$″ sanded gypsum 1:3 both faces; Design "C," Construction "S."	N/A	2 hrs.			2	1, 3, 4, 9, 10, 55	2
W-7-M-8a	$7^1/_4$″	6″ clay tile; Illinois surface clay; single cell thick; Plaster: $^5/_8$″ sanded gypsum 1:3 both faces; Design "C," Construction "E."	N/A	1 hr. 23 min			2	1-4, 9, 10, 55	$1^3/_4$
W-6-M-9	6″	Core: structural clay tile; see Notes 12, 16, 20; No facings.	N/A	20 min.		1		3, 5, 24	$^1/_3$
W-6-M-10	6″	Core: structural clay tile; see Notes 12, 17, 20; No facings.	N/A	25 min.		1		3, 5, 24	$^1/_3$
W-6-M-11	6″	Core: structural clay tile; see Notes 12, 16, 19; No facings.	N/A	15 min.		1		3, 5, 24	$^1/_4$
W-6-M-12	6″	Core: structural clay tile; see Notes 12, 17, 19; No facings.	N/A	20 min.		1		3, 5, 24	$^1/_3$
W-6-M-13	6″	Core: structural clay tile; see Notes 13, 16, 22; No facings.	N/A	45 min.		1		3, 5, 24	$^3/_4$
W-6-M-14	6″	Core: structural clay tile; see Notes 13, 17, 22; No facings.	N/A	1 hr.		1		3, 5, 24	1
W-6-M-15	6″	Core: structural clay tile; see Notes 15, 17, 19; No facings.	N/A	2 hrs.		1		3, 5, 24	2
W-6-M-16	6″	Core: structural clay tile; see Notes 15, 16, 19; No facings.	N/A	2 hrs.		1		3, 5, 24	2
W-6-M-17	6″	Cored concrete masonry; see Notes 12, 34, 36, 38, 41; No facings.	80 psi	3 hrs. 30 min.		1		5, 25	$3^1/_2$
W-6-M-18	6″	Cored concrete masonry; see Notes 12, 33, 36, 38, 41; No facings.	80 psi	3 hrs.		1		5, 25	3
W-6-M-19	$6^1/_2$″	Cored concrete masonry; see Notes 12, 34, 36, 38, 41; Facings: side 1; see Note 35.	80 psi	4 hrs.		1		5, 25	4
W-6-M-20	$6^1/_2$″	Cored concrete masonry; see Notes 12, 33, 36, 38, 41; Facings: side 1; see Note 35.	80 psi	4 hrs.		1		5, 25	4
W-6-M-21	$6^5/_8$″	Core: structural clay tile; see Notes 12, 16, 20; Facings: unexposed face only; see Note 18.	N/A	30 min.		1		3, 5, 24	$^1/_2$
W-6-M-22	$6^5/_8$″	Core: structural clay tile; see Notes 12, 17, 20; Facings: unexposed face only; see Note 18.	N/A	40 min.		1		3, 5, 24	$^2/_3$
W-6-M-23	$6^5/_8$″	Core: structural clay tile; see Notes 12, 16, 20; Facings: exposed face only; see Note 18.	N/A	1 hr.		1		3, 5, 24	1
W-6-M-24	$6^5/_8$″	Core: structural clay tile; see Notes 12, 17, 20; Facings: exposed face only; see Note 18.	N/A	1 hr. 5 min.		1		3, 5, 24	1
W-6-M-25	$6^5/_8$″	Core: structural clay tile; see Notes 12, 16, 19; Facings: unexposed side only; see Note 18.	N/A	25 min.		1		3, 5, 24	$^1/_3$

(Continued)

TABLE 1.1.3—MASONRY WALLS
6″ TO LESS THAN 8″ THICK—continued

ITEM CODE	THICKNESS	CONSTRUCTION DETAILS	PERFORMANCE		REFERENCE NUMBER			NOTES	REC. HOURS
			LOAD	TIME	PRE-BMS-92	BMS-92	POST-BMS-92		
W-6-M-26	$6^5/_8$″	Core: structural clay tile; see Notes 12, 7, 19; Facings: unexposed face only; see Note 18.	N/A	30 min.		1		3, 5, 24	$^1/_2$
W-6-M-27	$6^5/_8$″	Core: structural clay tile; see Notes 12, 16, 19; Facings: exposed side only; see Note 18.	N/A	1 hr.		1		3, 5, 24	1
W-6-M-28	$6^5/_8$″	Core: structural clay tile; see Notes 12, 17, 19; Facings: fire side only; see Note 18.	N/A	1 hr.		1		3, 5, 24	1
W-6-M-29	$6^5/_8$″	Core: structural clay tile; see Notes 13, 16, 22; Facings: unexposed side only; see Note 18.	N/A	1 hr.		1		3, 5, 24	1
W-6-M-30	$6^5/_8$″	Core: structural clay tile; see Notes 13, 17, 22; Facings: unexposed side only; see Note 18.	N/A	1 hr. 15 min.		1		3, 5, 24	$1^1/_4$
W-6-M-31	$6^5/_8$″	Core: structural clay tile; see Notes 13, 16, 22; Facings: fire side only; see Note 18.	N/A	1 hr. 15 min.		1		3, 5, 24	$1^1/_4$
W-6-M-32	$6^5/_8$″	Core: structural clay tile; see Notes 13, 17, 22; Facings: fire side only; see Note 18.	N/A	1 hr. 30 min.		1		3, 5, 24	$1^1/_2$
W-6-M-33	$6^5/_8$″	Core: structural clay tile; see Notes 15, 16, 19; Facings: unexposed side only; see Note 18.	N/A	2 hrs. 30 min.		1		3, 5, 24	$2^1/_2$
W-6-M-34	$6^5/_8$″	Core: structural clay tile; see Notes 15, 17, 19; Facings: unexposed side only; see Note 18.	N/A	2 hrs. 30 min.		1		3, 5, 24	$2^1/_2$
W-6-M-35	$6^5/_8$″	Core: structural clay tile; see Notes 15, 16, 19; Facings: fire side only; see Note 18.	N/A	2 hrs. 30 min.		1		3, 5, 24	$2^1/_2$
W-6-M-36	$6^5/_8$″	Core: structural clay tile; see Notes 15, 17, 19; Facings: fire side only; see Note 18.	N/A	2 hrs. 30 min.		1		3, 5, 24	$2^1/_2$
W-6-M-37	7″	Cored concrete masonry; see Notes 12, 34, 36, 38, 41; see Note 35 for facings on both sides.	80 psi	5 hrs.		1		5, 25	5
W-6-M-38	7″	Cored concrete masonry; see Notes 12, 33, 36, 38, 41; see Note 35 for facings.	80 psi	5 hrs.		1		5, 25	5
W-6-M-39	$7^1/_4$″	Core: structural clay tile; see Notes 12, 16, 20; Facings: both sides; see Note 18.	N/A	1 hr. 15 min.		1		3, 5, 24	$1^1/_4$
W-6-M-40	$7^1/_4$″	Core: structural clay tile; see Notes 12, 17, 20; Facings: both sides; see Note 18.	N/A	1 hr. 30 min.		1		3, 5, 24	$1^1/_2$
W-6-M-41	$7^1/_4$″	Core: structural clay tile; see Notes 12, 16, 19; Facings: both sides; see Note 18.	N/A	1 hr. 15 min.		1		3, 5, 24	$1^1/_4$
W-6-M-42	$7^1/_4$″	Core: structural clay tile; see Notes 12, 17, 19; Facings: both sides; see Note 18.	N/A	1 hr. 30 min.		1		3, 5, 24	$1^1/_2$

(Continued)

TABLE 1.1.3—MASONRY WALLS
6″ TO LESS THAN 8″ THICK—continued

ITEM CODE	THICKNESS	CONSTRUCTION DETAILS	PERFORMANCE		REFERENCE NUMBER			NOTES	REC. HOURS
			LOAD	TIME	PRE-BMS-92	BMS-92	POST-BMS-92		
W-7-M-43	$7^1/_4$″	Core: structural clay tile; see Notes 13, 16, 22; Facings: both sides of wall; see Note 18.	N/A	1 hr. 30 min.		1		3, 5, 24	$1^1/_2$
W-7-M-44	$7^1/_4$″	Core: structural clay tile; see Notes 13, 17, 22; Facings: both sides of wall; see Note 18.	N/A	2 hrs.		1		3, 5, 24	$1^1/_2$
W-7-M-45	$7^1/_4$″	Core: structural clay tile; see Notes 15, 16, 19; Facings: both sides; see Note 18.	N/A	3 hrs. 30 min.		1		3, 5, 24	$3^1/_2$
W-7-M-46	$7^1/_4$″	Core: structural clay tile; see Notes 15, 17, 19; Facings: both sides; see Note 18.	N/A	3 hrs. 30 min.		1		3, 5, 24	$3^1/_2$
W-6-M-47	6″	Core: 5″ thick solid gypsum blocks; see Note 45; Facings: both sides; see Note 45.	N/A	6 hrs.		1			6
W-6-M-48	6″	Core: hollow concrete units; see Notes 47, 50, 54; No facings.	N/A	1 hr. 15 min.		1			$1^1/_4$
W-6-M-49	6″	Core: hollow concrete units; see Notes 46, 50, 54; No facings.	N/A	1 hr. 30 min.		1			$1^1/_2$
W-6-M-50	6″	Core: hollow concrete units; see Notes 46, 41, 54; No facings.	N/A	2 hrs.		1			2
W-6-M-51	6″	Core: hollow concrete units; see Notes 46, 53, 54; No facings.	N/A	3 hrs.		1			3
W-6-M-52	6″	Core: hollow concrete units; see Notes 47, 53, 54; No facings.	N/A	2 hrs. 30 min.		1			$2^1/_2$
W-6-M-53	6″	Core: hollow concrete units; see Notes 47, 51, 54; No facings.	N/A	1 hr. 30 min.		1			$1^1/_2$
W-6-M-54	$6^1/_2$″	Core: hollow concrete units; see Notes 46, 50, 54; Facings: one side only; see Note 35.	N/A	2 hrs.		1			2
W-6-M-55	$6^1/_2$″	Core: hollow concrete units; see Notes 4, 51, 54; Facings: one side; see Note 35.	N/A	2 hrs. 30 min.		1			$2^1/_2$
W-6-M-56	$6^1/_2$″	Core: hollow concrete units; see Notes 46, 53, 54; Facings: one side; see Note 35.	N/A	4 hrs.		1			4
W-6-M-57	$6^1/_2$″	Core: hollow concrete units; see Notes 47, 53, 54; Facings: one side; see Note 35.	N/A	3 hrs.		1			3
W-6-M-58	$6^1/_2$″	Core: hollow concrete units; see Notes 47, 51, 54; Facings: one side; see Note 35.	N/A	2 hrs.		1			2
W-6-M-59	$6^1/_2$″	Core: hollow concrete units; see Notes 47, 50, 54; Facings: one side; see Note 35.	N/A	1 hr. 45 min.		1			$1^3/_4$
W-7-M-60	7″	Core: hollow concrete units; see Notes 46, 53, 54; Facings: both sides; see Note 35.	N/A	5 hrs.		1			5
W-7-M-61	7″	Core: hollow concrete units; see Notes 46, 51, 54; Facings: both sides; see Note 35.	N/A	3 hrs. 30 min.		1			$3^1/_2$

(Continued)

TABLE 1.1.3—MASONRY WALLS
6″ TO LESS THAN 8″ THICK—continued

ITEM CODE	THICKNESS	CONSTRUCTION DETAILS	PERFORMANCE		REFERENCE NUMBER			NOTES	REC. HOURS
			LOAD	TIME	PRE-BMS-92	BMS-92	POST-BMS-92		
W-7-M-62	7″	Core: hollow concrete units; see Notes 46, 50, 54; Facings: both sides; see Note 35.	N/A	2 hrs. 30 min.		1			$2^1/_2$
W-7-M-63	7″	Core: hollow concrete units; see Notes 47, 53, 54; Facings: both sides; see Note 35.	N/A	4 hrs.		1			4
W-7-M-64	7″	Core: hollow concrete units; see Notes 47, 51, 54; Facings: both sides; see Note 35.	N/A	2 hrs. 30 min.		1			$2^1/_2$
W-7-M-65	7″	Core: hollow concrete units; see Notes 47, 50, 54; Facings: both sides; see Note 35.	N/A	2 hrs.		1			2
W-6-M-66	6″	Concrete wall with 4″ × 4″ No. 6 wire fabric (welded) near wall center for reinforcement.	N/A	2 hrs. 30 min.			43	2	$2^1/_2$

For SI: 1 inch = 25.4 mm, 1 pound per square inch = 0.00689 MPa.

Notes:

1. Tested at NBS under ASA Spec. No. 43-1934 (ASTM C 19-53) except that hose stream testing where carried out was run on test specimens exposed for full test duration, not for a reduced period as is contemporarily done.
2. Failure by thermal criteria - maximum temperature rise.
3. For clay tile walls, unless the source or density of the clay can be positively identified or determined, it is suggested that the lowest hourly rating for the fire endurance of a clay tile partition of that thickness be followed. Identified sources of clay showing longer fire endurance can lead to longer time recommendations.
4. See Note 55 for construction and design details for clay tile walls.
5. Tested at NBS under ASA Spec. No. A2-1934.
6. Failure mode - collapse.
7. Collapsed on removal from furnace at 1 hour 9 minutes.
8. Hose stream - failed.
9. Hose stream - passed.
10. No end point met in test.
11. Wall collapsed at 1 hour 28 minutes.
12. One cell in wall thickness.
13. Two cells in wall thickness.
14. Double shells plus one cell in wall thickness.
15. One cell in wall thickness, cells filled with broken tile, crushed stone, slag, cinders or sand mixed with mortar.
16. Dense hard-burned clay or shale tile.
17. Medium-burned clay tile.
18. Not less than $^5/_8$ inch thickness of 1:3 sanded gypsum plaster.
19. Units of not less than 30 percent solid material.
20. Units of not less than 40 percent solid material.
21. Units of not less than 50 percent solid material.
22. Units of not less than 45 percent solid material.
23. Units of not less than 60 percent solid material.
24. All tiles laid in portland cement-lime mortar.
25. Load: 80 psi for gross cross sectional area of wall.
26. Three cells in wall thickness.
27. Minimum percent of solid material in concrete units = 52.
28. Minimum percent of solid material in concrete units = 54.
29. Minimum percent of solid material in concrete units = 55.
30. Minimum percent of solid material in concrete units = 57.
31. Minimum percent of solid material in concrete units = 62.
32. Minimum percent of solid material in concrete units = 65.
33. Minimum percent of solid material in concrete units = 70.
34. Minimum percent of solid material in concrete units = 76.
35. Not less than $^1/_2$ inch of 1:3 sanded gypsum plaster.
36. Noncombustible or no members framed into wall.
37. Combustible members framed into wall.
38. One unit in wall thickness.
39. Two units in wall thickness.

(Continued)

TABLE 1.1.3—MASONRY WALLS
6″ TO LESS THAN 8″ THICK—continued

40. Three units in wall thickness.
41. Concrete units made with expanded slag or pumice aggregates.
42. Concrete units made with expanded burned clay or shale, crushed limestone, air cooled slag or cinders.
43. Concrete units made with calcareous sand and gravel. Coarse aggregate, 60 percent or more calcite and dolomite.
44. Concrete units made with siliceous sand and gravel. Ninety percent or more quartz, chert or flint.
45. Laid in 1:3 sanded gypsum mortar.
46. Units of expanded slag or pumice aggregate.
47. Units of crushed limestone, blast furnace, slag, cinder and expanded clay or shale.
48. Units of calcareous sand and gravel. Coarse aggregate, 60 percent or more calcite and dolomite.
49. Units of siliceous sand and gravel. Ninety percent or more quartz, chert or flint.
50. Unit minimum 49 percent solid.
51. Unit minimum 62 percent solid.
52. Unit minimum 65 percent solid.
53. Unit minimum 73 percent solid.
54. Ratings based on one unit and one cell in wall section.
55. See Clay Tile Partition Design Construction drawings, below.

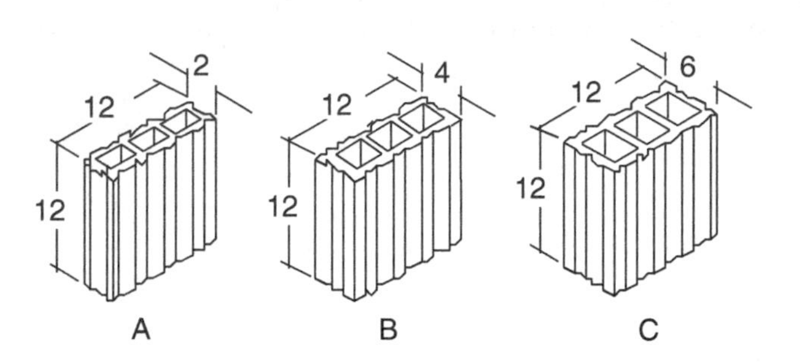

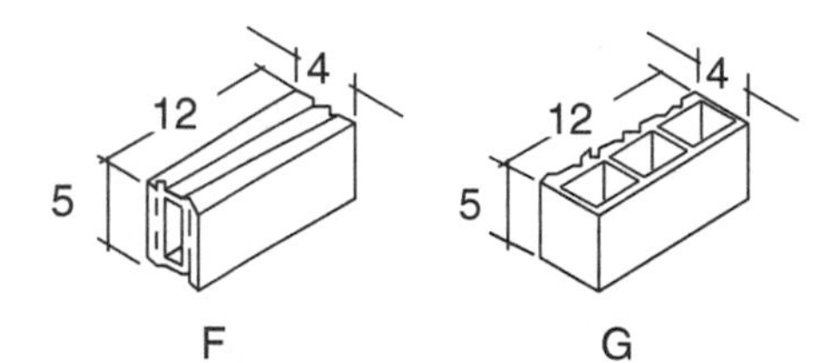

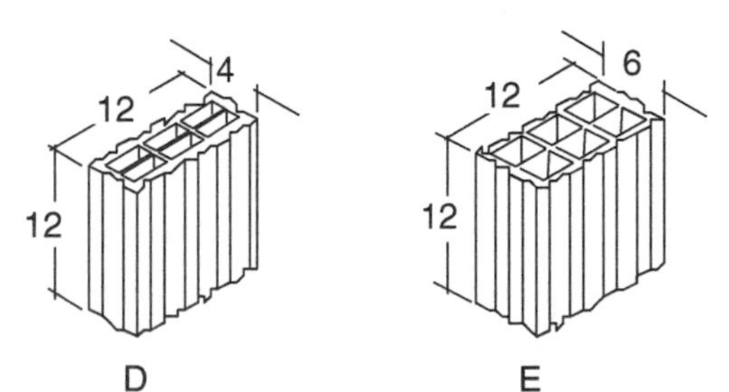

DESIGNS OF TILES USED IN FIRE-TEST PARTITIONS

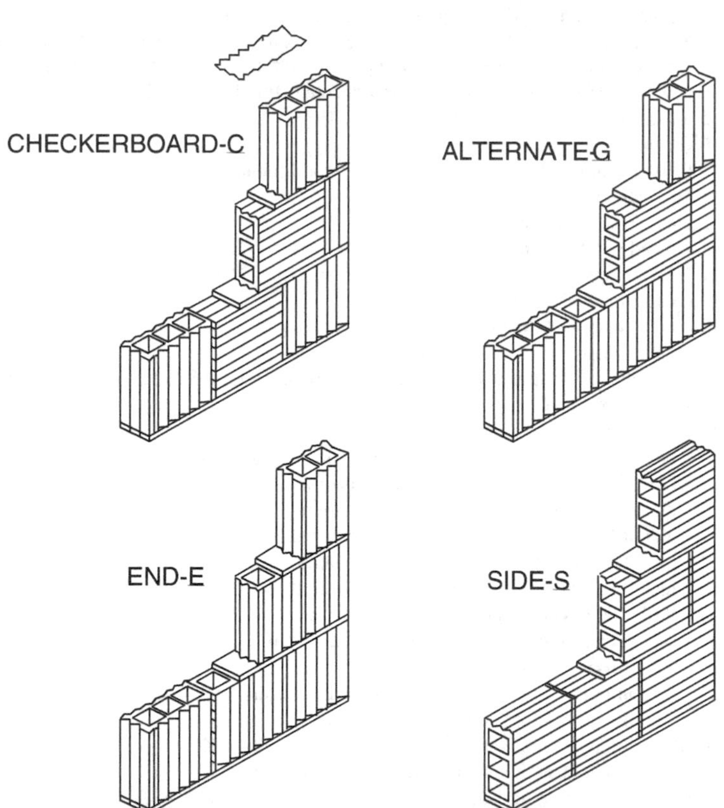

THE FOUR TYPES OF CONSTRUCTION USED IN FIRE-TEST PARTITIONS

FIGURE 1.1.4—MASONRY WALLS
8″ TO LESS THAN 10″ THICK

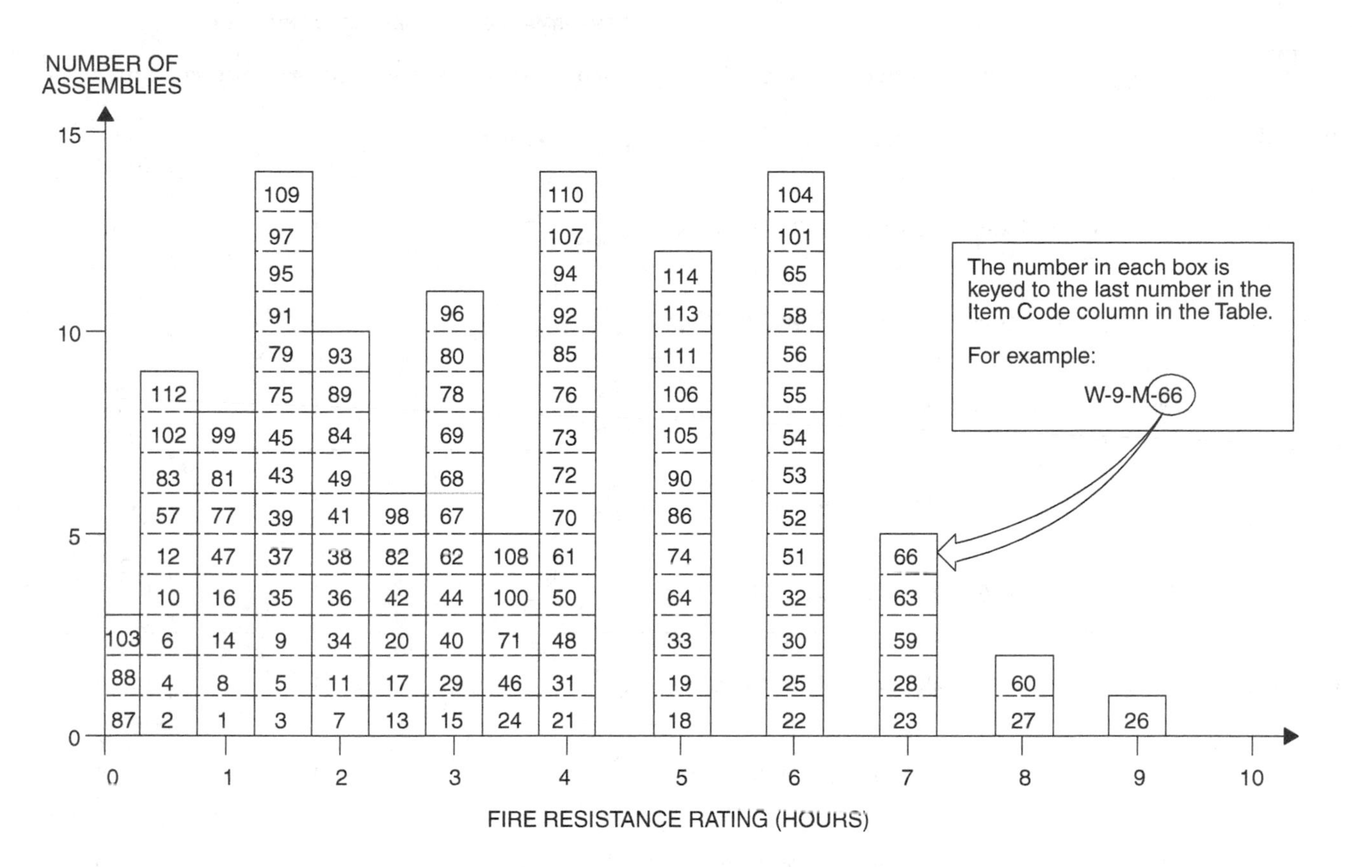

TABLE 1.1.4—MASONRY WALLS
8″ TO LESS THAN 10″ THICK

ITEM CODE	THICKNESS	CONSTRUCTION DETAILS	PERFORMANCE		REFERENCE NUMBER			NOTES	REC. HOURS
			LOAD	TIME	PRE-BMS-92	BMS-92	POST-BMS-92		
W-8-M-1	8″	Core: clay or shale structural tile; Units in wall thickness: 1; Cells in wall thickness: 2; Minimum % solids in units: 40.	80 psi	1 hr. 15 min.		1		1, 20	$1^{1}/_{4}$
W-8-M-2	8″	Core: clay or shale structural tile; Units in wall thickness: 1; Cells in wall thickness: 2; Minimum % solids in units: 40; No facings; Result for wall with combustible members framed into interior.	80 psi	45 min.		1		1, 20	$^{3}/_{4}$
W-8-M-3	8″	Core: clay or shale structural tile; Units in wall thickness: 1; Cells in wall thickness: 2; Minimum % solids in units: 43.	80 psi	1 hr. 30 min.		1		1, 20	$1^{1}/_{2}$
W-8-M-4	8″	Core: clay or shale structural tile; Units in wall thickness: 1; Cells in wall thickness: 2; Minimum % solids in units: 43; No facings; Combustible members framed into wall.	80 psi	45 min.		1		1, 20	$^{3}/_{4}$
W-8-M-5	8″	Core: clay or shale structural tile; No facings.	See Notes	1 hr. 30 min.		1		1, 2, 5, 10, 18, 20, 21	$1^{1}/_{2}$

(Continued)

TABLE 1.1.4—MASONRY WALLS
8″ TO LESS THAN 10″ THICK—continued

ITEM CODE	THICKNESS	CONSTRUCTION DETAILS	PERFORMANCE		REFERENCE NUMBER			NOTES	REC. HOURS
			LOAD	TIME	PRE-BMS-92	BMS-92	POST-BMS-92		
W-8-M-6	8″	Core: clay or shale structural tile; No facings.	See Notes	45 min.		1		1, 2, 5, 10,19, 20, 21	$^3/_4$
W-8-M-7	8″	Core: clay or shale structural tile; No facings	See Notes	2 hrs.		1		1, 2, 5, 13, 18, 20, 21	2
W-8-M-8	8″	Core: clay or shale structural tile; No facings.	See Notes	1 hr. 45 min.		1		1, 2, 5, 13, 19, 20, 21	$1^1/_4$
W-8-M-9	8″	Core: clay or shale structural tile; No facings.	See Notes	1 hr. 15 min.		1		1, 2, 6, 9, 18, 20, 21	$1^3/_4$
W-8-M-10	8″	Core: clay or shale structural tile; No facings.	See Notes	45 min.		1		1, 2, 6, 9, 19, 20, 21	$^3/_4$
W-8-M-11	8″	Core: clay or shale structural tile; No facings.	See Notes	2 hrs.		1		1, 2, 6, 10, 18, 20, 21	2
W-8-M-12	8″	Core: clay or shale structural tile; No facings.	See Notes	45 min.		1		1, 2, 6, 10, 19, 20, 21	$^3/_4$
W-8-M-13	8″	Core: clay or shale structural tile; No facings.	See Notes	2 hrs. 30 min.		1		1, 3, 6, 12, 18, 20, 21	$2^1/_2$
W-8-M-14	8″	Core: clay or shale structural tile; No facings.	See Notes	1 hr.		1		1, 2, 6, 12, 19, 20, 21	1
W-8-M-15	8″	Core: clay or shale structural tile; No facings.	See Notes	3 hrs.		1		1, 2, 6, 16, 18, 20, 21	3
W-8-M-16	8″	Core: clay or shale structural tile; No facings.	See Notes	1 hr. 15 min.		1		1, 2, 6, 16, 19, 20, 21	$1^1/_4$
W-8-M-17	8″	Cored clay or shale brick; Units in wall thickness: 1; Cells in wall thickness: 1; Minimum % solids: 70; No facings.	See Notes	2 hrs. 30 min.		1		1, 44	$2^1/_2$
W-8-M-18	8″	Cored clay or shale brick; Units in wall thickness: 2; Cells in wall thickness: 2; Minimum % solids: 87; No facings.	See Notes	5 hrs.		1		1, 45	5
W-8-M-19	8″	Core: solid clay or shale brick; No facings.	See Notes	5 hrs.		1		1, 22, 45	5
W-8-M-20	8″	Core: hollow rolok of clay or shale.	See Notes	2 hrs. 30 min.		1		1, 22, 45	$2^1/_2$
W-8-M-21	8″	Core: hollow rolok bak of clay or shale; No facings.	See Notes	4 hrs.		1		1, 45	4
W-8-M-22	8″	Core: concrete brick; No facings.	See Notes	6 hrs.		1		1, 45	6
W-8-M-23	8″	Core: sand-lime brick; No facings.	See Notes	7 hrs.		1		1, 45	7

(Continued)

TABLE 1.1.4—MASONRY WALLS
8″ TO LESS THAN 10″ THICK—continued

ITEM CODE	THICKNESS	CONSTRUCTION DETAILS	PERFORMANCE		REFERENCE NUMBER			NOTES	REC. HOURS
			LOAD	TIME	PRE-BMS-92	BMS-92	POST-BMS-92		
W-8-M-24	8″	Core: 4″, 40% solid clay or shale structural tile; 1 side 4″ brick facing.	See Notes	3 hrs. 30 min.		1		1, 20	$3^1/_2$
W-8-M-25	8″	Concrete wall (3220 psi); Reinforcing vertical rods 1″ from each face and 1″ diameter; horizontal rods $^5/_8$″ diameter.	22,200 lbs./ft.	6 hrs.			7		6
W-8-M-26	8″	Core: sand-line brick; $^1/_2$″ of 1:3 sanded gypsum plaster facings on one side.	See Notes	9 hrs.		1		1, 45	9
W-8-M-27	$8^1/_2$″	Core: sand-line brick; $^1/_2$″ of 1:3 sanded gypsum plaster facings on one side.	See Notes	8 hrs.		1		1, 45	8
W-8-M-28	$8^1/_2$″	Core: concrete; $^1/_2$″ of 1:3 sanded gypsum plaster facings on one side.	See Notes	7 hrs.		1		1, 45	7
W-8-M-29	$8^1/_2$″	Core: hollow rolok of clay or shale; $^1/_2$″ of 1:3 sanded gypsum plaster facings on one side.	See Notes	3 hrs.		1		1, 45	3
W-8-M-30	$8^1/_2$″	Core: solid clay or shale brick $^1/_2$″ thick, 1:3 sanded gypsum plaster facings on one side.	See Notes	6 hrs.		1		1, 22, 45,	6
W-8-M-31	$8^1/_2$″	Core: cored clay or shale brick; Units in wall thickness: 1; Cells in wall thickness: 1; Minimum % solids: 70; $^1/_2$″ of 1:3 sanded gypsum plaster facings on both sides.	See Notes	4 hrs.		1		1, 44	4
W-8-M-32	$8^1/_2$″	Core: cored clay or shale brick; Units in wall thickness: 2; Cells in wall thickness: 2; Minimum % solids: 87; $^1/_2$″ of 1:3 sanded gypsum plaster facings on one side.	See Notes	6 hrs.		1		1, 45	6
W-8-M-33	$8^1/_2$″	Core: hollow rolok bak of clay or shale; $^1/_2$″ of 1:3 sanded gypsum plaster facings on one side.	See Notes	5 hrs.		1		1, 45	5
W-8-M-34	$8^5/_8$″	Core: clay or shale structural tile; Units in wall thickness: 1; Cells in wall thickness: 2; Minimum % solids in units: 40; $^5/_8$″ of 1:3 sanded gypsum plaster facings on one side.	See Notes	2 hrs.		1		1, 20 21	2
W-8-M-35	$8^5/_8$″	Core: clay or shale structural tile; Units in wall thickness: 1; Cells in wall thickness: 2; Minimum % solids in units: 40; Exposed face: $^5/_8$″ of 1:3 sanded gypsum plaster.	See Notes	1 hr. 30 min.		1		1, 20, 21	$1^1/_2$

(Continued)

TABLE 1.1.4—MASONRY WALLS
8″ TO LESS THAN 10″ THICK—continued

ITEM CODE	THICKNESS	CONSTRUCTION DETAILS	PERFORMANCE		REFERENCE NUMBER			NOTES	REC. HOURS
			LOAD	TIME	PRE-BMS-92	BMS-92	POST-BMS-92		
W-8-M-36	$8^5/_8''$	Core: clay or shale structural tile; Units in wall thickness: 1; Cells in wall thickness: 2; Minimum % solids in units: 43; $^5/_8''$ of 1:3 sanded gypsum plaster facings on one side.	See Notes	2 hrs.				1, 20, 21	2
W-8-M-37	$8^5/_8''$	Core: clay or shale structural tile; Units in wall thickness: 1; Cells in wall thickness: 2; Minimum % solids in units: 43; $^5/_8''$ of 1:3 sanded gypsum plaster of the exposed face only.	See Notes	1 hr. 30 min.		1		1, 20, 21	$1^1/_2$
W-8-M-38	$8^5/_8''$	Core: clay or shale structural tile; Facings: side 1; see Note 17.	See Notes	2 hrs.		1		1, 2, 5, 10, 18, 20, 21	2
W-8-M-39	$8^5/_8''$	Core: clay or shale structural tile; Facings: exposed side only; see Note 17.	See Notes	1 hr. 30 min.		1		1, 2, 5, 10, 19, 20, 21	$1^1/_2$
W-8-M-40	$8^5/_8''$	Core: clay or shale structural tile; Facings: exposed side only; see Note 17.	See Notes	3 hrs.		1		1, 2, 5, 13, 18, 20, 21	3
W-8-M-41	$8^5/_8''$	Core: clay or shale structural tile; Facings: exposed side only; see Note 17.	See Notes	2 hrs.		1		1, 2, 5, 13, 19, 20, 21	2
W-8-M-42	$8^5/_8''$	Core: clay or shale structural tile; Facings: side 1; see Note 17.	See Notes	2 hrs. 30 min.		1		1, 2, 9, 18, 20, 21	$2^1/_2$
W-8-M-43	$8^5/_8''$	Core: clay or shale structural tile; Facings: exposed side only; see Note 17.	See Notes	1 hr. 30 min.		1		1, 2, 6, 9, 19, 20, 21	$1^1/_2$
W-8-M-44	$8^5/_8''$	Core: clay or shale structural tile; Facings: side 1, see Note 17; side 2, none.	See Notes	3 hrs.		1		1, 2, 10, 18, 20, 21	3
W-8-M-45	$8^5/_8''$	Core: clay or shale structural tile; Facings: fire side only; see Note 17.	See Notes	1 hr. 30 min.		1		1, 2, 6, 10, 19, 20, 21	$1^1/_2$
W-8-M-46	$8^5/_8''$	Core: clay or shale structural tile; Facings: side 1, see Note 17; side 2, none.	See Notes	3 hrs. 30 min.		1		1, 2, 6, 12, 18, 20, 21	$3^1/_2$
W-8-M-47	$8^5/_8''$	Core: clay or shale structural tile; Facings: exposed side only; see Note 17.	See Notes	1 hr. 45 min.		1		1, 2, 6, 12, 19, 20, 21	$1^3/_4$
W-8-M-48	$8^5/_8''$	Core: clay or shale structural tile; Facings: side 1, see Note 17; side 2, none.	See Notes	4 hrs.		1		1, 2, 6, 16, 18, 20, 21	4
W-8-M-49	$8^5/_8''$	Core: clay or shale structural tile; Facings: fire side only; see Note 17.	See Notes	2 hrs.		1		1, 2, 6, 16, 19, 20, 21	2

(Continued)

TABLE 1.1.4—MASONRY WALLS 8″ TO LESS THAN 10″ THICK—continued

ITEM CODE	THICKNESS	CONSTRUCTION DETAILS	PERFORMANCE		REFERENCE NUMBER			NOTES	REC. HOURS
			LOAD	TIME	PRE-BMS-92	BMS-92	POST-BMS-92		
W-8-M-50	$8^5/_8$″	Core: 4″, 40% solid clay or shale clay structural tile; 4″ brick plus $^5/_8$″ of 1:3 sanded gypsum plaster facings on one side.	See Notes	4 hrs.		1		1, 20	4
W-8-M-51	$8^3/_4$″	$8^3/_4$″ × $2^1/_2$″ and 4″ × $2^1/_2$″ cellular fletton (1873 psi) single and triple cell hollow brick set in $^1/_2$″ sand mortar in alternate courses.	3.6 tons/ft.	6 hrs.			7	23, 29	6
W-8-M-52	$8^3/_4$″	$8^3/_4$″ thick cement brick (2527 psi) with P.C. and sand mortar.	3.6 tons/ft.	6 hrs.			7	23, 24	6
W-8-M-53	$8^3/_4$″	$8^3/_4$″ × $2^1/_2$″ fletton brick (1831 psi) in $^1/_2$″ sand mortar.	3.6 tons/ft.	6 hrs.			7	23, 24	6
W-8-M-54	$8^3/_4$″	$8^3/_4$″ × $2^1/_2$″ London stock brick (683 psi) in $^1/_2$″ P.C. - sand mortar.	7.2 tons/ft.	6 hrs.			7	23, 24	6
W-9-M-55	9″	9″ × $2^1/_2$″ Leicester red wire-cut brick (4465 psi) in $^1/_2$″ P.C. - sand mortar.	6.0 tons/ft.	6 hrs.			7	23, 24	6
W-9-M-56	9″	9″ × 3″ sand-lime brick (2603 psi) in $^1/_2$″ P.C. - sand mortar.	3.6 tons/ft.	6 hrs.			7	23, 24	6
W-9-M-57	9″	2 layers $2^7/_8$″ fletton brick (1910 psi) with $3^1/_4$″ air space; Cement and sand mortar.	1.5 tons/ft.	32 min.			7	23, 25	$^1/_3$
W-9-M-58	9″	9″ × 3″ stairfoot brick (7527 psi) in $^1/_2$″ sand-cement mortar.	7.2 tons/ft.	6 hrs.			7	23, 24	6
W-9-M-59	9"	Core: solid clay or shale brick; $^1/_2$″ thick; 1:3 sanded gypsum plaster facings on both sides.	See Notes	7 hrs.		1		1, 22, 45	7
W-9-M-60	9″	Core: concrete brick; $^1/_2$″ of 1:3 sanded gypsum plaster facings on both sides.	See Notes	8 hrs.		1		1, 45	8
W-9-M-61	9″	Core: hollow rolok of clay or shale; $^1/_2$″ of 1:3 sanded gypsum plaster facings on both sides.	See Notes	4 hrs.		1		1, 45	4
W-9-M-62	9″	Cored clay or shale brick; Units in wall thickness: 1; Cells in wall thickness: 1; Minimum % solids: 70; $^1/_2$″ of 1:3 sanded gypsum plaster facings on one side.	See Notes	3 hrs.		1		1, 44	3
W-9-M-63	9″	Cored clay or shale brick; Units in wall thickness: 2; Cells in wall thickness: 2; Minimum % solids: 87; $^1/_2$″ of 1:3 sanded gypsum plaster facings on both sides.	See Notes	7 hrs.		1		1, 45	7
W-9-M-64	9-10″	Core: cavity wall of clay or shale brick; No facings.	See Notes	5 hrs.		1		1, 45	5
W-9-M-65	9-10″	Core: cavity construction of clay or shale brick; $^1/_2$″ of 1:3 sanded gypsum plaster facings on one side.	See Notes	6 hrs.		1		1, 45	6

(Continued)

TABLE 1.1.4—MASONRY WALLS
8″ TO LESS THAN 10″ THICK—continued

ITEM CODE	THICKNESS	CONSTRUCTION DETAILS	PERFORMANCE		REFERENCE NUMBER			NOTES	REC. HOURS
			LOAD	TIME	PRE-BMS-92	BMS-92	POST-BMS-92		
W-9-M-66	9-10″	Core: cavity construction of clay or shale brick; $^1/_2$″ of 1:3 sanded gypsum plaster facings on both sides.	See Notes	7 hrs.		1		1, 45	7
W-9-M-67	$9^1/_4$″	Core: clay or shale structural tile; Units in wall thickness: 1; Cells in wall thickness: 2; Minimum % solids in units: 40; $^5/_8$″ of 1:3 sanded gypsum plaster facings on both sides.	See Notes	3 hrs.		1		1, 20, 21	3
W-9-M-68	$9^1/_4$″	Core: clay or shale structural tile; Units in wall thickness: 1; Cells in wall thickness: 2; Minimum % solids in units: 43; $^5/_8$″ of 1:3 sanded gypsum plaster facings on both sides.	See Notes	3 hrs.		1		1, 20, 21	3
W-9-M-69	$9^1/_4$″	Core: clay or shale structural tile; Facings: sides 1 and 2; see Note 17.	See Notes	3 hrs.		1		1, 2, 5, 10, 18, 20, 21	3
W-9-M-70	$9^1/_4$″	Core: clay or shale structural tile; Facings: sides 1 and 2; see Note 17.	See Notes	4 hrs.		1		1, 2, 5, 13, 18, 20, 21	4
W-9-M-71	$9^1/_4$″	Core: clay or shale structural tile; Facings: sides 1 and 2; see Note 17.	See Notes	3 hrs. 30 min.		1		1, 2, 6, 9, 18, 20, 21	$3^1/_2$
W-9-M-72	$9^1/_4$″	Core: clay or shale structural tile; Facings: sides 1 and 2; see Note 17.	See Notes	4 hrs.		1		1, 2, 6, 10, 18, 20, 21	4
W-9-M-73	$9^1/_4$″	Core: clay or shale structural tile; Facings: sides 1 and 2; see Note 17.	See Notes	4 hrs.		1		1, 2, 6, 12, 18, 20, 21	4
W-9-M-74	$9^1/_4$″	Core: clay or shale structural tile; Facings: sides 1 and 2; see Note 17.	See Notes	5 hrs.		1		1, 2, 6 16, 18, 20, 21	5
W-9-M-75	8″	Cored concrete masonry; see Notes 2, 19, 26, 34, 40; No facings.	80 psi	1 hr. 30 min.		1		1, 20	$1^1/_2$
W-8-M-76	8″	Cored concrete masonry; see Notes 2, 18, 26, 34, 40; No facings	80 psi	4 hrs.		1		1, 20	4
W-8-M-77	8″	Cored concrete masonry; see Notes 2, 19, 26, 31, 40; No facings.	80 psi	1 hr. 15 min.		1		1, 20	$1^1/_4$
W-8-M-78	8″	Cored concrete masonry; see Notes 2, 18, 26, 31, 40; No facings.	80 psi	3 hrs.		1		1, 20	3
W-8-M-79	8″	Cored concrete masonry; see Notes 2, 19, 26, 36, 42; No facings.	80 psi	1 hr. 30 min.		1		1, 20	$1^1/_2$
W-8-M-80	8″	Cored concrete masonry; see Notes 2, 18, 26, 36, 41; No facings.	80 psi	3 hrs.		1		1, 20	3

(Continued)

TABLE 1.1.4—MASONRY WALLS
8″ TO LESS THAN 10″ THICK—continued

ITEM CODE	THICKNESS	CONSTRUCTION DETAILS	PERFORMANCE		REFERENCE NUMBER			NOTES	REC. HOURS
			LOAD	TIME	PRE-BMS-92	BMS-92	POST-BMS-92		
W-8-M-81	8″	Cored concrete masonry; see Notes 2, 19, 26, 34, 41; No facings.	80 psi	1 hr.		1		1, 20	1
W-8-M-82	8″	Cored concrete masonry; see Notes 2, 18, 26, 34, 41; No facings.	80 psi	2 hrs. 30 min.		1		1, 20	$2^1/_2$
W-8-M-83	8″	Cored concrete masonry; see Notes 2, 19, 26, 29, 41; No facings.	80 psi	45 min.		1		1, 20	$^3/_4$
W-8-M-84	8″	Cored concrete masonry; see Notes 2, 18, 26, 29, 41; No facings.	80 psi	2 hrs.		1		1, 20	2
W-8-M-85	$8^1/_2$″	Cored concrete masonry; see Notes 3, 18, 26, 34, 41; Facings: $2^1/_4$″ brick.	80 psi	4 hrs.		1		1, 20	4
W-8-M-86	8″	Cored concrete masonry; see Notes 3, 18, 26, 34, 41; Facings: $3^3/_4$″ brick face.	80 psi	5 hrs.		1		1, 20	5
W-8-M-87	8″	Cored concrete masonry; see Notes 2, 19, 26, 30, 43; No facings.	80 psi	12 min.		1		1, 20	$^1/_5$
W-8-M-88	8″	Cored concrete masonry; see Notes 2, 18, 26, 30, 43; No facings.	80 psi	12 min.		1		1, 20	$^1/_5$
W-8-M-89	$8^1/_2$″	Cored concrete masonry; see Notes 2, 19, 26, 34, 40; Facings: fire side only; see Note 38.	80 psi	2 hrs.		1		1, 20	2
W-8-M 90	$8^1/_2$″	Cored concrete masonry; see Notes 2, 18, 26, 34, 40; Facings: side 1, see Note 38.	80 psi	5 hrs.		1		1, 20	5
W-8-M-91	$8^1/_2$″	Cored concrete masonry; see Notes 2, 19, 26, 31, 40; Facings: fire side only; see Note 38.	80 psi	1 hr. 45 min.		1		1, 20	$1^3/_4$
W-8-M-92	$8^1/_2$″	Cored concrete masonry; see Notes 2, 18, 26, 31, 40; Facings: one side; see Note 38.	80 psi	4 hrs.		1		1, 20	4
W-8-M-93	$8^1/_2$″	Cored concrete masonry; see Notes 2, 19, 26, 36, 41; Facings: fire side only; see Note 38.	80 psi	2 hrs.		1		1, 20	2
W-8-M-94	$8^1/_2$″	Cored concrete masonry; see Notes 2, 18, 26, 36, 41; Facings: fire side only; see Note 38.	80 psi	4 hrs.		1		1, 20	4
W-8-M-95	$8^1/_2$″	Cored concrete masonry; see Notes 2, 19, 26, 34, 41; Facings: fire side only; see Note 38.	80 psi	1 hr. 30 min.		1		1, 20	$1^1/_2$
W-8-M-96	$8^1/_2$″	Cored concrete masonry; see Notes 2, 18, 26, 34, 41; Facings: one side; see Note 38.	80 psi	3 hrs.				1, 20	3
W-8-M-97	$8^1/_2$″	Cored concrete masonry; see Notes 2, 19, 26, 29, 41; Facings: fire side only; see Note 38.	80 psi	1 hr. 30 min.		1		1, 20	$1^1/_2$

(Continued)

TABLE 1.1.4—MASONRY WALLS
8″ TO LESS THAN 10″ THICK—continued

ITEM CODE	THICKNESS	CONSTRUCTION DETAILS	PERFORMANCE		REFERENCE NUMBER			NOTES	REC. HOURS
			LOAD	TIME	PRE-BMS-92	BMS-92	POST-BMS-92		
W-8-M-98	$8^1/_2$″	Cored concrete masonry; see Notes 2, 18, 26, 29, 41; Facings: one side; see Note 38.	80 psi	2 hrs. 30 min.		1		1, 20	$2^1/_2$
W-8-M-99	$8^1/_2$″	Cored concrete masonry; see Notes 3, 19, 23, 27, 41; No facings.	80 psi	1 hr. 15 min.		1		1, 20	$1^1/_4$
W-8-M-100	$8^1/_2$″	Cored concrete masonry; see Notes 3, 18, 23, 27, 41; No facings.	80 psi	3 hrs. 30 min.		1		1, 20	$3^1/_2$
W-8-M-101	$8^1/_2$″	Cored concrete masonry; see Notes 3, 18, 26, 34, 41; Facings: $3^3/_4$″ brick face; one side only; see Note 38.	80 psi	6 hrs.		1		1, 20	6
W-8-M-102	$8^1/_2$″	Cored concrete masonry; see Notes 2, 19, 26, 30, 43; Facings: fire side only; see Note 38.	80 psi	30 min.		1		1, 20	$^1/_2$
W-8-M-103	$8^1/_2$″	Cored concrete masonry; see Notes 2, 18, 26, 30, 43; Facings: one side only; see Note 38.	80 psi	12 min.		1		1, 20	$^1/_5$
W-8-M-104	9″	Cored concrete masonry; see Notes 2, 18, 26, 34, 40; Facings: both sides; see Note 38.	80 psi	6 hrs.		1		1, 20	6
W-8-M-105	9″	Cored concrete masonry; see Notes 2, 18, 26, 31, 40; Facings: both sides; see Note 38.	80 psi	5 hrs.		1		1, 20	5
W-8-M-106	9″	Cored concrete masonry; see Notes 2, 18, 26, 36, 41; Facings: both sides of wall; see Note 38.	80 psi	5 hrs.		1		1, 20	5
W-8-M-107	9″	Cored concrete masonry; see Notes 2, 18, 26, 34, 41; Facings: both sides; see Note 38.	80 psi	4 hrs.		1		1, 20	4
W-8-M-108	9″	Cored concrete masonry; see Notes 2, 18, 26, 29, 41; Facings: both sides; see Note 38.	80 psi	3 hrs. 30 min.		1		1, 20	$3^1/_2$
W-8-M-109	9″	Cored concrete masonry; see Notes 3, 19, 23, 27, 40; Facings: fire side only; see Note 38.	80 psi	1 hr. 45 min.		1		1, 20	$1^3/_4$
W-8-M-110	9″	Cored concrete masonry; see Notes 3, 18, 23, 27, 41; Facings: one side only; see Note 38.	80 psi	4 hrs.		1		1, 20	4
W-8-M-111	9″	Cored concrete masonry; see Notes 3, 18, 26, 34, 41; $2^1/_4$″ brick face on one side only; see Note 38.	80 psi	5 hrs.		1		1, 20	5
W-8-M-112	9″	Cored concrete masonry; see Notes 2, 18, 26, 30, 43; Facings: both sides; see Note 38.	80 psi	30 min.		1		1, 20	$^1/_2$

(Continued)

TABLE 1.1.4—MASONRY WALLS
8″ TO LESS THAN 10″ THICK—continued

ITEM CODE	THICKNESS	CONSTRUCTION DETAILS	PERFORMANCE		REFERENCE NUMBER			NOTES	REC. HOURS
			LOAD	TIME	PRE-BMS-92	BMS-92	POST-BMS-92		
W-9-M-113	$9^1/_2$″	Cored concrete masonry; see Notes 3, 18, 23, 27, 41; Facings: both sides; see Note 38.	80 psi	5 hrs.		1		1, 20	5
W-8-M-114	8″		200 psi	5 hrs.			43	22	5

For SI: 1 inch = 25.4 mm, 1 pound per square inch = 0.00689 MPa.

Notes:

1. Tested at NBS under ASA Spec. No. 43-1934 (ASTM C 19-53).
2. One unit in wall thickness.
3. Two units in wall thickness.
4. Two or three units in wall thickness.
5. Two cells in wall thickness.
6. Three or four cells in wall thickness.
7. Four or five cells in wall thickness.
8. Five or six cells in wall thickness.
9. Minimum percent of solid materials in units = 40%.
10. Minimum percent of solid materials in units = 43%.
11. Minimum percent of solid materials in units = 46%.
12. Minimum percent of solid materials in units = 48%.
13. Minimum percent of solid materials in units = 49%.
14. Minimum percent of solid materials in units = 45%.
15. Minimum percent of solid materials in units = 51%.
16. Minimum percent of solid materials in units = 53%.
17. Not less than $^5/_8$ inch thickness of 1:3 sanded gypsum plaster.
18. Noncombustible or no members framed into wall.
19. Combustible members framed into wall.
20. Load: 80 psi for gross cross-sectional area of wall.
21. Portland cement-lime mortar.
22. Failure mode thermal.
23. British test.
24. Passed all criteria.
25. Failed by sudden collapse with no preceding signs of impending failure.
26. One cell in wall thickness.
27. Two cells in wall thickness.
28. Three cells in wall thickness.
29. Minimum percent of solid material in concrete units = 52.
30. Minimum percent of solid material in concrete units = 54.
31. Minimum percent of solid material in concrete units = 55.
32. Minimum percent of solid material in concrete units = 57.
33. Minimum percent of solid material in concrete units = 60.
34. Minimum percent of solid material in concrete units = 62.
35. Minimum percent of solid material in concrete units = 65.
36. Minimum percent of solid material in concrete units = 70.
37. Minimum percent of solid material in concrete units = 76.
38. Not less than $^1/_2$ inch of 1:3 sanded gypsum plaster.
39. Three units in wall thickness.
40. Concrete units made with expanded slag or pumice aggregates.
41. Concrete units made with expanded burned clay or shale, crushed limestone, air cooled slag or cinders.
42. Concrete units made with calcareous sand and gravel. Coarse aggregate, 60 percent or more calcite and dolomite.
43. Concrete units made with siliceous sand and gravel. Ninety percent or more quartz, chert and dolomite.
44. Load: 120 psi for gross cross-sectional area of wall.
45. Load: 160 psi for gross cross-sectional area of wall.

FIGURE 1.1.5—MASONRY WALLS 10″ TO LESS THAN 12″ THICK

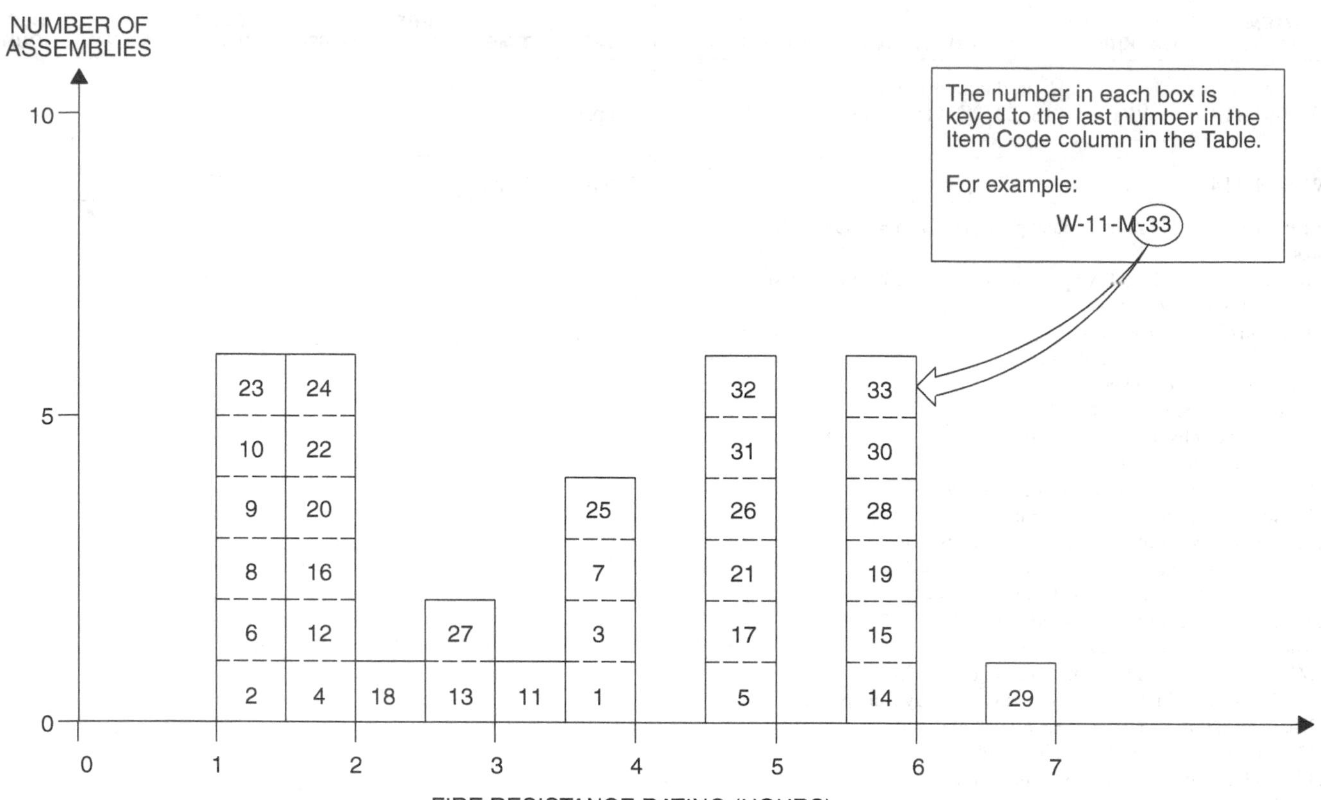

TABLE 1.1.5—MASONRY WALLS 10″ TO LESS THAN 12″ THICK

ITEM CODE	THICKNESS	CONSTRUCTION DETAILS	PERFORMANCE		REFERENCE NUMBER			NOTES	REC. HOURS
			LOAD	TIME	PRE-BMS-92	BMS-92	POST-BMS-92		
W-10-M-1	10″	Core: two 3³/₄″, 40% solid clay or shale structural tiles with 2″ air space between; Facings: ³/₄″ portland cement plaster on stucco on both sides.	80 psi	4 hrs.		1		1, 20	4
W-10-M-2	10″	Core: cored concrete masonry, 2″ air cavity; see Notes 3, 19, 27, 34, 40; No facings.	80 psi	1 hr. 30 min.		1		1, 20	1¹/₂
W-10-M-3	10″	Cored concrete masonry; see Notes 3, 18, 27, 34, 40; No facings.	80 psi	4 hrs.		1		1, 20	4
W-10-M-4	10″	Cored concrete masonry; see Notes 2, 19, 26, 34, 40; No facings.	80 psi	2 hrs.		1		1, 20	2
W-10-M-5	10″	Cored concrete masonry; see Notes 2, 18, 26, 33, 40; No facings.	80 psi	5 hrs.		1		1, 20	5
W-10-M-6	10″	Cored concrete masonry; see Notes 2, 19, 26, 33, 41; No facings.	80 psi	1 hr. 30 min.		1		1, 20	1¹/₂
W-10-M-7	10″	Cored concrete masonry; see Notes 2, 18, 26, 33, 41; No facings.	80 psi	4 hrs.		1		1, 20	4

(Continued)

TABLE 1.1.5—MASONRY WALLS
10″ TO LESS THAN 12″ THICK—continued

ITEM CODE	THICKNESS	CONSTRUCTION DETAILS	PERFORMANCE		REFERENCE NUMBER			NOTES	REC. HOURS
			LOAD	TIME	PRE-BMS-92	BMS-92	POST-BMS-92		
W-10-M-8	10″	Cored concrete masonry (cavity type 2″ air space); see Notes 3, 19, 27, 34, 42; No facings.	80 psi	1 hr. 15min.		1		1, 20	$1^1/_4$
W-10-M-9	10″	Cored concrete masonry (cavity type 2″ air space); see Notes 3, 18, 27, 34, 42; No facings.	80 psi	1 hr. 15 min.		1		1, 20	$1^1/_4$
W-10-M-10	10″	Cored concrete masonry (cavity type 2″ air space); see Notes 3, 19, 27, 34, 41; No facings.	80 psi	1 hr. 15 min.		1		1, 20	$1^1/_4$
W-10-M-11	10″	Cored concrete masonry (cavity type 2″ air space); see Notes 3, 18, 27, 34, 41; No facings.	80 psi	3 hrs. 30 min.		1		1, 20	$3^1/_2$
W-10-M-12	10″	9″ thick concrete block ($11^3/_4$″ × 9″ × $4^1/_4$″) with two 2″ thick voids included; $^3/_8$″ P.C. plaster $^1/_8$″ neat gypsum.	N/A	1 hr. 53 min.			7	23, 44	$1^3/_4$
W-10-M-13	10″	Holly clay tile block wall - $8^1/_2$″ block with two 3″ voids in each $8^1/_2$″ section; $^3/_4$″ gypsum plaster - each face.	N/A	2 hrs. 42 min.			7	23, 25	$2^1/_2$
W-10-M-14	10″	Two layers $4^1/_4$″ brick with $1^1/_2$″ air space; No ties sand cement mortar. (Fletton brick - 1910 psi).	N/A	6 hrs.			7	23, 24	6
W-10-M-15	10″	Two layers $4^1/_4$″ thick Fletton brick (1910 psi); $1^1/_2$″ air space; Ties: 18″ o.c. vertical; 3′ o.c. horizontal.	N/A	6 hrs.			7	23, 24	6
W-10-M-16	$10^1/_2$″	Cored concrete masonry; 2″ air cavity; see Notes 3, 19, 27, 34, 40; Facings: fire side only; see Note 38.	80 psi	2 hrs.		1		1, 20	2
W-10-M-17	$10^1/_2$″	Cored concrete masonry; see Notes 3, 18, 27, 34, 40; Facings: side 1 only; see Note 38.	80 psi	5 hrs.		1		1, 20	5
W-10-M-18	$10^1/_2$″	Cored concrete masonry; see Notes 2, 19, 26, 33, 40; Facings: fire side only; see Note 38.	80 psi	2 hrs. 30 min.		1		1, 20	$2^1/_2$
W-10-M-19	$10^1/_2$″	Cored concrete masonry; see Notes 2, 18, 26, 33, 40; Facings: one side; see Note 38.	80 psi	6 hrs.		1		1, 20	6
W-10-M-20	$10^1/_2$″	Cored concrete masonry; see Notes 2, 19, 26, 33, 41; Facings: fire side of wall only; see Note 38.	80 psi	2 hrs.		1		1, 20	2
W-10-M-21	$10^1/_2$″	Cored concrete masonry; see Notes 2, 18, 26, 33, 41; Facings: one side only; see Note 38.	80 psi	5 hrs.		1		1, 20	5
W-10-M-22	$10^1/_2$″	Cored concrete masonry (cavity type 2″ air space); see Notes 3,19, 27, 34, 42; Facings: fire side only; see Note 38.	80 psi	1 hr. 45 min.		1		1, 20	$1^3/_4$

(Continued)

TABLE 1.1.5—MASONRY WALLS
10″ TO LESS THAN 12″ THICK—continued

ITEM CODE	THICKNESS	CONSTRUCTION DETAILS	PERFORMANCE		REFERENCE NUMBER			NOTES	REC. HOURS
			LOAD	TIME	PRE-BMS-92	BMS-92	POST-BMS-92		
W-10-M-23	$10^1/_2$″	Cored concrete masonry (cavity type 2″ air space); see Notes 3, 18, 27, 34, 42; Facings: one side only; see Note 38.	80 psi	1 hr. 15 min.		1		1, 20	$1^1/_4$
W-10-M-24	$10^1/_2$″	Cored concrete masonry (cavity type 2″ air space); see Notes 3, 19, 27, 34, 41; Facings: fire side only; see Note 38.	80 psi	2 hrs.		1		1, 20	2
W-10-M-25	$10^1/_2$″	Cored concrete masonry (cavity type 2″ air space); see Notes 3, 18, 27, 34, 41; Facings: one side only; see Note 38.	80 psi	4 hrs.		1		1, 20	4
W-10-M-26	$10^5/_8$″	Core: 8″, 40% solid tile plus 2″ furring tile; $^5/_8$″ sanded gypsum plaster between tile types; Facings: both sides $^3/_4$″ portland cement plaster or stucco.	80 psi	5 hrs.		1		1, 20	5
W-10-M-27	$10^5/_8$″	Core: 8″, 40% solid tile plus 2″ furring tile; $^5/_8$″ sanded gypsum plaster between tile types; Facings: one side $^3/_4$″ portland cement plaster or stucco.	80 psi	3 hrs. 30 min.		1		1, 20	$3^1/_2$
W-11-M-28	11″	Cored concrete masonry; see Notes 3, 18, 27, 34, 40; Facings: both sides; see Note 38.	80 psi	6 hrs.		1		1, 20	6
W-11-M-29	11″	Cored concrete masonry; see Notes 2, 18, 26, 33, 40; Facings: both sides; see Note 38.	80 psi	7 hrs.		1		1, 20	7
W-11-M-30	11″	Cored concrete masonry; see Notes 2, 18, 26, 33, 41; Facings: both sides of wall; see Note 38.	80 psi	6 hrs.		1		1, 20	6
W-11-M-31	11″	Cored concrete masonry (cavity type 2″ air space); see Notes 3, 18, 27, 34, 42; Facings: both sides; see Note 38.	80 psi	5 hrs.		1		1, 20	5
W-11-M-32	11″	Cored concrete masonry (cavity type 2″ air space); see Notes 3, 18, 27, 34, 41; Facings: both sides; see Note 38.	80 psi	5 hrs.		1		1, 20	5

(Continued)

TABLE 1.1.5—MASONRY WALLS
10″ TO LESS THAN 12″ THICK—continued

ITEM CODE	THICKNESS	CONSTRUCTION DETAILS	PERFORMANCE		REFERENCE NUMBER			NOTES	REC. HOURS
			LOAD	TIME	PRE-BMS-92	BMS-92	POST-BMS-92		
W-11-M-33	11″	Two layers brick ($4^1/_2$″ Fletton, 2,428 psi) 2″ air space; galvanized ties; 18″ o.c. - horizontal; 3′ o.c. - vertical.	3 tons/ft.	6 hrs.			7	23, 24	6

For SI: 1 inch = 25.4 mm, 1 pound per square inch = 0.00689 MPa.

Notes:

1. Tested at NBS - ASA Spec. No. A2-1934.
2. One unit in wall thickness.
3. Two units in wall thickness.
4. Two or three units in wall thickness.
5. Two cells in wall thickness.
6. Three or four cells in wall thickness.
7. Four or five cells in wall thickness.
8. Five or six cells in wall thickness.
9. Minimum percent of solid materials in units = 40%.
10. Minimum percent of solid materials in units = 43%.
11. Minimum percent of solid materials in units = 46%.
12. Minimum percent of solid materials in units = 48%.
13. Minimum percent of solid materials in units = 49%.
14. Minimum percent of solid materials in units = 45%.
15. Minimum percent of solid materials in units = 51%.
16. Minimum percent of solid materials in units = 53%.
17. Not less than $^5/_8$ inch thickness of 1:3 sanded gypsum plaster.
18. Noncombustible or no members framed into wall.
19. Combustible members framed into wall.
20. Load: 80 psi for gross cross sectional area of wall.
21. Portland cement-lime mortar.
22. Failure mode - thermal.
23. British test.
24. Passed all criteria.
25. Failed by sudden collapse with no preceding signs of impending failure.
26. One cell in wall thickness.
27. Two cells in wall thickness.
28. Three cells in wall thickness.
29. Minimum percent of solid material in concrete units = 52%.
30. Minimum percent of solid material in concrete units = 54%.
31. Minimum percent of solid material in concrete units = 55%.
32. Minimum percent of solid material in concrete units = 57%.
33. Minimum percent of solid material in concrete units = 60%.
34. Minimum percent of solid material in concrete units = 62%.
35. Minimum percent of solid material in concrete units = 65%.
36. Minimum percent of solid material in concrete units = 70%.
37. Minimum percent of solid material in concrete units = 76%.
38. Not less than $^1/_2$ inch of 1:3 sanded gypsum plaster.
39. Three units in wall thickness.
40. Concrete units made with expanded slag or pumice aggregates.
41. Concrete units made with expanded burned clay or shale, crushed limestone, air cooled slag or cinders.
42. Concrete units made with calcareous sand and gravel. Coarse aggregate, 60 percent or more calcite and dolomite.

FIGURE 1.1.6—MASONRY WALLS 12″ TO LESS THAN 14″ THICK

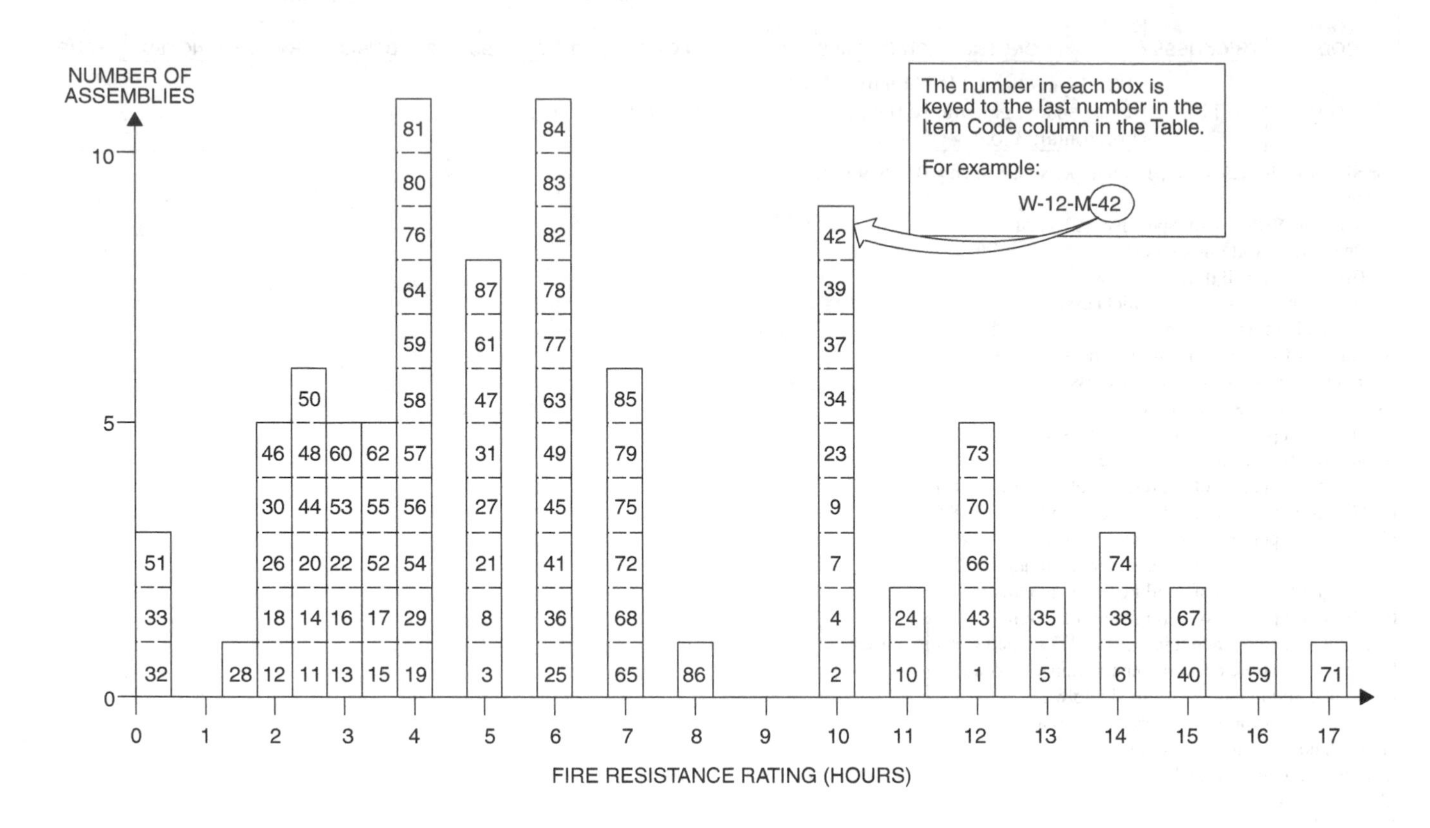

TABLE 1.1.6—MASONRY WALLS 12″ TO LESS THAN 14″ THICK

ITEM CODE	THICKNESS	CONSTRUCTION DETAILS	PERFORMANCE		REFERENCE NUMBER			NOTES	REC. HOURS
			LOAD	TIME	PRE-BMS-92	BMS-92	POST-BMS-92		
W-12-M-1	12″	Core: solid clay or shale brick; No facings.	N/A	12 hrs.		1		1	12
W-12-M-2	12″	Core: solid clay or shale brick; No facings.	160 psi	10 hrs.		1		1, 44	10
W-12-M-3	12″	Core: hollow rolok of clay or shale; No facings.	160 psi	5 hrs.		1		1, 44	5
W-12-M-4	12″	Core: hollow rolok bak of clay or shale; No facings.	160 psi	10 hrs.		1		1, 44	10
W-12-M-5	12″	Core: concrete brick; No facings.	160 psi	13 hrs.		1		1, 44	13
W-12-M-6	12″	Core: sand-lime brick; No facings.	N/A	14 hrs.		1		1	14
W-12-M-7	12″	Core: sand-lime brick; No facings.	160 psi	10 hrs.		1		1, 44	10
W-12-M-8	12″	Cored clay or shale brick; Units in wall thickness: 1; Cells in wall thickness: 2; Minimum % solids: 70; No facings.	120 psi	5 hrs.		1		1, 45	5
W-12-M-9	12″	Cored clay or shale brick; Units in wall thickness: 3; Cells in wall thickness: 3; Minimum % solids: 87; No facings.	160 psi	10 hrs.		1		1, 44	10

(Continued)

TABLE 1.1.6—MASONRY WALLS
12″ TO LESS THAN 14″ THICK—continued

ITEM CODE	THICKNESS	CONSTRUCTION DETAILS	PERFORMANCE		REFERENCE NUMBER			NOTES	REC. HOURS
			LOAD	TIME	PRE-BMS-92	BMS-92	POST-BMS-92		
W-12-M-10	12″	Cored clay or shale brick; Units in wall thickness: 3; Cells in wall thickness: 3; Minimum % solids: 87; No facings.	N/A	11 hrs.		1		1	11
W-12-M-11	12″	Core: clay or shale structural tile; see Notes 2, 6, 9, 18; No facings.	80 psi	2 hrs.		1		1, 20	$2^1/_2$
W-12-M-12	12″	Core: clay or shale structural tile; see Notes 2, 4, 9, 19; No facings.	80 psi	2 hrs.		1		1, 20	2
W-12-M-13	12″	Core: clay or shale structural tile; see Notes 2, 6, 14, 19; No facings.	80 psi	3 hrs.		1		1, 20	3
W-12-M-14	12″	Core: clay or shale structural tile; see Notes 2, 6, 14, 18; No facings.	80 psi	2 hrs. 30 min.		1		1, 20	$2^1/_2$
W-12-M-15	12″	Core: clay or shale structural tile; see Notes 2, 4, 13, 18; No facings.	80 psi	3 hrs. 30 min.		1		1, 20	$3^1/_2$
W-12-M-16	12″	Core: clay or shale structural tile; see Notes 2, 4, 13, 19; No facings.	80 psi	3 hrs.		1		1, 20	3
W-12-M-17	12″	Core: clay or shale structural tile; see Notes 3, 6, 9, 18; No facings.	80 psi	3 hrs. 30 min.		1		1, 20	$3^1/_2$
W-12-M-18	12″	Core: clay or shale structural tile; see Notes 3, 6, 9, 19; No facings.	80 psi	2 hrs.		1		1, 20	2
W-12-M-19	12″	Core: clay or shale structural tile; see Notes 3, 6, 14, 18; No facings.	80 psi	4 hrs.		1		1, 20	4
W-12-M-20	12″	Core: clay or shale structural tile; see Notes 3, 6, 14, 19; No facings.	80 psi	2 hrs. 30 min		1		1, 20	$2^1/_2$
W-12-M-21	12″	Core: clay or shale structural tile; see Notes 3, 6, 16, 18; No facings.	80 psi	5 hrs.		1		1, 20	5
W-12-M-22	12″	Core: clay or shale structural tile; see Notes 3, 6, 16, 19; No facings.	80 psi	3 hrs.		1		1, 20	3
W-12-M-23	12″	Core: 8″, 70% solid clay or shale structural tile; 4″ brick facings on one side.	80 psi	10 hrs.		1		1, 20	10
W-12-M-24	12″	Core: 8″, 70% solid clay or shale structural tile; 4″ brick facings on one side.	N/A	11 hrs.		1		1	11
W-12-M-25	12″	Core: 8″, 40% solid clay or shale structural tile; 4″ brick facings on one side.	80 psi	6 hrs.		1		1, 20	6
W-12-M-26	12″	Cored concrete masonry; see Notes 1, 9, 15, 16, 20; No facings.	80 psi	2 hrs.		1		1, 20	2
W-12-M-27	12″	Cored concrete masonry; see Notes 2, 18, 26, 34, 41; No facings.	80 psi	5 hrs.		1		1, 20	5
W-12-M-28	12″	Cored concrete masonry; see Notes 2, 19, 26, 31, 41; No facings.	80 psi	1 hr. 30 min.		1		1, 20	$1^1/_2$
W-12-M-29	12″	Cored concrete masonry; see Notes 2, 18, 26, 31, 41; No facings.	80 psi	4 hrs.		1		1, 20	4

(Continued)

TABLE 1.1.6—MASONRY WALLS 12″ TO LESS THAN 14″ THICK—continued

ITEM CODE	THICKNESS	CONSTRUCTION DETAILS	PERFORMANCE		REFERENCE NUMBER			NOTES	REC. HOURS
			LOAD	TIME	PRE-BMS-92	BMS-92	POST-BMS-92		
W-12-M-30	12″	Cored concrete masonry; see Notes 3, 19, 27, 31, 43; No facings.	80 psi	2 hrs.		1		1, 20	2
W-12-M-31	12″	Cored concrete masonry; see Notes 3, 18, 27, 31, 43; No facings.	80 psi	5 hrs.		1		1, 20	5
W-12-M-32	12″	Cored concrete masonry; see Notes 2, 19, 26, 32, 43; No facings.	80 psi	25 min.		1		1, 20	$^1/_3$
W-12-M-33	12″	Cored concrete masonry; see Notes 2, 18, 26, 32, 43; No facings.	80 psi	25 min.		1		1, 20	$^1/_3$
W-12-M-34	$12^1/_2$″	Core: solid clay or shale brick; $^1/_2$″ of 1:3 sanded gypsum plaster facings on one side.	160 psi	10 hrs.		1		1, 44	10
W-12-M-35	$12^1/_2$″	Core: solid clay or shale brick; $^1/_2$″ of 1:3 sanded gypsum plaster facings on one side.	N/A	13 hrs.		1		1	13
W-12-M-36	$12^1/_2$″	Core: hollow rolok of clay or shale; $^1/_2$″ of 1:3 sanded gypsum plaster facings on one side.	160 psi	6 hrs.		1		1, 44	6
W-12-M-37	$12^1/_2$″	Core: hollow rolok bak of clay or shale; $^1/_2$″ of 1:3 sanded gypsum plaster facings on one side.	160 psi	10 hrs.		1		1, 44	10
W-12-M-38	$12^1/_2$″	Core: concrete; $^1/_2$″ of 1:3 sanded gypsum plaster facings on one side.	160 psi	14 hrs.		1		1, 44	14
W-12-M-39	$12^1/_2$″	Core: sand-lime brick; $^1/_2$″ of 1:3 sanded gypsum plaster facings on one side.	160 psi	10 hrs.		1		1, 44	10
W-12-M-40	$12^1/_2$″	Core: sand-lime brick; $^1/_2$″ of 1:3 sanded gypsum plaster facings on one side.	N/A	15 hrs.		1		1	15
W-12-M-41	$12^1/_2$″	Cored clay or shale brick; Units in wall thickness: 1; Cells in wall thickness: 2; Minimum % solids: 70; $^1/_2$″ of 1:3 sanded gypsum plaster facings on one side.	120 psi	6 hrs.		1		1, 45	6
W-12-M-42	$12^1/_2$″	Cored clay or shale brick; Units in wall thickness: 3; Cells in wall thickness: 3; Minimum % solids: 87; $^1/_2$″ of 1:3 sanded gypsum plaster facings on one side.	160 psi	10 hrs.		1		1, 44	10
W-12-M-43	$12^1/_2$″	Cored clay or shale brick; Units in wall thickness: 3; Cells in wall thickness: 3; Minimum % solids: 87; $^1/_2$″ of 1:3 sanded gypsum plaster facings on one side.	N/A	12 hrs.		1		1	12

(Continued)

TABLE 1.1.6—MASONRY WALLS 12″ TO LESS THAN 14″ THICK—continued

ITEM CODE	THICKNESS	CONSTRUCTION DETAILS	PERFORMANCE		REFERENCE NUMBER			NOTES	REC. HOURS
			LOAD	TIME	PRE-BMS-92	BMS-92	POST-BMS-92		
W-12-M-44	$12^1/_2$″	Cored concrete masonry; see Notes 2, 19, 26, 34, 41; Facings: fire side only; see Note 38.	80 psi	2 hrs. 30 min.		1		1, 20	$2^1/_2$
W-12-M-45	$12^1/_2$″	Cored concrete masonry; see Notes 2, 18, 26, 34, 39, 41; Facings: one side only; see Note 38.	80 psi	6 hrs.		1		1, 20	6
W-12-M-46	$12^1/_2$″	Cored concrete masonry; see Notes 2, 19, 26, 31, 41; Facings: fire side only; see Note 38.	80 psi	2 hrs.		1		1, 20	2
W-12-M-47	$12^1/_2$″	Cored concrete masonry; see Notes 2, 18, 26, 31, 41; Facings: one side of wall only; see Note 38.	80 psi	5 hrs.		1		1, 20	5
W-12-M-48	$12^1/_2$″	Cored concrete masonry; see Notes 3, 19, 27, 31, 43; Facings: fire side only; see Note 38.	80 psi	2 hrs. 30 min.		1		1, 20	$2^1/_2$
W-12-M-49	$12^1/_2$″	Cored concrete masonry; see Notes 3, 18, 27, 31, 43; Facings: one side only; see Note 38.	80 psi	6 hrs.		1		1, 20	6
W-12-M-50	$12^1/_2$″	Cored concrete masonry; see Notes 2, 19, 26, 32, 43; Facings: fire side only; see Note 38.	80 psi	2 hrs. 30 min.		1		1, 20	$2^1/_2$
W-12-M-51	$12^1/_2$″	Cored concrete masonry; see Notes 2, 18, 26, 32, 43; Facings: one side only; see Note 38.	80 psi	25 min.		1		1, 20	$^1/_3$
W-12-M-52	$12^5/_8$″	Clay or shale structural tile; see Notes 2, 6, 9, 18; Facings: side 1, see Note 17; side 2, none.	80 psi	3 hrs. 30 min.		1		1, 20	$3^1/_2$
W-12-M-53	$12^5/_8$″	Clay or shale structural tile; see Notes 2, 6, 9, 19; Facings: fire side only; see Note 17.	80 psi	3 hrs.		1		1, 20	3
W-12-M-54	$12^5/_8$″	Clay or shale structural tile; see Notes 2, 6, 14, 19; Facings: side 1, see Note 17; side 2, none.	80 psi	4 hrs.		1		1, 20	4
W-12-M-55	$12^5/_8$″	Clay or shale structural tile; see Notes 2, 6, 14, 18; Facings: exposed side only; see Note 17.	80 psi	3 hrs. 30 min.		1		1, 20	$3^1/_2$
W-12-M-56	$12^5/_8$″	Clay or shale structural tile; see Notes 2, 4, 13, 18; Facings: side 1, see Note 17; side 2, none.	80 psi	4 hrs.		1		1, 20	4
W-12-M-57	$12^5/_8$″	Clay or shale structural tile; see Notes 1, 4, 13, 19; Facings: fire side only; see Note 17.	80 psi	4 hrs.		1		1, 20	4
W-12-M-58	$12^5/_8$″	Clay or shale structural tile; see Notes 3, 6, 9, 18; Facings: side 1, see Note 17; side 2, none.	80 psi	4 hrs.		1		1, 20	4
W-12-M-59	$12^5/_8$″	Clay or shale structural tile; see Notes 3, 6, 9, 19; Facings: fire side only; see Note 17.	80 psi	3 hrs.		1		1, 20	3

(Continued)

TABLE 1.1.6—MASONRY WALLS
12″ TO LESS THAN 14″ THICK—continued

ITEM CODE	THICKNESS	CONSTRUCTION DETAILS	PERFORMANCE		REFERENCE NUMBER			NOTES	REC. HOURS
			LOAD	TIME	PRE-BMS-92	BMS-92	POST-BMS-92		
W-12-M-60	$12^5/_8$″	Clay or shale structural tile; see Notes 3, 6, 14, 18; Facings: side 1, see Note 17; side 2, none.	80 psi	5 hrs.		1		1, 20	5
W-12-M-61	$12^5/_8$″	Clay or shale structural tile; see Notes 3, 6, 14, 19; Facings: fire side only; see Note 17.	80 psi	3 hrs. 30 min.		1		1, 20	$3^1/_2$
W-12-M-62	$12^5/_8$″	Clay or shale structural tile; see Notes 3, 6, 16, 18; Facings: side 1, see Note 17; side 2, none.	80 psi	6 hrs.		1		1, 20	6
W-12-M-63	$12^5/_8$″	Clay or shale structural tile; see Notes 3, 6, 16, 19; Facings: fire side only; see Note 17.	80 psi	4 hrs.		1		1, 20	4
W-12-M-64	$12^5/_8$″	Core: 8″, 40% solid clay or shale structural tile; Facings: 4″ brick plus $^5/_8$″ of 1:3 sanded gypsum plaster on one side.	80 psi	7 hrs.		1		1, 20	7
W-13-M-65	13″	Core: solid clay or shale brick; $^1/_2$″ of 1:3 sanded gypsum plaster facings on both sides.	160 psi	12 hrs.		1		1, 44	12
W-13-M-66	13″	Core: solid clay or shale brick; $^1/_2$″ of 1:3 sanded gypsum plaster facings on both sides.	N/A	15 hrs.		1		1, 20	15
W-13-M-67	13″	Core: solid clay or shale brick; $^1/_2$″ of 1:3 sanded gypsum plaster facings on both sides.	N/A	15 hrs.		1		1	15
W-13-M-68	13″	Core: hollow rolok of clay or shale; $^1/_2$″ of 1:3 sanded gypsum plaster facings on both sides.	80 psi	7 hrs.		1		1, 20	7
W-13-M-69	13″	Core: concrete brick; $^1/_2$″ of 1:3 sanded gypsum plaster facings on both sides.	160 psi	16 hrs.		1		1, 44	16
W-13-M-70	13″	Core: sand-lime brick; $^1/_2$″ of 1:3 sanded gypsum plaster facings on both sides.	160 psi	12 hrs.		1		1, 44	12
W-13-M-71	13″	Core: sand-lime brick; $^1/_2$″ of 1:3 sanded gypsum plaster facings on both sides.	N/A	17 hrs.		1		1	17
W-13-M-72	13″	Cored clay or shale brick; Units in wall thickness: 1; Cells in wall thickness: 2; Minimum % solids: 70; $^1/_2$″ of 1:3 sanded gypsum plaster facings on both sides.	120 psi	7 hrs.		1		1, 45	7
W-13-M-73	13″	Cored clay or shale brick; Units in wall thickness: 3; Cells in wall thickness: 3; Minimum % solids: 87; $^1/_2$″ of 1:3 sanded gypsum plaster facings on both sides.	160 psi	12 hrs.		1		1, 44	12
W-13-M-74	13″	Cored clay or shale brick; Units in wall thickness: 3; Cells in wall thickness: 2; Minimum % solids: 87; $^1/_2$″ of 1:3 sanded gypsum plaster facings on both sides.	N/A	14 hrs.		1		1	14

(Continued)

TABLE 1.1.6—MASONRY WALLS
12″ TO LESS THAN 14″ THICK—continued

ITEM CODE	THICKNESS	CONSTRUCTION DETAILS	PERFORMANCE		REFERENCE NUMBER			NOTES	REC. HOURS
			LOAD	TIME	PRE-BMS-92	BMS-92	POST-BMS-92		
W-13-M-75	13″	Cored concrete masonry; see Notes 18, 23, 28, 39, 41; No facings.	80 psi	7 hrs.		1		1, 20	7
W-13-M-76	13″	Cored concrete masonry; see Notes 19, 23, 28, 39, 41; No facings.	80 psi	4 hrs.		1		1, 20	4
W-13-M-77	13″	Cored concrete masonry; see Notes 3, 18, 27, 31, 43; Facings: both sides; see Note 38.	80 psi	6 hrs.		1		1, 20	6
W-13-M-78	13″	Cored concrete masonry; see Notes 2, 18, 26, 31, 41; Facings: both sides; see Note 38.	80 psi	6 hrs.		1		1, 20	6
W-13-M-79	13″	Cored concrete masonry; see Notes 2, 18, 26, 34, 41; Facings: both sides of wall; see Note 38.	80 psi	7 hrs.		1		1, 20	7
W-13-M-80	$13^1/_4$″	Core: clay or shale structural tile; see Notes 2, 6, 9, 18; Facings: both sides; see Note 17.	80 psi	4 hrs.		1		1, 20	4
W-13-M-81	$13^1/_4$″	Core: clay or shale structural tile; see Notes 2, 6, 14, 19; Facings: both sides; see Note 17.	80 psi	4 hrs.		1		1, 20	4
W-13-M-82	$13^1/_4$″	Core: clay or shale structural tile; see Notes 2, 4, 13, 18; Facings: both sides; see Note 17.	80 psi	6 hrs.		1		1, 20	6
W-13-M-83	$13^1/_4$″	Core: clay or shale structural tile; see Notes 3, 6, 9, 18; Facings: both sides; see Note 17.	80 psi	6 hrs.		1		1, 20	6
W-13-M-84	$13^1/_4$″	Core: clay or shale structural tile; see Notes 3, 6, 14, 18; Facings: both sides; see Note 17.	80 psi	6 hrs.		1		1, 20	6
W-13-M-85	$13^1/_4$″	Core: clay or shale structural tile; see Notes 3, 6, 16, 18; Facings: both sides; see Note 17.	80 psi	7 hrs.		1		1, 20	7
W-13-M-86	$13^1/_2$″	Cored concrete masonry; see Notes 18, 23, 28, 39, 41; Facings: one side only; see Note 38.	80 psi	8 hrs.		1		1, 20	8

(Continued)

TABLE 1.1.6—MASONRY WALLS 12" TO LESS THAN 14" THICK—continued

ITEM CODE	THICKNESS	CONSTRUCTION DETAILS	PERFORMANCE		REFERENCE NUMBER			NOTES	REC. HOURS
			LOAD	TIME	PRE-BMS-92	BMS-92	POST-BMS-92		
W-13-M-87	$13^1/_2$"	Cored concrete masonry; see Notes 19, 23, 28, 39, 41; Facings: fire side only; see Note 38.	80 psi	5 hrs.		1		1, 20	5

For SI: 1 inch = 25.4 mm, 1 pound per square inch = 0.00689 MPa.

Notes:

1. Tested at NBS - ASA Spec. No. A2-1934.
2. One unit in wall thickness.
3. Two units in wall thickness.
4. Two or three units in wall thickness.
5. Two cells in wall thickness.
6. Three or four cells in wall thickness.
7. Four or five cells in wall thickness.
8. Five or six cells in wall thickness.
9. Minimum percent of solid materials in units = 40%.
10. Minimum percent of solid materials in units = 43%.
11. Minimum percent of solid materials in units = 46%.
12. Minimum percent of solid materials in units = 48%.
13. Minimum percent of solid materials in units = 49%.
14. Minimum percent of solid materials in units = 45%.
15. Minimum percent of solid materials in units = 51%.
16. Minimum percent of solid materials in units = 53%.
17. Not less than $^5/_8$ inch thickness of 1:3 sanded gypsum plaster.
18. Noncombustible or no members framed into wall.
19. Combustible members framed into wall.
20. Load: 80 psi for gross area.
21. Portland cement-lime mortar.
22. Failure mode-thermal.
23. British test.
24. Passed all criteria.
25. Failed by sudden collapse with no preceding signs of impending failure.
26. One cell in wall thickness.
27. Two cells in wall thickness.
28. Three cells in wall thickness.
29. Minimum percent of solid material in concrete units = 52%.
30. Minimum percent of solid material in concrete units = 54%.
31. Minimum percent of solid material in concrete units = 55%.
32. Minimum percent of solid material in concrete units = 57%.
33. Minimum percent of solid material in concrete units = 60%.
34. Minimum percent of solid material in concrete units = 62%.
35. Minimum percent of solid material in concrete units = 65%.
36. Minimum percent of solid material in concrete units = 70%.
37. Minimum percent of solid material in concrete units = 76%.
38. Not less than $^1/_2$ inch of 1:3 sanded gypsum plaster.
39. Three units in wall thickness.
40. Concrete units made with expanded slag or pumice aggregates.
41. Concrete units made with expanded burned clay or shale, crushed limestone, air cooled slag or cinders.
42. Concrete units made with calcareous sand and gravel. Coarse aggregate, 60 percent or more calcite and dolomite.
43. Concrete units made with siliceous sand and gravel. Ninety percent or more quartz, chert or flint.
44. Load: 160 psi of gross wall cross sectional area.
45. Load: 120 psi of gross wall cross sectional area.

FIGURE 1.1.7—MASONRY WALLS 14″ OR MORE THICK

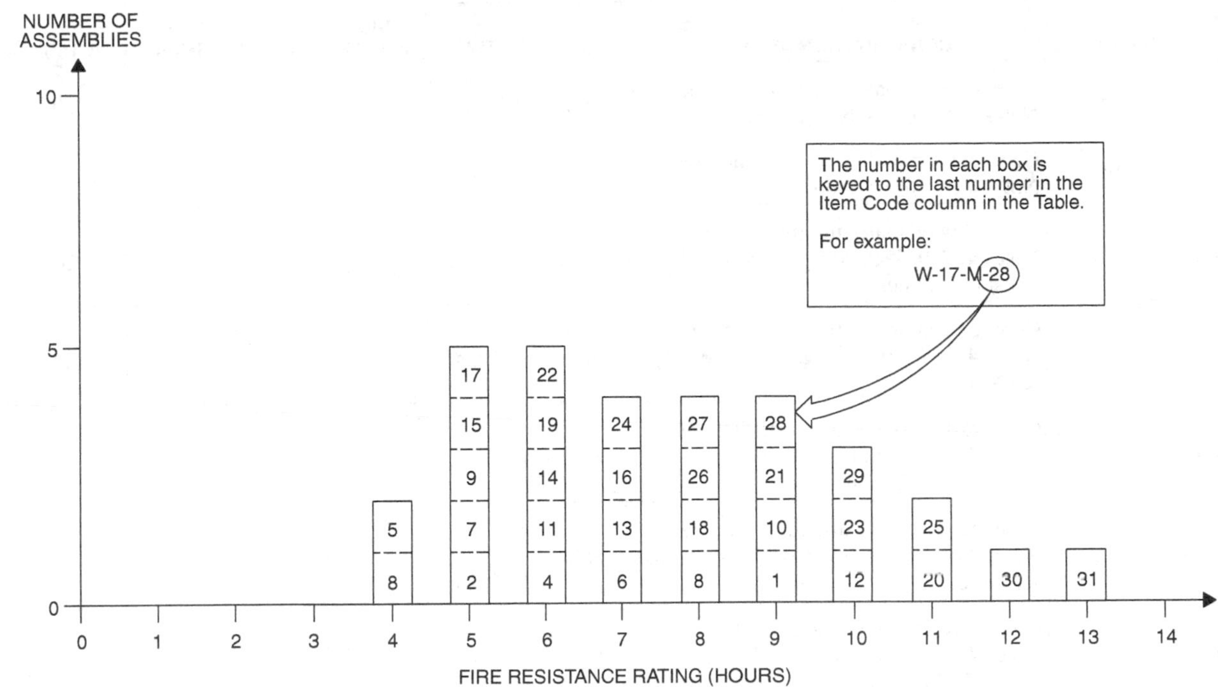

TABLE 1.1.7—MASONRY WALLS 14″ OR MORE THICK

ITEM CODE	THICKNESS	CONSTRUCTION DETAILS	PERFORMANCE		REFERENCE NUMBER			NOTES	REC. HOURS
			LOAD	TIME	PRE-BMS-92	BMS-92	POST-BMS-92		
W-14-M-1	14″	Core: cored masonry; see Notes 18, 28, 33, 39, 41; Facings: both sides; see Note 38.	80 psi	9 hrs.		1		1, 20	9
W-16-M-2	16″	Core: clay or shale structural tile; see Notes 4, 7, 9, 19; No facings.	80 psi	5 hrs.		1		1, 20	5
W-16-M-3	16″	Core: clay or shale structural tile; see Notes 4, 7, 9, 19; No facings.	80 psi	4 hrs.		1		1, 20	4
W-16-M-4	16″	Core: clay or shale structural tile; see Notes 4, 7, 10, 18; No facings.	80 psi	6 hrs.		1		1, 20	6
W-16-M-5	16″	Core: clay or shale structural tile; see Notes 4, 7, 10, 19; No facings.	80 psi	4 hrs.		1		1, 20	4
W-16-M-6	16″	Core: clay or shale structural tile; see Notes 4, 7, 11, 18; No facings.	80 psi	7 hrs.		1		1, 20	7
W-16-M-7	16″	Core: clay or shale structural tile; see Notes 4, 7, 11, 19; No facings.	80 psi	5 hrs.		1		1, 20	5
W-16-M-8	16″	Core: clay or shale structural tile; see Notes 4, 8, 13, 18; No facings.	80 psi	8 hrs.		1		1, 20	8
W-16-M-9	16″	Core: clay or shale structural tile; see Notes 4, 8, 13, 19; No facings.	80 psi	5 hrs.		1		1, 20	5
W-16-M-10	16″	Core: clay or shale structural tile; see Notes 4, 8, 15, 18; No facings.	80 psi	9 hrs.		1		1, 20	9
W-16-M-11	16″	Core: clay or shale structural tile; see Notes 3, 7, 14, 18; No facings.	80 psi	6 hrs.		1		1, 20	6

(Continued)

TABLE 1.1.7—MASONRY WALLS 14″ OR MORE THICK—continued

ITEM CODE	THICKNESS	CONSTRUCTION DETAILS	PERFORMANCE		REFERENCE NUMBER			NOTES	REC. HOURS
			LOAD	TIME	PRE-BMS-92	BMS-92	POST-BMS-92		
W-16-M-12	16″	Core: clay or shale structural tile; see Notes 4, 8, 16, 18; No facings.	80 psi	10 hrs.		1		1, 20	10
W-16-M-13	16″	Core: clay or shale structural tile; see Notes 4, 6, 16, 19; No facings.	80 psi	7 hrs.		1		1, 20	7
W-16-M-14	16$^5/_8$″	Core: clay or shale structural tile; see Notes 4, 7, 9, 18; Facings: side 1, see Note 17; side 2, none.	80 psi	6 hrs.		1		1, 20	6
W-16-M-15	16$^5/_8$″	Core: clay or shale structural tile; see Notes 4, 7, 9, 19, Facings: fire side only; see Note 17.	80 psi	5 hrs.		1		1, 20	5
W-16-M-16	16$^5/_8$″	Core: clay or shale structural tile; see Notes 4, 7, 10, 18; Facings: side 1, see Note 17; side 2, none.	80 psi	7 hrs.		1		1, 20	7
W-16-M-17	16$^5/_8$″	Core: clay or shale structural tile; see Notes 4, 7, 10, 19; Facings: fire side only; see Note 17.	80 psi	5 hrs.		1		1, 20	5
W-16-M-18	16$^5/_8$″	Core: clay or shale structural tile; see Notes 4, 7, 11, 18; Facings: side 1, see Note 17; side 2, none.	80 psi	5 hrs.		1		1, 20	5
W-16-M-19	16$^5/_8$″	Core: clay or shale structural tile; see Notes 4, 7, 11, 19; Facings: fire side only; see Note 17.	80 psi	6 hrs.		1		1, 20	6
W-16-M-20	16$^5/_8$″	Core: clay or shale structural tile; see Notes 4, 8, 13, 18; Facings: sides 1 and 2; see Note 17.	80 psi	11 hrs.		1		1, 20	11
W-16-M-21	16$^5/_8$″	Core: clay or shale structural tile; see Notes 4, 8, 13 18; Facings: side 1, see Note 17; side 2, none.	80 psi	9 hrs.		1		1, 20	9
W-16-M-22	16$^5/_8$″	Core: clay or shale structural tile; see Notes 4, 8, 13, 19; Facings: fire side only; see Note 17.	80 psi	6 hrs.		1		1, 20	6
W-16-M-23	16$^5/_8$″	Core: clay or shale structural tile; see Notes 4, 8, 15, 18; Facings: side 1, see Note 17; side 2, none.	80 psi	10 hrs.		1		1, 20	10
W-16-M-24	16$^5/_8$″	Core: clay or shale structural tile; see Notes 4, 8, 15, 19; Facings: fire side only; see Note 17.	80 psi	7 hrs.		1		1, 20	7
W-16-M-25	16$^5/_8$″	Core: clay or shale structural tile; see Notes 4, 6, 16, 18; Facings: side 1, see Note 17; side 2, none.	80 psi	11 hrs.		1		1, 20	11
W-16-M-26	16$^5/_8$″	Core: clay or shale structural tile; see Notes 4, 6, 16, 19; Facings: fire side only; see Note 17.	80 psi	8 hrs.		1		1, 20	8
W-17-M-27	17$^1/_4$″	Core: clay or shale structural tile; see Notes 4, 7, 9, 18; Facings: sides 1 and 2; see Note 17.	80 psi	8 hrs.		1		1, 20	8
W-17-M-28	17$^1/_4$″	Core: clay or shale structural tile; see Notes 4, 7, 10, 18; Facings: sides 1 and 2; see Note 17.	80 psi	9 hrs.		1		1, 20	9

(Continued)

TABLE 1.1.7—MASONRY WALLS 14″ OR MORE THICK—continued

ITEM CODE	THICKNESS	CONSTRUCTION DETAILS	PERFORMANCE		REFERENCE NUMBER			NOTES	REC. HOURS
			LOAD	TIME	PRE-BMS-92	BMS-92	POST-BMS-92		
W-17-M-29	$17^{1}/_{4}''$	Core: clay or shale structural tile; see Notes 4, 7, 11, 18; Facings: sides 1 and 2; see Note 17.	80 psi	10 hrs.		1		1, 20	10
W-17-M-30	$17^{1}/_{4}''$	Core: clay or shale structural tile; see Notes 4, 8, 15, 18; Facings: sides 1 and 2; see Note 17.	80 psi	12 hrs.		1		1, 20	12
W-17-M-31	$17^{1}/_{4}''$	Core: clay or shale structural tile; see Notes 4, 6, 16, 18; Facings: sides 1 and 2; see Note 17.	80 psi	13 hrs.		1		1, 20	13

For SI: 1 inch = 25.4 mm, 1 pound per square inch = 0.00689 MPa.

Notes:

1. Tested at NBS - ASA Spec. No. A2-1934.
2. One unit in wall thickness.
3. Two units in wall thickness.
4. Two or three units in wall thickness.
5. Two cells in wall thickness.
6. Three or four cells in wall thickness.
7. Four or five cells in wall thickness.
8. Five or six cells in wall thickness.
9. Minimum percent of solid materials in units = 40%.
10. Minimum percent of solid materials in units = 43%.
11. Minimum percent of solid materials in units = 46%.
12. Minimum percent of solid materials in units = 48%.
13. Minimum percent of solid materials in units = 49%.
14. Minimum percent of solid materials in units = 45%.
15. Minimum percent of solid materials in units = 51%.
16. Minimum percent of solid materials in units = 53%.
17. Not less than $^{5}/_{8}$ inch thickness of 1:3 sanded gypsum plaster.
18. Noncombustible or no members framed into wall.
19. Combustible members framed into wall.
20. Load: 80 psi for gross area.
21. Portland cement-lime mortar.
22. Failure mode - thermal.
23. British test.
24. Passed all criteria.
25. Failed by sudden collapse with no preceding signs of impending failure.
26. One cell in wall thickness.
27. Two cells in wall thickness.
28. Three cells in wall thickness.
29. Minimum percent of solid material in concrete units = 52%.
30. Minimum percent of solid material in concrete units = 54%.
31. Minimum percent of solid material in concrete units = 55%.
32. Minimum percent of solid material in concrete units = 57%.
33. Minimum percent of solid material in concrete units = 60%.
34. Minimum percent of solid material in concrete units = 62%.
35. Minimum percent of solid material in concrete units = 65%.
36. Minimum percent of solid material in concrete units = 70%.
37. Minimum percent of solid material in concrete units = 76%.
38. Not less than $^{1}/_{2}$ inch of 1:3 sanded gypsum plaster.
39. Three units in wall thickness.
40. Concrete units made with expanded slag or pumice aggregates.
41. Concrete units made with expanded burned clay or shale, crushed limestone, air cooled slag or cinders.
42. Concrete units made with calcareous sand and gravel. Coarse aggregate, 60 percent or more calcite and dolomite.
43. Concrete units made with siliceous sand and gravel. Ninety percent or more quartz, chert or flint.

FIGURE 1.2.1—METAL FRAME WALLS
0″ TO LESS THAN 4″ THICK

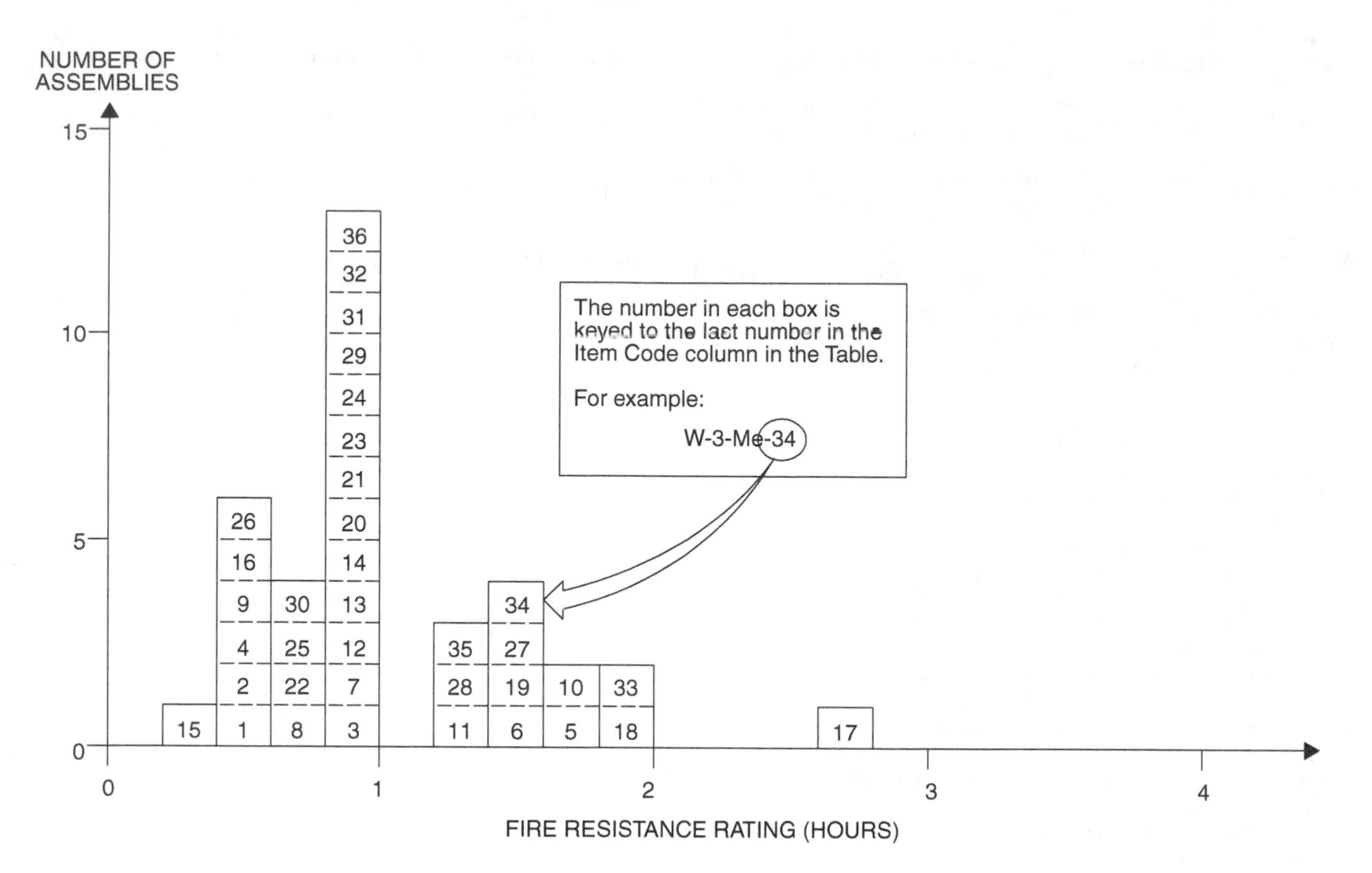

TABLE 1.2.1—METAL FRAME WALLS
0″ TO LESS THAN 4″ THICK

ITEM CODE	THICKNESS	CONSTRUCTION DETAILS	PERFORMANCE		REFERENCE NUMBER			NOTES	REC. HOURS
			LOAD	TIME	PRE-BMS-92	BMS-92	POST-BMS-92		
W-3-Me-1	3″	Core: steel channels having three rows of 4″ × $^1/_8$″ staggered slots in web; core filled with heat expanded vermiculite weighing 1.5 lbs./ft.^{2} of wall area; Facings: sides 1 and 2, 18 gage steel, spot welded to core.	N/A	25 min.		1			$^1/_3$
W-3-Me-2	3″	Core: steel channels having three rows of 4″ × $^1/_8$″ staggered slots in web; core filled with heat expanded vermiculite weighing 2 lbs./ft.^{2} of wall area; Facings: sides 1 and 2, 18 gage steel, spot welded to core.	N/A	30 min.		1			$^1/_2$
W-3-Me-3	$2^1/_2$″	Solid partition: $^3/_8$″ tension rods (vertical) 3′ o.c. with metal lath; Scratch coat: cement/sand/lime plaster; Float coats: cement/sand/lime plaster; Finish coats: neat gypsum plaster.	N/A	1 hr.			7	1	1
W-2-Me-4	2″	Solid wall: steel channel per Note 1; 2″ thickness of 1:2; 1:3 portland cement on metal lath.	N/A	30 min.		1			$^1/_2$

(Continued)

TABLE 1.2.1—METAL FRAME WALLS 0″ TO LESS THAN 4″ THICK—continued

ITEM CODE	THICKNESS	CONSTRUCTION DETAILS	PERFORMANCE		REFERENCE NUMBER			NOTES	REC. HOURS
			LOAD	TIME	PRE-BMS-92	BMS-92	POST-BMS-92		
W-2-Me-5	2″	Solid wall: steel channel per Note 1; 2″ thickness of neat gypsum plaster on metal lath.	N/A	1 hr. 45 min.		1			$1^3/_4$
W-2-Me-6	2″	Solid wall: steel channel per Note 1; 2″ thickness of 1:$1^1/_2$; 1:$1^1/_2$ gypsum plaster on metal lath.	N/A	1 hr. 30 min.		1			$1^1/_2$
W-2-Me-7	2″	Solid wall: steel channel per Note 2; 2″ thickness of 1:1; 1:1 gypsum plaster on metal lath.	N/A	1 hr.		1			1
W-2-Me-8	2″	Solid wall: steel channel per Note 1; 2″ thickness of 1:2; 1:2 gypsum plaster on metal lath.	N/A	45 min.		1			$^3/_4$
W-2-Me-9	$2^1/_4$″	Solid wall: steel channel per Note 2; $2^1/_4$″ thickness of 1:2; 1:3 portland cement on metal lath.	N/A	30 min.		1			$^1/_2$
W-2-Me-10	$2^1/_4$″	Solid wall: steel channel per Note 2; $2^1/_4$″ thickness of neat gypsum plaster on metal lath.	N/A	2 hrs.		1			2
W-2-Me-11	$2^1/_4$″	Solid wall: steel channel per Note 2; $2^1/_4$″ thickness of 1:$^1/_2$; 1:$^1/_2$ gypsum plaster on metal lath.	N/A	1 hr. 45 min.		1			$1^3/_4$
W-2-Me-12	$2^1/_4$″	Solid wall: steel channel per Note 2; $2^1/_4$″ thickness of 1:1; 1:1 gypsum plaster on metal lath.	N/A	1 hr. 15 min.		1			$1^1/_4$
W-2-Me-13	$2^1/_4$″	Solid wall: steel channel per Note 2; $2^1/_4$″ thickness of 1:2; 1:2 gypsum plaster on metal lath.	N/A	1 hr.		1			1
W-2-Me-14	$2^1/_2$″	Solid wall: steel channel per Note 1; $2^1/_2$″ thickness of 4.5:1:7; 4.5:1:7 portland cement, sawdust and sand sprayed on wire mesh; see Note 3.	N/A	1 hr.		1			1
W-2-Me-15	$2^1/_2$″	Solid wall: steel channel per Note 2; $2^1/_2$″ thickness of 1:4; 1:4 portland cement sprayed on wire mesh; see Note 3.	N/A	20 min.		1			$^1/_3$
W-2-Me-16	$2^1/_2$″	Solid wall: steel channel per Note 2; $2^1/_2$″ thickness of 1:2; 1:3 portland cement on metal lath.	N/A	30 min.		1			$^1/_2$
W-2-Me-17	$2^1/_2$″	Solid wall: steel channel per Note 2; $2^1/_2$″ thickness of neat gypsum plaster on metal lath.	N/A	2 hrs. 30 min.		1			$2^1/_2$
W-2-Me-18	$2^1/_2$″	Solid wall: steel channel per Note 2; $2^1/_2$″ thickness of 1:$^1/_2$; 1:$^1/_2$ gypsum plaster on metal lath.	N/A	2 hrs.		1			2
W-2-Me-19	$2^1/_2$″	Solid wall: steel channel per Note 2; $2^1/_2$″ thickness of 1:1; 1:1 gypsum plaster on metal lath.	N/A	1 hr. 30 min.		1			$1^1/_2$
W-2-Me-20	$2^1/_2$″	Solid wall: steel channel per Note 2; $2^1/_2$″ thickness of 1:2; 1:2 gypsum plaster on metal lath.	N/A	1 hr.		1			1

(Continued)

TABLE 1.2.1—METAL FRAME WALLS
0″ TO LESS THAN 4″ THICK—continued

ITEM CODE	THICKNESS	CONSTRUCTION DETAILS	PERFORMANCE		REFERENCE NUMBER			NOTES	REC. HOURS
			LOAD	TIME	PRE-BMS-92	BMS-92	POST-BMS-92		
W-2-Me-21	$2^1/_2$″	Solid wall: steel channel per Note 2; $2^1/_2$″ thickness of 1:2; 1:3 gypsum plaster on metal lath.	N/A	1 hr.		1			1
W-3-Me-22	3″	Core: steel channel per Note 2; 1:2; 1:2 gypsum plaster on $^3/_4$″ soft asbestos lath; plaster thickness 2″.	N/A	45 min.		1			$^3/4$
W-3-Me-23	$3^1/_2$″	Solid wall: steel channel per Note 2; $2^1/_2$″ thickness of 1:2; 1:2 gypsum plaster on $^3/_4$″ asbestos lath.	N/A	1 hr.		1			1
W-3-Me-24	$3^1/_2$″	Solid wall: steel channel per Note 2; lath over and $1{:}2^1/_2$; $1{:}2^1/_2$ gypsum plaster on 1″ magnesium oxysulfate wood fiberboard; plaster thickness $2^1/_2$″.	N/A	1 hr.		1			1
W-3-Me-25	$3^1/_2$″	Core: steel studs; see Note 4; Facings: $^3/_4$″ thickness of $1{:}^1/_{30}{:}2$; $1{:}^1/_{30}{:}3$ portland cement and asbestos fiber plaster.	N/A	45 min.		1			$^3/4$
W-3-Me-26	$3^1/_2$″	Core: steel studs; see Note 4; Facings: both sides $^3/_4$″ thickness of 1:2; 1:3 portland cement.	N/A	30 min.		1			$^1/_2$
W-3-Me-27	$3^1/_2$″	Core: steel studs; see Note 4; Facings: both sides $^3/_4$″ thickness of neat gypsum plaster.	N/A	1 hr. 30 min.		1			$1^1/2$
W-3-Me-28	$3^1/_2$″	Core: steel studs; see Note 4; Facings: both sides $^3/_4$″ thickness of $1{:}^1/_2$; $1{:}^1/_2$ gypsum plaster.	N/A	1 hr. 15 min.		1			$1^1/4$
W-3-Me-29	$3^1/_2$″	Core: steel studs; see Note 4; Facings: both sides $^3/_4$″ thickness of 1:2; 1:2 gypsum plaster.	N/A	1 hr.		1			1
W-3-Me-30	$3^1/_2$″	Core: steel studs; see Note 4; Facings: both sides $^3/_4$″ thickness of 1:2; 1:3 gypsum plaster.	N/A	45 min.		1			$^3/_4$
W-3-Me-31	$3^3/_4$″	Core: steel studs; see Note 4; Facings: both sides $^7/_8$″ thickness of $1{:}^1/_{30}{:}2$; $1{:}^1/_{30}{:}$ 3 portland cement and asbestos fiber plaster.	N/A	1 hr.		1			1
W-3-Me-32	$3^3/_4$″	Core: steel studs; see Note 4; Facings: both sides $^7/_8$″ thickness of 1:2; 1:3 portland cement.	N/A	45 min.		1			$^3/_4$
W-3-Me-33	$3^3/_4$″	Core: steel studs; see Note 4; Facings: both sides $^7/_8$″ thickness of neat gypsum plaster.	N/A	2 hrs.		1			2
W-3-Me-34	$3^3/_4$″	Core: steel studs; see Note 4; Facings: both sides $^7/_8$″ thickness of $1{:}^1/_2$; $1{:}^1/_2$ gypsum plaster.	N/A	1 hr. 30 min.		1			$1^1/_2$
W-3-Me-35	$3^3/_4$″	Core: steel studs; see Note 4; Facings: both sides $^7/_8$″ thickness of 1:2; 1:2 gypsum plaster.	N/A	1 hr. 15 min.		1			$1^1/4$

(Continued)

TABLE 1.2.1—METAL FRAME WALLS
0″ TO LESS THAN 4″ THICK—continued

ITEM CODE	THICKNESS	CONSTRUCTION DETAILS	PERFORMANCE		REFERENCE NUMBER			NOTES	REC. HOURS
			LOAD	TIME	PRE-BMS-92	BMS-92	POST-BMS-92		
W-3-Me-36	$3^{3}/_{4}''$	Core: steel; see Note 4; Facings: $^{7}/_{8}''$ thickness of 1:2; 1:3 gypsum plaster on both sides.	N/A	1 hr.		1			1

For SI: 1 inch = 25.4 mm.

Notes:

1. Failure mode - local temperature rise - back face.
2. Three-fourths inch or 1 inch channel framing - hot-rolled or strip-steel channels.
3. Reinforcement is 4-inch square mesh of No. 6 wire welded at intersections (no channels).
4. Ratings are for any usual type of nonload-bearing metal framing providing 2 inches (or more) air space.

General Note:

The construction details of the wall assemblies are as complete as the source documentation will permit. Data on the method of attachment of facings and the gauge of steel studs was provided when known. The cross-sectional area of the steel stud can be computed, thereby permitting a reasoned estimate of actual loading conditions. For load-bearing assemblies, the maximum allowable stress for the steel studs has been provided in the table "Notes." More often, it is the thermal properties of the facing materials, rather than the specific gauge of the steel, that will determine the degree of fire resistance. This is particularly true for nonbearing wall assemblies.

FIGURE 1.2.2—METAL FRAME WALLS
4″ TO LESS THAN 6″ THICK

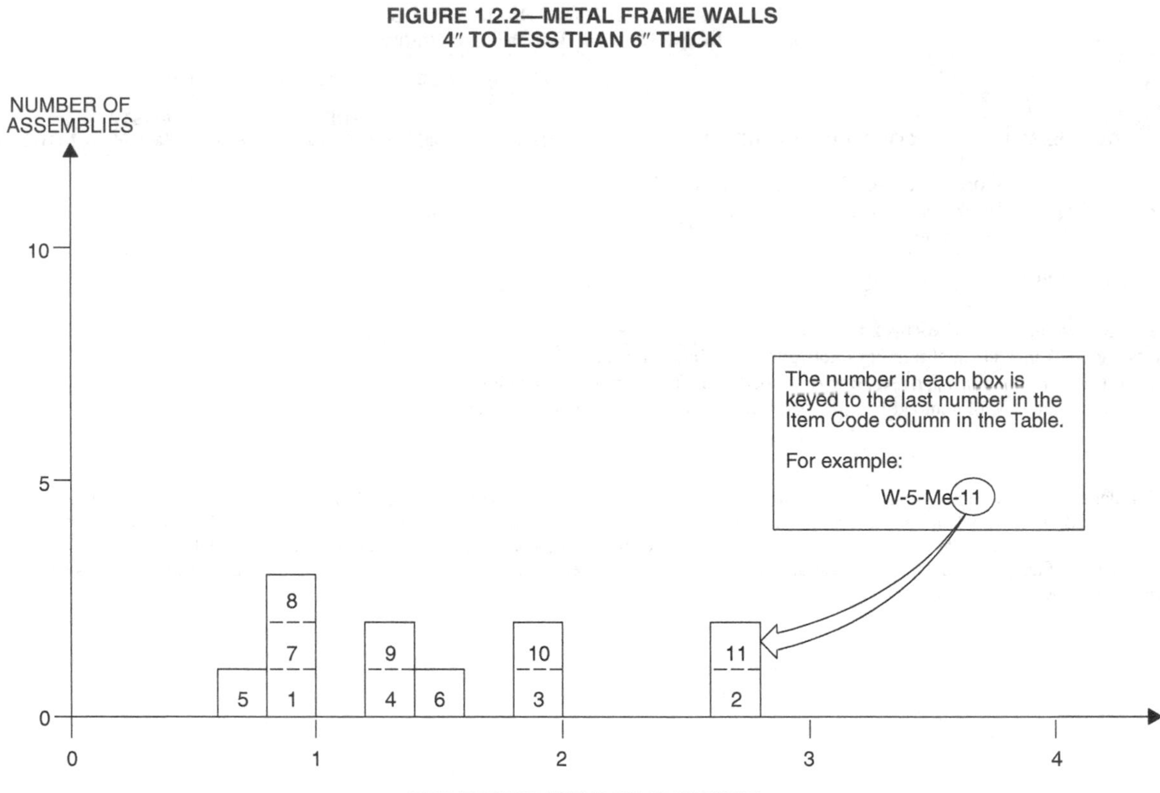

TABLE 1.2.2—METAL FRAME WALLS
4″ TO LESS THAN 6″ THICK

ITEM CODE	THICKNESS	CONSTRUCTION DETAILS	PERFORMANCE		REFERENCE NUMBER			NOTES	REC. HOURS
			LOAD	TIME	PRE-BMS-92	BMS-92	POST-BMS-92		
W-5-Me-1	5$^1/_2$″	3″ cavity with 16 ga. channel studs (3$^1/_2$″ o.c.) of $^1/_2$″ × $^1/_2$″ channel and 3″ spacer; Metal lath on ribs with plaster (three coats) $^3/_4$″ over face of lath; Plaster (each side): scratch coat, cement/lime/sand with hair; float coat, cement/lime/sand; finish coat, neat gypsum.	N/A	1 hr. 11 min.			7	1	1
W-4-Me-2	4″	Core: steel studs; see Note 2; Facings: both sides 1″ thickness of neat gypsum plaster.	N/A	2 hrs. 30 min.		1			2$^1/_2$
W-4-Me-3	4″	Core: steel studs; see Note 2; Facings: both sides 1″ thickness of 1:$^1/_2$; 1:$^1/_2$ gypsum plaster.	N/A	2 hrs.		1			2
W-4-Me-4	4″	Core: steel; see Note 2; Facings: both sides 1″ thickness of 1:2; 1:3 gypsum plaster.	N/A	1 hr. 15 min.		1			1$^1/_4$
W-4-Me-5	4$^1/_2$″	Core: lightweight steel studs 3″ in depth; Facings: both sides $^3/_4$″ thick sanded gypsum plaster, 1:2 scratch coat, 1:3 brown coat applied on metal lath.	See Note 4	45 min.		1		5	$^3/4$
W-4-Me-6	4$^1/_2$″	Core: lightweight steel studs 3″ in depth; Facings: both sides $^3/_4$″ thick neat gypsum plaster on metal lath.	See Note 4	1 hr. 30 min.		1		5	1$^1/_2$

(Continued)

TABLE 1.2.2—METAL FRAME WALLS
4″ TO LESS THAN 6″ THICK—continued

ITEM CODE	THICKNESS	CONSTRUCTION DETAILS	PERFORMANCE		REFERENCE NUMBER			NOTES	REC. HOURS
			LOAD	TIME	PRE-BMS-92	BMS-92	POST-BMS-92		
W-4-Me-7	$4^{1}/_{2}$″	Core: lightweight steel studs 3″ in depth; Facings: both sides $^{3}/_{4}$″ thick sanded gypsum plaster, 1:2 scratch and brown coats applied on metal lath.	See Note 4	1 hr.		1		5	1
W-4-Me-8	$4^{3}/_{4}$″	Core: lightweight steel studs 3″ in depth; Facings: both sides $^{7}/_{8}$″ thick sanded gypsum plaster, 1:2 scratch coat, 1:3 brown coat, applied on metal lath.	See Note 4	1 hr.		1		5	1
W-4-Me-9	$4^{3}/_{4}$″	Core: lightweight steel studs 3″ in depth; Facings: both sides $^{7}/_{8}$″ thick sanded gypsum plaster, 1:2 scratch and 1:3 brown coats applied on metal lath.	See Note 4	1 hr. 15 min.		1		5	$1^{1}/_{4}$
W-5-Me-10	5″	Core: lightweight steel studs 3″ in depth; Facings: both sides 1″ thick neat gypsum plaster on metal lath.	See Note 4	2 hrs.		1		5	2
W-5-Me-11	5″	Core: lightweight steel studs 3″ in depth; Facings: both sides 1″ thick neat gypsum plaster on metal lath.	See Note 4	2 hrs. 30 min.		1		5, 6	$2^{1}/_{2}$

For SI: 1 inch = 25.4 mm, 1 pound per square inch = 0.00689 MPa.

Notes:

1. Failure mode - local back face temperature rise.
2. Ratings are for any usual type of nonbearing metal framing providing a minimum 2 inches air space.
3. Facing materials secured to lightweight steel studs not less than 3 inches deep.
4. Rating based on loading to develop a maximum stress of 7270 psi for net area of each stud.
5. Spacing of steel studs must be sufficient to develop adequate rigidity in the metal-lath or gypsum-plaster base.
6. As per Note 4 but load/stud not to exceed 5120 psi.

General Note:

The construction details of the wall assemblies are as complete as the source documentation will permit. Data on the method of attachment of facings and the gauge of steel studs was provided when known. The cross sectional area of the steel stud can be computed, thereby permitting a reasoned estimate of actual loading conditions. For load-bearing assemblies, the maximum allowable stress for the steel studs has been provided in the table "Notes." More often, it is the thermal properties of the facing materials, rather than the specific gauge of the steel, that will determine the degree of fire resistance. This is particularly true for nonbearing wall assemblies.

TABLE 1.2.3—METAL FRAME WALLS
6″ TO LESS THAN 8″ THICK

ITEM CODE	THICKNESS	CONSTRUCTION DETAILS	PERFORMANCE		REFERENCE NUMBER			NOTES	REC. HOURS
			LOAD	TIME	PRE-BMS-92	BMS-92	POST-BMS-92		
W-6-Me-1	$6^5/_8$″	On one side of 1″ magnesium oxysulfate wood fiberboard sheathing attached to steel studs (see Notes 1 and 2), 1″ air space, $3^3/_4$″ brick secured with metal ties to steel frame every fifth course; Inside facing of $^7/_8$″ 1:2 sanded gypsum plaster on metal lath secured directly to studs; Plaster side exposed to fire.	See Note 2	1 hr. 45 min.		1		1	$1^3/_4$
W-6-Me-2	$6^5/_8$″	On one side of 1″ magnesium oxysulfate wood fiberboard sheathing attached to steel studs (see Notes 1 and 2), 1″ air space, $3^3/_4$″ brick secured with metal ties to steel frame every fifth course; Inside facing of $^7/_8$″ 1:2 sanded gypsum plaster on metal lath secured directly to studs; Brick face exposed to fire.	See Note 2	4 hrs.		1		1	4
W-6-Me-3	$6^5/_8$″	On one side of 1″ magnesium oxysulfate wood fiberboard sheathing attached to steel studs (see Notes 1 and 2), 1″ air space, $3^3/_4$″ brick secured with metal ties to steel frame every fifth course; Inside facing of $^7/_8$″ vermiculite plaster on metal lath secured directly to studs; Plaster side exposed to fire.	See Note 2	2 hrs.		1		1	2

For SI: 1 inch = 25.4 mm, 1 pound per square inch = 0.00689 MPa.

Notes:

1. Lightweight steel studs (minimum 3 inches deep) used. Stud spacing dependent on loading, but in each case, spacing is to be such that adequate rigidity is provided to the metal lath plaster base.
2. Load is such that stress developed in studs is not greater than 5120 psi calculated from net stud area.

General Note:

The construction details of the wall assemblies are as complete as the source documentation will permit. Data on the method of attachment of facings and the gauge of steel studs was provided when known. The cross sectional area of the steel stud can be computed, thereby permitting a reasoned estimate of actual loading conditions. For load-bearing assemblies, the maximum allowable stress for the steel studs has been provided in the table "Notes." More often, it is the thermal properties of the facing materials, rather than the specific gauge of the steel, that will determine the degree of fire resistance. This is particularly true for nonbearing wall assemblies.

TABLE 1.2.4—METAL FRAME WALLS
8″ TO LESS THAN 10″ THICK

ITEM CODE	THICKNESS	CONSTRUCTION DETAILS	PERFORMANCE		REFERENCE NUMBER			NOTES	REC. HOURS
			LOAD	TIME	PRE-BMS-92	BMS-92	POST-BMS-92		
W-9-Me-1	$9^1/_{16}$″	On one side of $^1/_2$″ wood fiberboard sheathing next to studs, $^3/_4$″ air space formed with $^3/_4$″ × $1^5/_8$″ wood strips placed over the fiberboard and secured to the studs, paper backed wire lath nailed to strips $3^3/_4$″ brick veneer held in place by filling a $^3/_4$″ space between the brick and paper backed lath with mortar; Inside facing of $^3/_4$″ neat gypsum plaster on metal lath attached to $^5/_{16}$″ plywood strips secured to edges of steel studs; Rated as combustible because of the sheathing; See Notes 1 and 2; Plaster exposed.	See Note 2	1 hr. 45 min.		1		1	$1^3/_4$
W-9-Me-2	$9^1/_{16}$″	Same as above with brick exposed.	See Note 2	4 hrs.		1		1	4
W-8-Me-3	$8^1/_2$″	On one side of paper backed wire lath attached to studs and $3^3/_4$″ brick veneer held in place by filling a 1″ space between the brick and lath with mortar; Inside facing of 1″ paper-enclosed mineral wool blanket weighing 0.6 lb./ft.^{2} attached to studs, metal lath or paper backed wire lath laid over the blanket and attached to the studs, $^3/_4$″ sanded gypsum plaster 1:2 for the scratch coat and 1:3 for the brown coat; See Notes 1 and 2; Plaster face exposed.	See Note 2	4 hrs.		1		1	4
W-8-Me-4	$8^1/_2$″	Same as above with brick exposed.	See Note 2	5 hrs.		1		1	5

For SI: 1 inch = 25.4 mm, 1 pound per square inch = 0.00689 MPa.

Notes:

1. Lightweight steel studs ≥ 3 inches in depth. Stud spacing dependent on loading, but in any case, the spacing is to be such that adequate rigidity is provided to the metal-lath plaster base.
2. Load is such that stress developed in studs is ≤ 5120 psi calculated from the net area of the stud.

General Note:

The construction details of the wall assemblies are as complete as the source documentation will permit. Data on the method of attachment of facings and the gauge of steel studs was provided when known. The cross sectional area of the steel stud can be computed, thereby permitting a reasoned estimate of actual loading conditions. For load-bearing assemblies, the maximum allowable stress for the steel studs has been provided in the table "Notes." More often, it is the thermal properties of the facing materials, rather than the specific gauge of the steel, that will determine the degree of fire resistance. This is particularly true for nonbearing wall assemblies.

TABLE 1.3.1—WOOD FRAME WALLS
0″ TO LESS THAN 4″ THICK

ITEM CODE	THICKNESS	CONSTRUCTION DETAILS	PERFORMANCE		REFERENCE NUMBER			NOTES	REC. HOURS
			LOAD	TIME	PRE-BMS-92	BMS-92	POST-BMS-92		
W-3-W-1	$3^3/_4$″	Solid wall: $2^1/_4$″ wood-wool slab core; $^3/_4$″ gypsum plaster each side.	N/A	2 hrs.			7	1, 6	2
W-3-W-2	$3^7/_8$″	2 × 4 stud wall; $^3/_{16}$″ thick cement asbestos board on both sides of wall.	360 psi net area	10 min.		1		2-5	$^1/_6$
W-3-W-3	$3^7/_8$″	Same as W-3-W-2 but stud cavities filled with 1 lb./ft.^{2} mineral wool batts.	360 psi net area	40 min.		1		2-5	$^2/_3$

For SI: 1 inch = 25.4 mm, 1 pound per square inch = 0.00689 MPa.

Notes:

1. Achieved "Grade C" fire resistance (British).
2. Nominal 2′4 wood studs of No. 1 common or better lumber set edgewise, 2′4 plates at top and bottom and blocking at mid height of wall.
3. All horizontal joints in facing material backed by 2′4 blocking in wall.
4. Load: 360 psi of net stud cross sectional area.
5. Facings secured with 6d casing nails. Nail holes predrilled and 0.02 inch to 0.03 inch smaller than nail diameter.
6. The wood-wool core is a pressed excelsior slab which possesses insulating properties similar to cellulosic insulation.

FIGURE 1.3.2—WOOD FRAME WALLS
4″ TO LESS THAN 6″ THICK

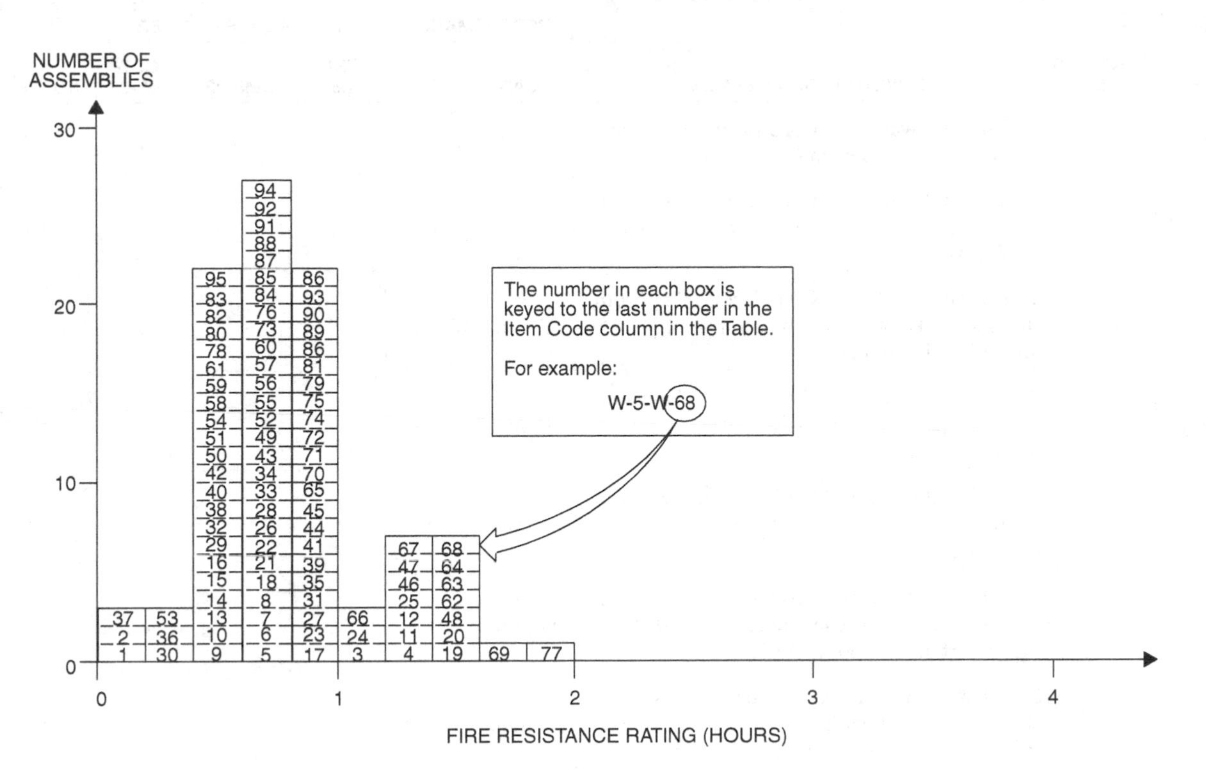

TABLE 1.3.2—WOOD FRAME WALLS
4″ TO LESS THAN 6″ THICK

ITEM CODE	THICKNESS	CONSTRUCTION DETAILS	PERFORMANCE		REFERENCE NUMBER			NOTES	REC. HOURS
			LOAD	TIME	PRE-BMS-92	BMS-92	POST-BMS-92		
W-4-W-1	4″	2″ × 4″ stud wall; $^3/_{16}$″ CAB; no insulation; Design A.	35 min.	10 min.			4	1-10	$^1/_6$
W-4-W-2	$4^1/_8$″	2″ × 4″ stud wall; $^3/_{16}$″ CAB; no insulation; Design A.	38 min.	9 min.			4	1-10	$^1/_6$
W-4-W-3	$4^3/_4$″	2″ × 4″ stud wall; $^3/_{16}$″ CAB and $^3/_8$″ gypsum board face (both sides); Design B.	62 min.	64 min.			4	1-10	1
W-5-W-4	5″	2″ × 4″ stud wall; $^3/_{16}$″ CAB and $^1/_2$″ gypsum board (both sides); Design B.	79 min.	Greater than 90 min.			4	1-10	1
W-4-W-5	$4^3/_4$″	2″ × 4″ stud wall; $^3/_{16}$″ CAB and $^3/_8$″ gypsum board (both sides); Design B.	45 min.	45 min.			4	1-12	—
W-5-W-6	5″	2″ × 4″ stud wall; $^3/_{16}$″ CAB and $^1/_2$″ gypsum board face (both sides); Design B.	45 min.	45 min.			4	1-10, 12, 13	—
W-4-W-7	4″	2″ × 4″ stud wall; $^3/_{16}$″ CAB face; $3^1/_2$″ mineral wool insulation; Design C.	40 min.	42 min.			4	1-10	$^2/_3$
W-4-W-8	4″	2″ × 4″ stud wall; $^3/_{16}$″ CAB face; $3^1/_2$″ mineral wool insulation; Design C.	46 min.	46 min.			4	1-10, 43	$^2/_3$
W-4-W-9	4″	2″ × 4″ stud wall; $^3/_{16}$″ CAB face; $3^1/_2$″ mineral wool insulation; Design C.	30 min.	30 min.			4	1-10, 12, 14	—

(Continued)

TABLE 1.3.2—WOOD FRAME WALLS
4″ TO LESS THAN 6″ THICK—continued

ITEM CODE	THICKNESS	CONSTRUCTION DETAILS	PERFORMANCE		REFERENCE NUMBER			NOTES	REC. HOURS
			LOAD	TIME	PRE-BMS-92	BMS-92	POST-BMS-92		
W-4-W-10	$4^1/_8$″	2″ × 4″ stud wall; $^3/_{16}$″ CAB face; $3^1/_2$″ mineral wool insulation; Design C.	—	30 min.			4	1-8, 12, 14	—
W-4-W-11	$4^3/_4$″	2″ × 4″ stud wall; $^3/_{16}$″ CAB face; $^3/_8$″ gypsum strips over studs; $5^1/_2$″ mineral wool insulation; Design D.	79 min.	79 min.			4	1-10	1
W-4-W-12	$4^3/_4$″	2″ × 4″ stud wall; $^3/_{16}$″ CAB face; $^3/_8$″ gypsum strips at stud edges; $7^1/_2$″ mineral wool insulation; Design D.	82 min.	82 min.			4	1-10	1
W-4-W-13	$4^3/_4$″	2″ × 4″ stud wall; $^3/_{16}$″ CAB face; $^3/_8$″ gypsum board strips over studs; $5^1/_2$″ mineral wool insulation; Design D.	30 min.	30 min.			4	1-12	—
W-4-W-14	$4^3/_4$″	2″ × 4″ stud wall; $^3/_{16}$″ CAB face; $^3/_8$″ gypsum board strips over studs; 7″ mineral wool insulation; Design D.	30 min.	30 min.			4	1-12	—
W-5-W-15	$5^1/_2$″	2″ × 4″ stud wall; Exposed face: CAB shingles over 1″ × 6″; Unexposed face: $^1/_8$″ CAB sheet; $^7/_{16}$″ fiberboard (wood); Design E.	34 min.	—			4	1-10	$^1/_2$
W-5-W-16	$5^1/_2$″	2″ × 4″ stud wall; Exposed face: $^1/_8$″ CAB sheet; $^7/_{16}$″ fiberboard; Unexposed face: CAB shingles over 1″ × 6″; Design E.	32 min.	33 min.			4	1-10	$^1/_2$
W-5-W-17	$5^1/_2$″	2″ × 4″ stud wall; Exposed face: CAB shingles over 1″ × 6″; Unexposed face: $^1/_8$″ CAB sheet; gypsum at stud edges; $3^1/_2$″ mineral wood insulation; Design F.	51 min.	—			4	1-10	$^3/_4$
W-5-W-18	$5^1/_2$″	2″ × 4″ stud wall; Exposed face: $^1/_8$″ CAB sheet; gypsum board at stud edges; Unexposed face: CAB shingles over 1″ × 6″; $3^1/_2$″ mineral wool insulation; Design F.	42 min.	—			4	1-10	$^2/_3$
W-5-W-19	$5^5/_8$″	2″ × 4″ stud wall; Exposed face: CAB shingles over 1″ × 6″; Unexposed face: $^1/_8$″ CAB sheet; gypsum board at stud edges; $5^1/_2$″ mineral wool insulation; Design G.	74 min.	85 min.			4	1-10	1
W-5-W-20	$5^5/_8$″	2″ × 4″ stud wall; Exposed face: $^1/_8$″ CAB sheet; gypsum board at $^3/_{16}$″ stud edges; $^7/_{16}$″ fiberboard; Unexposed face: CAB shingles over 1″ × 6″; $5^1/_2$″ mineral wool insulation; Design G.	79 min.	85 min.			4	1-10	$1^1/_4$
W-5-W-21	$5^5/_8$″	2″ × 4″ stud wall; Exposed face: CAB shingles 1″ × 6″ sheathing; Unexposed face: CAB sheet; gypsum board at stud edges; $5^1/_2$″ mineral wool insulation; Design G.	38 min.	38 min.			4	1-10, 12, 14	—

(Continued)

TABLE 1.3.2—WOOD FRAME WALLS
4″ TO LESS THAN 6″ THICK—continued

ITEM CODE	THICKNESS	CONSTRUCTION DETAILS	PERFORMANCE		REFERENCE NUMBER			NOTES	REC. HOURS
			LOAD	TIME	PRE-BMS-92	BMS-92	POST-BMS-92		
W-5-W-22	$5^5/_8$″	2″ × 4″ stud wall; Exposed face: CAB sheet; gypsum board at stud edges; Unexposed face: CAB shingles 1″ × 6″ sheathing; $5^1/_2$″ mineral wool insulation; Design G.	38 min.	38 min.			4	1-12	—
W-6-W-23	6″	2″ × 4″ stud wall; 16″ o.c.; $^1/_2$″ gypsum board each side; $^1/_2$″ gypsum plaster each side.	N/A	60 min.			7	15	1
W-6-W-24	6″	2″ × 4″ stud wall; 16″ o.c.; $^1/_2$″ gypsum board each side; $^1/_2$″ gypsum plaster each side.	N/A	68 min.			7	16	1
W-6-W-25	$6^7/_8$″	2″ × 4″ stud wall; 18″ o.c.; $^3/_4$″ gypsum plank each side; $^3/_{16}$″ gypsum plaster each side.	N/A	80 min.			7	15	$1^1/_3$
W-5-W-26	$5^1/_8$″	2″ × 4″ stud wall; 16″ o.c.; $^3/_8$″ gypsum board each side; $^3/_{16}$″ gypsum plaster each side.	N/A	37 min.			7	15	$^1/_2$
W-5-W-27	$5^3/_4$″	2″ × 4″ stud wall; 16″ o.c.; $^3/_8$″ gypsum lath each side; $^1/_2$″ gypsum plaster each side.	N/A	52 min.			7	15	$^3/_4$
W-5-W-28	5″	2″ × 4″ stud wall; 16″ o.c.; $^1/_2$″ gypsum board each side.	N/A	37 min.			7	16	$^1/_2$
W-5-W-29	5″	2″ × 4″ stud wall; $^1/_2$″ fiberboard both sides 14% M.C. with F.R. paint at 35 gm./ft.^{2}.	N/A	28 min.			7	15	$^1/_3$
W-4-W-30	$4^3/_4$″	2″ × 4″ stud wall; Fire side: $^1/_2$″ (wood) fiberboard; Back side: $^1/_4$″ CAB; 16″ o.c.	N/A	17 min.			7	15, 16	$^1/_4$
W-5-W-31	$5^1/_8$″	2″ × 4″ stud wall; 16″ o.c.; $^1/_2$″ fiberboard insulation with $^1/_{32}$″ asbestos (both sides of each board).	N/A	50 min.			7	16	$^3/_4$
W-4-W-32	$4^1/_4$″	2″ × 4″ stud wall; $^3/_8$″ thick gypsum wallboard on both faces; insulated cavities.	See Note 23	25 min.		1		17, 18, 23	$^1/_3$
W-4-W-33	$4^1/_2$″	2″ × 4″ stud wall; $^1/_2$″ thick gypsum wallboard on both faces.	See Note 17	40 min.		1		17, 23	$^1/_3$
W-4-W-34	$4^1/_2$″	2″ × 4″ stud wall; $^1/_2$″ thick gypsum wallboard on both faces; insulated cavities.	See Note 17	45 min.		1		17, 18, 23	$^3/_4$
W-4-W-35	$4^1/_2$″	2″ × 4″ stud wall; $^1/_2$″ thick gypsum wallboard on both faces; insulated cavities.	N/A	1 hr.		1		17, 18, 24	1
W-4-W-36	$4^1/_2$″	2″ × 4″ stud wall; $^1/_2$″ thick, 1.1 lbs./ft.^{2} wood fiberboard sheathing on both faces.	See Note 23	15 min.		1		17, 23	$^1/_4$
W-4-W-37	$4^1/_2$″	2″ × 4″ stud wall; $^1/_2$″ thick, 0.7 lb./ft.^{2} wood fiberboard sheathing on both faces.	See Note 23	10 min.		1		17, 23	$^1/_6$
W-4-W-38	$4^1/_2$″	2″ × 4″ stud wall; $^1/_2$″ thick, flameproofed 1.6 lbs./ft.^{2} wood fiberboard sheathing on both faces.	See Note 23	30 min.		1		17, 23	$^1/_2$

(Continued)

**TABLE 1.3.2—WOOD FRAME WALLS
4″ TO LESS THAN 6″ THICK—continued**

ITEM CODE	THICKNESS	CONSTRUCTION DETAILS	PERFORMANCE		REFERENCE NUMBER			NOTES	REC. HOURS
			LOAD	TIME	PRE-BMS-92	BMS-92	POST-BMS-92		
W-4-W-39	$4^1/_2$″	2″ × 4″ stud wall; $^1/_2$″ thick gypsum wallboard on both faces; insulated cavities.	See Note 23	1 hr.		1		17, 18, 23	1
W-4-W-40	$4^1/_2$″	2″ × 4″ stud wall; $^1/_2$″ thick, 1:2; 1:3 gypsum plaster on wood lath on both faces.	See Note 23	30 min.		1		17, 21, 23	$^1/_2$
W-4-W-41	$4^1/_2$″	2″ × 4″ stud wall; $^1/_2$″, 1:2; 1:3 gypsum plaster on wood lath on both faces; insulated cavities.	See Note 23	1 hr.		1		17, 18, 21, 24	1
W-4-W-42	$4^1/_2$″	2″ × 4″ stud wall; $^1/_2$″, 1:5; 1:7.5 lime plaster on wood lath on both wall faces.	See Note 23	30 min.		1		17, 21, 23	$^1/_2$
W-4-W-43	$4^1/_2$″	2″ × 4″ stud wall; $^1/_2$″ thick 1:5; 1:7.5 lime plaster on wood lath on both faces; insulated cavities.	See Note 23	45 min.		1		17, 18, 21, 23	$^3/_4$
W-4-W-44	$4^5/_8$″	2″ × 4″ stud wall; $^3/_{16}$″ thick cement-asbestos over $^3/_8$″ thick gypsum board on both faces.	See Note 23	1 hr.		1		23, 25, 26, 27	1
W-4-W-45	$4^5/_8$″	2″ × 4″ stud wall; studs faced with 4″ wide strips of $^3/_8$″ thick gypsum board; $^3/_{16}$″ thick gypsum cement-asbestos board on both faces; insulated cavities.	See Note 23	1 hr.		1		23, 25, 27, 28	1
W-4-W-46	$4^5/_8$″	Same as W-4-W-45 but nonload bearing.	N/A	1 hr. 15 min.		1		24, 28	$1^1/_4$
W-4-W-47	$4^7/_8$″	2″ × 4″ stud wall; $^3/_{16}$″ thick cement-asbestos board over $^1/_2$″ thick gypsum sheathing on both faces.	See Note 23	1 hr. 15 min.		1		23, 25, 26, 27	$1^1/_4$
W-4-W-48	$4^7/_8$″	Same as W-4-W-47 but nonload bearing.	N/A	1 hr. 30 min.		1		24, 27	$1^1/_2$
W-5-W-49	5″	2″ × 4″ stud wall; Exterior face: $^3/_4$″ wood sheathing; asbestos felt 14 lbs./100 ft.^{2} and $^5/_{32}$″ cement-asbestos shingles; Interior face: 4″ wide strips of $^3/_8$″ gypsum board over studs; wall faced with $^3/_{16}$″ thick cement-asbestos board.	See Note 23	40 min.		1		18, 23, 25, 26, 29	$^2/_3$
W-5-W-50	5″	2″ × 4″ stud wall; Exterior face: as per W-5-W-49; Interior face: $^9/_{16}$″ composite board consisting of $^7/_{16}$″ thick wood fiberboard faced with $^1/_8$″ thick cement-asbestos board; Exterior side exposed to fire.	See Note 23	30 min.		1		23, 25, 26, 30	$^1/_2$
W-5-W-51	5″	Same as W-5-W-50 but interior side exposed to fire.	See Note 23	30 min.		1		23, 25, 26	$^1/_2$
W-5-W-52	5″	Same as W-5-W-49 but exterior side exposed to fire.	See Note 23	45 min.		1		18, 23, 25, 26	$^3/_4$
W-5-W-53	5″	2″ × 4″ stud wall; $^3/_4$″ thick T&G wood boards on both sides.	See Note 23	20 min.		1		17, 23	$^1/_3$

(Continued)

TABLE 1.3.2—WOOD FRAME WALLS
4″ TO LESS THAN 6″ THICK—continued

ITEM CODE	THICKNESS	CONSTRUCTION DETAILS	PERFORMANCE		REFERENCE NUMBER			NOTES	REC. HOURS
			LOAD	TIME	PRE-BMS-92	BMS-92	POST-BMS-92		
W-5-W-54	5″	Same as W-5-W-53 but with insulated cavities.	See Note 23	35 min.		1		17, 18, 23	$^{1}/_{2}$
W-5-W-55	5″	2″ × 4″ stud wall; $^{3}/_{4}$″ thick T&G wood boards on both sides with 30 lbs./100 ft.^{2} asbestos; paper, between studs and boards.	See Note 23	45 min.		1		17, 23	$^{3}/_{4}$
W-5-W-56	5″	2″ × 4″ stud wall; $^{1}/_{2}$″ thick, 1:2; 1:3 gypsum plaster on metal lath on both sides of wall.	See Note 23	45 min.		1		17, 21, 34	$^{3}/_{4}$
W-5-W-57	5″	2″ × 4″ stud wall; $^{3}/_{4}$″ thick 2:1:8; 2:1:12 lime and Keene's cement plaster over metal lath on both sides of wall.	See Note 23	45 min.		1		17, 21, 23	$^{1}/_{2}$
W-5-W-58	5″	2″ × 4″ stud wall; $^{3}/_{4}$″ thick 2:1:8; 2:1:10 lime portland cement plaster over metal lath on both sides of wall.	See Note 23	30 min.		1		17, 21, 23	$^{1}/_{2}$
W-5-W-59	5″	2″ × 4″ stud wall; $^{3}/_{4}$″ thick 1:5; 1:7.5 lime plaster on metal lath on both sides of wall.	See Note 23	30 min.		1		17, 21, 23	$^{1}/_{2}$
W-5-W-60	5″	2″ × 4″ stud wall; $^{3}/_{4}$″ thick 1:$^{1}/_{30}$:2; 1:$^{1}/_{30}$:3 portland cement, asbestos fiber plaster on metal lath on both sides of wall.	See Note 23	45 min.		1		17, 21, 23	$^{3}/_{4}$
W-5-W-61	5″	2″ × 4″ stud wall; $^{3}/_{4}$″ thick 1:2; 1:3 portland cement plaster on metal lath on both sides of wall.	See Note 23	30 min.		1		17, 21, 23	$^{1}/_{2}$
W-5-W-62	5″	2″ × 4″ stud wall; $^{3}/_{4}$″ thick neat gypsum plaster on metal lath on both sides of wall.	N/A	1 hr. 30 min.		1		17, 22, 24	1$^{1}/_{2}$
W-5-W-63	5″	2″ × 4″ stud wall; $^{3}/_{4}$″ thick neat gypsum plaster on metal lath on both sides of wall.	See Note 23	1 hr. 30 min.		1		17, 21, 23	1$^{1}/_{2}$
W-5-W-64	5″	2″ × 4″ stud wall; $^{3}/_{4}$″ thick 1:2; 1:2 gypsum plaster on metal lath on both sides of wall; insulated cavities.	See Note 23	1 hr. 30 min.		1		17, 18, 21, 23	1$^{1}/_{2}$
W-5-W-65	5″	2″ × 4″ stud wall; same as W-5-W-64 but cavities not insulated.	See Note 23	1 hr.		1		17, 21, 23	1
W-5-W-66	5″	2″ × 4″ stud wall; $^{3}/_{4}$″ thick 1:2; 1:3 gypsum plaster on metal lath on both sides of wall; insulated cavities.	See Note 23	1 hr. 15 min.		1		17, 18, 21, 23	1$^{1}/_{4}$
W-5-W-67	5$^{1}/_{16}$″	Same as W-5-W-49 except cavity insulation of 1.75 lbs./ft.^{2} mineral wool bats; rating applies when either wall side exposed to fire.	See Note 23	1 hr. 15 min.		1		23, 26, 25	1$^{1}/_{4}$
W-5-W-68	5$^{1}/_{4}$″	2″ × 4″ stud wall, $^{7}/_{8}$″ thick 1:2; 1:3 gypsum plaster on metal lath on both sides of wall; insulated cavities.	See Note 23	1 hr. 30 min.		1		17, 18, 21, 23	1$^{1}/_{2}$

(Continued)

TABLE 1.3.2—WOOD FRAME WALLS 4″ TO LESS THAN 6″ THICK—continued

ITEM CODE	THICKNESS	CONSTRUCTION DETAILS	PERFORMANCE		REFERENCE NUMBER			NOTES	REC. HOURS
			LOAD	TIME	PRE-BMS-92	BMS-92	POST-BMS-92		
W-5-W-69	$5^1/_4$″	2″ × 4″ stud wall; $^7/_8$″ thick neat gypsum plaster applied on metal lath on both sides of wall.	N/A	1 hr. 45 min.		1		17, 22, 24	$1^3/_4$
W-5-W-70	$5^1/_4$″	2″ × 4″ stud wall; $^1/_2$″ thick neat gypsum plaster on $^3/_8$″ plain gypsum lath on both sides of wall.	See Note 23	1 hr.		1		17, 22, 23	1
W-5-W-71	$5^1/_4$″	2″ × 4″ stud wall; $^1/_2$″ thick of 1:2; 1:2 gypsum plaster on $^3/_8$″ thick plain gypsum lath with $1^3/_4$″ × $1^3/_4$″ metal lath pads nailed 8″ o.c. vertically and 16″ o.c. horizontally on both sides of wall.	See Note 23	1 hr.		1		17, 21, 23	1
W-5-W-72	$5^1/_4$″	2″ × 4″ stud wall; $^1/_2$″ thick of 1:2; 1:2 gypsum plaster on $^3/_8$″ perforated gypsum lath, one $^3/_4$″ diameter hole or larger per 16″ square of lath surface, on both sides of wall.	See Note 23	1 hr.		1		17, 21, 23	1
W-5-W-73	$5^1/_4$″	2″ × 4″ stud wall; $^1/_2$″ thick of 1:2; 1:2 gypsum plaster on $^3/_8$″ gypsum lath (plain, indented or perforated) on both sides of wall.	See Note 23	45 min.		1		17, 21, 23	$^3/_4$
W-5-W-74	$5^1/_4$″	2″ × 4″ stud wall; $^7/_8$″ thick of 1:2; 1:3 gypsum plaster over metal lath on both sides of wall.	See Note 23	1 hr.		1		17, 21, 23	1
W-5-W-75	$5^1/_4$″	2″ × 4″ stud wall; $^7/_8$″ thick of 1:$^1/_{30}$:2; 1:$^1/_{30}$:3 portland cement, asbestos plaster applied over metal lath on both sides of wall.	See Note 23	1 hr.		1		17, 21, 23	1
W-5-W-76	$5^1/_4$″	2″ × 4″ stud wall; $^7/_8$″ thick of 1:2; 1:3 portland cement plaster over metal lath on both sides of wall.	See Note 23	45 min.		1		17, 21, 23	$^3/_4$
W-5-W-77	$5^1/_2$″	2″ × 4″ stud wall; 1″ thick neat gypsum plaster over metal lath on both sides of wall; nonload bearing.	N/A	2 hrs.		1		17, 22, 24	2
W-5-W-78	$5^1/_2$″	2″ × 4″ stud wall; $^1/_2$″ thick of 1:2; 1:2 gypsum plaster on $^1/_2$″ thick, 0.7 lb./ft.^{2} wood fiberboard on both sides of wall.	See Note 23	35 min.		1		17, 21, 23	$^1/_2$
W-4-W-79	$4^3/_4$″	2″ × 4″ wood stud wall; $^1/_2$″ thick of 1:2; 1:2 gypsum plaster over wood lath on both sides of wall; mineral wool insulation.	N/A	1 hr.			43	21, 31, 35, 38	1
W-4-W-80	$4^3/_4$″	Same as W-4-W-79 but uninsulated.	N/A	35 min.			43	21, 31, 35	$^1/_2$
W-4-W-81	$4^3/_4$″	2″ × 4″ wood stud wall; $^1/_2$″ thick of 3:1:8; 3:1:12 lime, Keene's cement, sand plaster over wood lath on both sides of wall; mineral wool insulation.	N/A	1 hr.			43	21, 31, 35, 40	1

(Continued)

TABLE 1.3.2—WOOD FRAME WALLS
4″ TO LESS THAN 6″ THICK—continued

ITEM CODE	THICKNESS	CONSTRUCTION DETAILS	PERFORMANCE		REFERENCE NUMBER			NOTES	REC. HOURS
			LOAD	TIME	PRE-BMS-92	BMS-92	POST-BMS-92		
W-4-W-82	$4^3/_4$″	2″ × 4″ wood stud wall; $^1/_2$″ thick of 1:6$^1/_4$; 1:6$^1/_4$ lime Keene's cement plaster over wood lath on both sides of wall; mineral wool insulation.	N/A	30 min.			43	21, 31, 35, 40	$^1/_2$
W-4-W-83	$4^3/_4$″	2″ × 4″ wood stud wall; $^1/_2$″ thick of 1:5; 1:7.5 lime plaster over wood lath on both sides of wall.	N/A	30 min.			43	21, 31, 35	$^1/_2$
W-5-W-84	$5^1/_8$″	2″ × 4″ wood stud wall; $^{11}/_{16}$″ thick of 1:5; 1:7.5 lime plaster over wood lath on both sides of wall; mineral wool insulation.	N/A	45 min.			43	21, 31, 35, 39	$^3/_4$
W-5-W-85	$5^1/_4$″	2″ × 4″ wood stud wall; $^3/_4$″ thick of 1:5; 1:7 lime plaster over wood lath on both sides of wall; mineral wool insulation.	N/A	40 min.			43	21, 31, 35, 40	$^2/_3$
W-5-W-86	$5^1/_4$″	2″ × 4″ wood stud wall; $^1/_2$″ thick of 2:1:12 lime, Keene's cement and sand scratch coat; $^1/_2$″ thick 2:1:18 lime, Keene's cement and sand brown coat over wood lath on both sides of wall; mineral wool insulation.	N/A	1 hr.			43	21, 31, 35, 40	1
W-5-W-87	$5^1/_4$″	2″ × 4″ wood stud wall; $^1/_2$″ thick of 1:2; 1:2 gypsum plaster over $^3/_8$″ plaster board on both sides of wall.	N/A	45 min.			43	21, 31	$^3/_4$
W-5-W-88	$5^1/_4$″	2″ × 4″ wood stud wall; $^1/_2$″ thick of 1:2; 1:2 gypsum plaster over $^3/_8$″ gypsum lath on both sides of wall.	N/A	45 min.			43	21, 31	$^3/_4$
W-5-W-89	$5^1/_4$″	2″ × 4″ wood stud wall; $^1/_2$″ thick of 1:2; 1:2 gypsum plaster over $^3/_8$″ gypsum lath on both sides of wall.	N/A	1 hr.			43	21, 31, 33	1
W-5-W-90	$5^1/_4$″	2″ × 4″ wood stud wall; $^1/_2$″ thick neat plaster over $^3/_8$″ thick gypsum lath on both sides of wall.	N/A	1 hr.			43	21, 22, 31	1
W-5-W-91	$5^1/_4$″	2″ × 4″ wood stud wall; $^1/_2$″ thick of 1:2; 1:2 gypsum plaster over $^3/_8$″ thick indented gypsum lath on both sides of wall.	N/A	45 min.			43	21, 31	$^3/_4$
W-5-W-92	$5^1/_4$″	2″ × 4″ wood stud wall; $^1/_2$″ thick of 1:2; 1:2 gypsum plaster over $^3/_8$″ thick perforated gypsum lath on both sides of wall.	N/A	45 min.			43	21, 31, 34	$^3/_4$
W-5-W-93	$5^1/_4$″	2″ × 4″ wood stud wall; $^1/_2$″ thick of 1:2; 1:2 gypsum plaster over $^3/_8$″ perforated gypsum lath on both sides of wall.	N/A	1 hr.			43	21, 31	1
W-5-W-94	$5^1/_4$″	2″ × 4″ wood stud wall; $^1/_2$″ thick of 1:2; 1:2 gypsum plaster over $^3/_8$″ thick perforated gypsum lath on both sides of wall.	N/A	45 min.			43	21, 31, 34	$^3/_4$

(Continued)

TABLE 1.3.2—WOOD FRAME WALLS
4″ TO LESS THAN 6″ THICK—continued

ITEM CODE	THICKNESS	CONSTRUCTION DETAILS	PERFORMANCE		REFERENCE NUMBER			NOTES	REC. HOURS
			LOAD	TIME	PRE-BMS-92	BMS-92	POST-BMS-92		
W-5-W-95	5[1]/2″	2″ × 4″ wood stud wall; 1/2″ thick of 1:2; 1:2 gypsum plaster over 1/2″ thick wood fiberboard plaster base on both sides of wall.	N/A	35 min.			43	21, 31, 36	1/2
W-5-W-96	5[3]/4″	2″ × 4″ wood stud wall; 1/2″ thick of 1:2; 1:2 gypsum plaster over 7/8″ thick flameproofed wood fiberboard on both sides of wall.	N/A	1 hr.			43	21, 31, 37	1

For SI: 1 inch = 25.4 mm, 1 foot = 305 mm, 1 pound = 0.004448 kN, 1 pound per square inch = 0.00689 MPa, 1 pound per square foot = 47.9 N/m^2.

Notes:

1. All specimens 8 feet or 8 feet 8 inches by 10 feet, 4 inches, i.e. one half of furnace size. See Note 42 for design cross section.
2. Specimens tested in tandem (two per exposure).
3. Test per ASA No. A2-1934 except where unloaded. Also, panels were of "half" size of furnace opening. Time value signifies a thermal failure time.
4. Two-inch by 4-inch studs: 16 inches on center.; where 10 feet 4 inches, blocking at 2-foot 4-inch height.
5. Facing 4 feet by 8 feet, cement-asbestos board sheets, 3/16 inch thick.
6. Sheathing (diagonal): 25/22 inch by 5[1]/2 inch, 1 inch by 6 inches pine.
7. Facing shingles: 24 inches by 12 inches by 5/32 inch where used.
8. Asbestos felt: asphalt sat between sheathing and shingles.
9. Load: 30,500 pounds or 360 psi/stud where load was tested.
10. Walls were tested beyond achievement of first test end point. A load-bearing time in excess of performance time indicates that although thermal criteria were exceeded, load-bearing ability continued.
11. Wall was rated for one hour combustible use in original source.
12. Hose steam test specimen. See table entry of similar design above for recommended rating.
13. Rated one and one-fourth hour load bearing. Rated one and one-half hournonload bearing.
14. Failed hose stream.
15. Test terminated due to flame penetration.
16. Test terminated - local back face temperature rise.
17. Nominal 2-inch by 4-inch wood studs of No. 1 common or better lumber set edgewise. Two-inch by four-inch plates at top and bottom and blocking at mid height of wall.
18. Cavity insulation consists of rock wool bats 1.0 lb./ft.^{2} of filled cavity area.
19. Cavity insulation consists of glass wool bats 0.6 lb./ft.^{2} of filled cavity area.
20. Cavity insulation consists of blown-in forck wool 2.0 lbs./ft.^{2} of filled cavity area
21. Mix proportions for plastered walls as follows: first ratio indicates scratch coat mix, weight of dry plaster: dry sand; second ratio indicates brown coat mix.
22. "Neat" plaster is taken to mean unsanded wood-fiber gypsum plaster.
23. Load: 360 psi of net stud cross sectional area.
24. Rated as nonload bearing.
25. Nominal 2-inch by 4-inch studs per Note 17, spaced at 16 inches on center.
26. Horizontal joints in facing material supported by 2-inch by 4-inch blocking within wall.
27. Facings secured with 6d casing nails. Nail holes predrilled and were 0.02 to 0.03 inch smaller than nail diameter.
28. Cavity insulation consists of mineral wool bats weighing 2 lbs./ft.^{2} of filled cavity area.
29. Interior wall face exposed to fire.
30. Exterior wall faced exposed to fire.
31. Nominal 2-inch by 4-inch studs of yellow pine or Douglas-fir spaced 16 inches on center in a single row.
32. Studs as in Note 31 except double row, with studs in rows staggered.
33. Six roofing nails with metal-lath pads around heats to each 16-inch by 48-inch lath.
34. Areas of holes less than 2[3]/4 percent of area of lath.
35. Wood laths were nailed with either 3d or 4d nails, one nail to each bearing, and the end joining broken every seventh course.
36. One-half-inch thick fiberboard plaster base nailed with 3d or 4d common wire nails spaced 4 to 6 inches on center.
37. Seven-eighths-inch thick fiberboard plaster base nailed with 5d common wire nails spaced 4 to 6 inches on center.
38. Mineral wood bats 1.05 to 1.25 lbs./ft.^{2} with waterproofed-paper backing.
39. Blown-in mineral wool insulation, 2.2 lbs./ft.^{2}.
40. Mineral wool bats, 1.4 lbs./ft.^{2} with waterproofed-paper backing.
41. Mineral wood bats, 0.9 lb./ft.^{2}.
42. See wall design diagram, below.

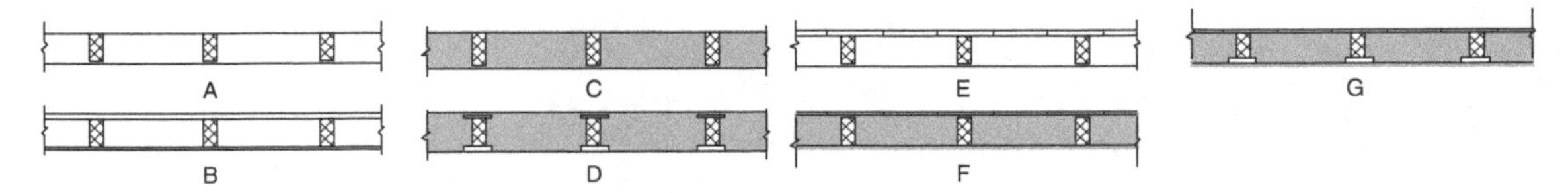

43. Duplicate specimen of W-4-W-7, tested simultaneously with W-4-W-7 in 18-foot test furnace.

TABLE 1.3.3—WOOD FRAME WALLS
6″ TO LESS THAN 8″ THICK

ITEM CODE	THICKNESS	CONSTRUCTION DETAILS	PERFORMANCE		REFERENCE NUMBER			NOTES	REC. HOURS
			LOAD	TIME	PRE-BMS-92	BMS-92	POST-BMS-92		
W-6-W-1	$6^1/_4$″	2 × 4 stud wall; $^1/_2$″ thick, 1:2; 1:2 gypsum plaster on $^7/_8$″ flameproofed wood fiberboard weighing 2.8 lbs./ft.^{2} on both sides of wall.	See Note 3	1 hr.		1		1-3	1
W-6-W-2	$6^1/_2$″	2 × 4 stud wall; $^1/_2$″ thick, 1:3; 1:3 gypsum plaster on 1″ thick magnesium oxysulfate wood fiberboard on both sides of wall.	See Note 3	45 min.		1		1-3	$^3/_4$
W-7-W-3	$7^1/_4$″	Double row of 2 × 4 studs, $^1/_2$″ thick of 1:2; 1:2 gypsum plaster applied over $^3/_8$″ thick perforated gypsum lath on both sides of wall; mineral wool insulation.	N/A	1 hr.			43	2, 4, 5	1
W-7-W-4	$7^1/_2$″	Double row of 2 × 4 studs, $^5/_8$″ thick of 1:2; 1:2 gypsum plaster applied over $^3/_8$″ thick perforated gypsum lath over laid with 2″ × 2″, 16 gage wire fabric, on both sides of wall.	N/A	1 hr. 15 min.			43	2, 4	$1^1/_4$

For SI: 1 inch = 25.4 mm, 1 pound = 0.004448 kN, 1 pound per square inch = 0.00689 MPa, 1 pound per square foot = 47.9 N/m^2.

Notes:

1. Nominal 2-inch by 4-inch wood studs of No. 1 common or better lumber set edgewise. Two-inch by 4-inch plates at top and bottom and blocking at mid height of wall.
2. Mix proportions for plastered walls as follows: first ratio indicates scratch coat mix, weight of dry plaster:dry sand; second ratio indicates brown coat mix.
3. Load: 360 psi of net stud cross sectional area.
4. Nominal 2-inch by 4-inch studs of yellow pine of Douglas-fir spaced 16 inches in a double row, with studs in rows staggered.
5. Mineral wool bats, 0.19 lb./ft.^{2}.

TABLE 1.4.1—MISCELLANEOUS MATERIALS WALLS
0″ TO LESS THAN 4″ THICK

ITEM CODE	THICKNESS	CONSTRUCTION DETAILS	PERFORMANCE		REFERENCE NUMBER			NOTES	REC. HOURS
			LOAD	TIME	PRE-BMS-92	BMS-92	POST-BMS-92		
W-3-Mi-1	$3^7/_8$″	Glass brick wall: (bricks $5^3/_4$″ × $5^3/_4$″ × $3^7/_8$″) $^1/_4$″ mortar bed, cement/lime/sand; mounted in brick (9″) wall with mastic and $^1/_2$″ asbestos rope.	N/A	1 hr.			7	1, 2	1
W-3-Mi-2	3″	Core: 2″ magnesium oxysulfate wood-fiber blocks; laid in portland cement-lime mortar; Facings: on both sides; see Note 3.	N/A	1 hr.		1		3	1
W-3-Mi-3	$3^7/_8$″	Core: 8″ × $4^7/_8$″ glass blocks $3^7/_8$″ thick weighing 4 lbs. each; laid in portland cement-lime mortar; horizontal mortar joints reinforced with metal lath.	N/A	15 min.		1			$^1/_4$

For SI: 1 inch = 25.4 mm, 1 pound = 0.004448 kN.

Notes:

1. No failure reached at 1 hour.
2. These glass blocks are assumed to be solid based on other test data available for similar but hollow units which show significantly reduced fire endurance.
3. Minimum of $^1/_2$ inch of 1:3 sanded gypsum plaster required to develop this rating.

**TABLE 1.4.2—MISCELLANEOUS MATERIALS WALLS
4″ TO LESS THAN 6 ″THICK**

ITEM CODE	THICKNESS	CONSTRUCTION DETAILS	PERFORMANCE		REFERENCE NUMBER			NOTES	REC. HOURS
			LOAD	TIME	PRE-BMS-92	BMS-92	POST-BMS-92		
W-4-Mi-1	4″	Core: 3″ magnesium oxysulfate wood-fiber blocks; laid in portland cement mortar; Facings: both sides; see Note 1.	N/A	2 hrs.		1			2

For SI: 1 inch = 25.4 mm.

Notes:

1. One-half inch sanded gypsum plaster. Voids in hollow blocks to be not more than 30 percent.

FIGURE 1.5.1—FINISH RATINGS—INORGANIC MATERIALS

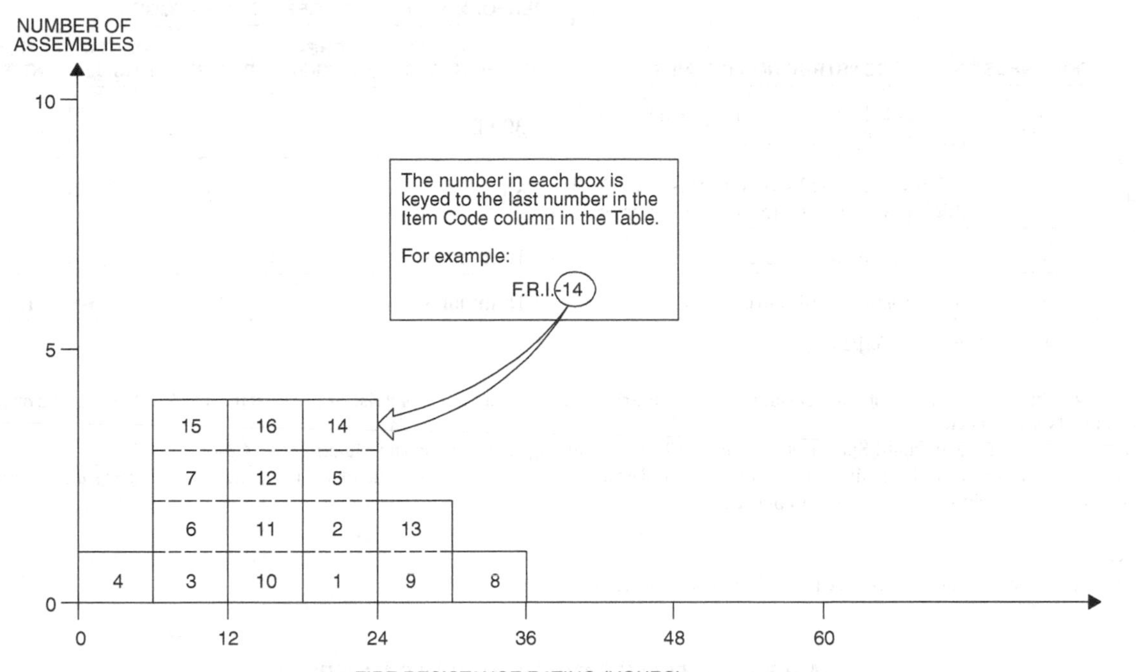

TABLE 1.5.1—FINISH RATINGS—INORGANIC MATERIALS

ITEM CODE	THICKNESS	CONSTRUCTION DETAILS	PERFORMANCE	REFERENCE NUMBER			NOTES	REC. F.R. (MIN.)
			FINISH RATING	PRE-BMS-92	BMS-92	POST-BMS-92		
F.R.-I-1	$^{9}/_{16}$″	$^{3}/_{8}$″ gypsum wallboard faced with $^{3}/_{16}$″ cement-asbestos board.	20 minutes		1		1, 2	15
F.R.-I-2	$^{11}/_{16}$″	$^{1}/_{2}$″ gypsum sheathing faced with $^{3}/_{16}$″ cement-asbestos board.	20 minutes		1		1, 2	20
F.R.-I-3	$^{3}/_{16}$″	$^{3}/_{16}$″ cement-asbestos board over uninsulated cavity.	10 minutes		1		1, 2	5
F.R.-I-4	$^{3}/_{16}$″	$^{3}/_{16}$″ cement-asbestos board over insulated cavities.	5 minutes		1		1, 2	5
F.R.-I-5	$^{3}/_{4}$″	$^{3}/_{4}$″ thick 1:2; 1:3 gypsum plaster over paper backed metal lath.	20 minutes		1		1, 2, 3	20
F.R.-I-6	$^{3}/_{4}$″	$^{3}/_{4}$″ thick portland cement plaster on metal lath.	10 minutes		1		1, 2	10
F.R.-I-7	$^{3}/_{4}$″	$^{3}/_{4}$″ thick 1:5; 1:7.5 lime plaster on metal lath.	10 minutes		1		1, 2	10
F.R.-I-8	1″	1″ thick neat gypsum plaster on metal lath.	35 minutes		1		1, 2, 4	35
F.R.-I-9	$^{3}/_{4}$″	$^{3}/_{4}$″ thick neat gypsum plaster on metal lath.	30 minutes		1		1, 2, 4	30
F.R.-I-10	$^{3}/_{4}$″	$^{3}/_{4}$″ thick 1:2; 1:2 gypsum plaster on metal lath.	15 minutes		1		1, 2, 3	15
F.R.-I-11	$^{1}/_{2}$″	Same as F.R.-1-7, except $^{1}/_{2}$″ thick on wood lath.	15 minutes		1		1, 2, 3	15
F.R.-I-12	$^{1}/_{2}$″	$^{1}/_{2}$″ thick 1:2; 1:3 gypsum plaster on wood lath.	15 minutes		1		1, 2, 3	15

(Continued)

TABLE 1.5.1—FINISH RATINGS—INORGANIC MATERIALS—continued

ITEM CODE	THICKNESS	CONSTRUCTION DETAILS	PERFORMANCE	REFERENCE NUMBER			NOTES	REC. F.R. (MIN.)
			FINISH RATING	PRE-BMS-92	BMS-92	POST-BMS-92		
F.R.-I-13	$^7/_8$″	$^1/_2$″ thick 1:2; 1:2 gypsum plaster on $^3/_8$″ perforated gypsum lath.	30 minutes		1		1, 2, 3	30
F.R.-I-14	$^7/_8$″	$^1/_2$″ thick 1:2; 1:2 gypsum plaster on $^3/_8$″ thick plain or indented gypsum plaster.	20 minutes		1		1, 2, 3	20
F.R.-I-15	$^3/_8$″	$^3/_8$″ gypsum wallboard.	10 minutes		1		1, 2	10
F.R.-I-16	$^1/_2$″	$^1/_2$″ gypsum wallboard.	15 minutes		1		1, 2	15

For SI: 1 inch = 25.4 mm, °C = [(°F) - 32]/1.8.

Notes:

1. The finish rating is the time required to obtain an average temperature rise of 250°F, or a single point rise of 325°F, at the interface between the material being rated and the substrate being protected.
2. Tested in accordance with the Standard Specifications for Fire Tests of Building Construction and Materials, ASA No. A2-1932.
3. Mix proportions for plasters as follows: first ratio, dry weight of plaster: dry weight of sand for scratch coat; second ratio, plaster: sand for brown coat.
4. Neat plaster means unsanded wood-fiber gypsum plaster.

General Note:

The finish rating of modern building materials can be found in the current literature.

TABLE 1.5.2—FINISH RATINGS—ORGANIC MATERIALS

ITEM CODE	THICKNESS	CONSTRUCTION DETAILS	PERFORMANCE	REFERENCE NUMBER			NOTES	REC. F.R. (MIN.)
			FINISH RATING	PRE-BMS-92	BMS-92	POST-BMS-92		
F.R.-O-1	$^9/_{16}$″	$^7/_{16}$″ wood fiberboard faced with $^1/_8$″ cement-asbestos board.	15 minutes		1		1, 2	15
F.R.-O-2	$^{29}/_{32}$″	$^3/_4$″ wood sheathing, asbestos felt weighing 14 lbs./100 ft.^{2} and $^5/_{32}$″ cement-asbestos shingles.	20 minutes		1		1, 2	20
F.R.-O-3	$1^1/_2$″	1″ thick magnesium oxysulfate wood fiberboard faced with 1:3; 1:3 gypsum plaster, $^1/_2$″ thick.	20 minutes		1		1, 2, 3	20
F.R.-O-4	$^1/_2$″	$^1/_2$″ thick wood fiberboard.	5 minutes		1		1, 2	5
F.R.-O-5	$^1/_2$″	$^1/_2$″ thick flameproofed wood fiberboard.	10 minutes		1		1, 2	10
F.R.-O-6	1″	$^1/_2$″ thick wood fiberboard faced with $^1/_2$″ thick 1:2; 1:2 gypsum plaster.	15 minutes		1		1, 2, 3	30
F.R.-O-7	$1^3/_8$″	$^7/_8$″ thick flameproofed wood fiberboard faced with $^1/_2$″ thick 1:2; 1:2 gypsum plaster.	30 minutes		1		1, 2, 3	30
F.R.-O-8	$1^1/_4$″	$1^1/_4$″ thick plywood.	30 minutes			35		30

For SI: 1 inch = 25.4 mm, 1 pound = 0.004448 kN, 1 pound per square foot = 47.9 N/m^2, °C = [(°F) - 32]/1.8.

Notes:

1. The finish rating is the time required to obtain an average temperature rise of 250°F, or a single point rise of 325°F, at the interface between the material being rated and he substrate being protected.
2. Tested in accordance with the Standard Specifications for Fire Tests of Building Construction and Materials, ASA No. A2-1932.
3. Plaster ratios as follows: first ratio is for scratch coat, weight of dry plaster: weight of dry sand; second ratio is for the brown coat.

General Note:

The finish rating of thinner materials, particularly thinner woods, have not been listed because the possible effects of shrinkage, warpage and aging cannot be predicted.

SECTION II—COLUMNS

TABLE 2.1.1—REINFORCED CONCRETE COLUMNS MINIMUM DIMENSION 0″ TO LESS THAN 6″

ITEM CODE	MINIMUM DIMENSION	CONSTRUCTION DETAILS	PERFORMANCE		REFERENCE NUMBER			NOTES	REC. HOURS
			LOAD	TIME	PRE-BMS-92	BMS-92	POST-BMS-92		
C-6-RC-1	6″	6″ × 6″ square columns; gravel aggregate concrete (4030 psi); Reinforcement: vertical, four $^7/_8$″ rebars; horizontal, $^5/_{16}$″ ties at 6″ pitch; Cover: 1″.	34.7 tons	62 min.			7	1, 2	1
C-6-RC-2	6″	6″ × 6″ square columns; gravel aggregate concrete (4200 psi); Reinforcement: vertical, four $^1/_2$″ rebars; horizontal, $^5/_{16}$″ ties at 6″ pitch; Cover: 1″.	21 tons	69 min.			7	1, 2	1

Notes:
1. Collapse.
2. British Test.

FIGURE 2.1.2—REINFORCED CONCRETE COLUMNS MINIMUM DIMENSION 10″ TO LESS THAN 12″

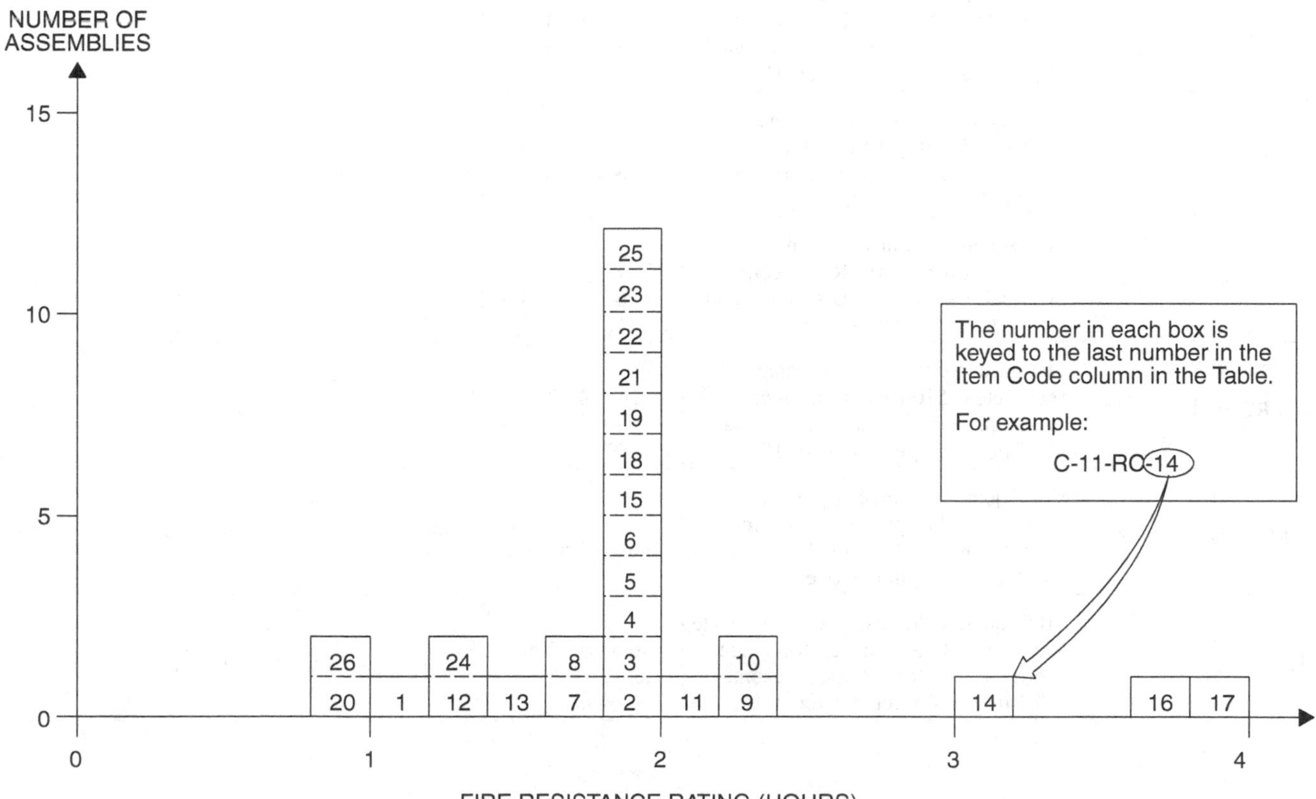

TABLE 2.1.2—REINFORCED CONCRETE COLUMNS MINIMUM DIMENSION 10″ TO LESS THAN 12″

ITEM CODE	MINIMUM DIMENSION	CONSTRUCTION DETAILS	PERFORMANCE		REFERENCE NUMBER			NOTES	REC. HOURS
			LOAD	TIME	PRE-BMS-92	BMS-92	POST-BMS-92		
C-10-RC-1	10″	10″ square columns; aggregate concrete (4260 psi); Reinforcement: vertical, four $1^1/_4$″ rebars; horizontal, $^3/_8$″ ties at 6″ pitch; Cover: $1^1/_4$″.	92.2 tons	1 hr. 2 min.			7	1	1
C-10-RC-2	10″	10″ square columns; aggregate concrete (2325 psi); Reinforcement: vertical, four $^1/_2$″ rebars; horizontal, $^5/_{16}$″ ties at 6″ pitch; Cover: 1″.	46.7 tons	1 hr. 52 min.			7	1	$1^3/_4$
C-10-RC-3	10″	10″ square columns; aggregate concrete (5370 psi); Reinforcement: vertical, four $^1/_2$″ rebars; horizontal, $^5/_{16}$″ ties at 6″ pitch; Cover: 1″.	46.5 tons	2 hrs.			7	2, 3, 11	2
C-10-RC-4	10″	10″ square columns; aggregate concrete (5206 psi); Reinforcement: vertical, four $^1/_2$″ rebars; horizontal, $^5/_{16}$″ ties at 6″ pitch; Cover: 1″.	46.5 tons	2 hrs.			7	2, 7	2
C-10-RC-5	10″	10″ square columns; aggregate concrete (5674 psi); Reinforcement: vertical, four $^1/_2$″ rebars; horizontal, $^5/_{16}$″ ties at 6″ pitch; Cover: 1″.	46.7 tons	2 hrs.			7	1	2
C-10-RC-6	10″	10″ square columns; aggregate concrete (5150 psi); Reinforcement: vertical, four $1^1/_2$″ rebars; horizontal, $^5/_{16}$″ ties at 6″ pitch; Cover: 1″.	66 tons	1 hr. 43 min.			7	1	$1^3/_4$
C-10-RC-7	10″	10″ square columns; aggregate concrete (5580 psi); Reinforcement: vertical, four $^1/_2$″ rebars; horizontal, $^5/_{16}$″ ties at 6″ pitch; Cover: $1^1/_8$″.	62.5 tons	1 hr. 38 min.			7	1	$1^1/_2$
C-10-RC-8	10″	10″ square columns; aggregate concrete (4080 psi); Reinforcement: vertical, four $1^1/_8$″ rebars; horizontal, $^5/_{16}$″ ties at 6″ pitch; Cover: $1^1/_8$″.	72.8 tons	1 hr. 48 min.			7	1	$1^3/_4$
C-10-RC-9	10″	10″ square columns; aggregate concrete (2510 psi); Reinforcement: vertical, four $^1/_2$″ rebars; horizontal, $^5/_{16}$″ ties at 6″ pitch; Cover: 1″.	51 tons	2 hrs. 16 min.			7	1	$2^1/_4$
C-10-RC-10	10″	10″ square columns; aggregate concrete (2170 psi); Reinforcement: vertical, four $^1/_2$″ rebars; horizontal, $^5/_{16}$″ ties at 6″ pitch; Cover: 1″.	45 tons	2 hrs. 14 min.			7	12	$2^1/_4$
C-10-RC-11	10″	10″ square columns; gravel aggregate concrete (4015 psi); Reinforcement: vertical, four $^1/_2$″ rebars; horizontal, $^5/_{16}$″ ties at 6″ pitch; Cover: $1^1/_8$″.	46.5 tons	2 hrs. 6 min.			7	1	2

(Continued)

TABLE 2.1.2—REINFORCED CONCRETE COLUMNS MINIMUM DIMENSION 10″ TO LESS THAN 12″—continued

ITEM CODE	MINIMUM DIMENSION	CONSTRUCTION DETAILS	PERFORMANCE		REFERENCE NUMBER			NOTES	REC. HOURS
			LOAD	TIME	PRE-BMS-92	BMS-92	POST-BMS-92		
C-11-RC-12	11″	11″ square columns; gravel aggregate concrete (4150 psi); Reinforcement: vertical, four $1^1/_4$″ rebars; horizontal, $^3/_8$″ ties at $7^1/_2$″ pitch; Cover: $1^1/_2$″.	61 tons	1 hr. 23 min.			7	1	$1^1/_4$
C-11-RC-13	11″	11″ square columns; gravel aggregate concrete (4380 psi); Reinforcement: vertical, four $1^1/_4$″ rebars; horizontal, $^3/_8$″ ties at $7^1/_2$″ pitch; Cover: $1^1/_2$″.	61 tons	1 hr. 26 min.			7	1	$1^1/_4$
C-11-RC-14	11″	11″ square columns; gravel aggregate concrete (4140 psi); Reinforcement: vertical, four $1^1/_4$″ rebars; horizontal, $^3/_8$″ ties at $7^1/_2$″ pitch; steel mesh around reinforcement; Cover: $1^1/_2$″.	61 tons	3 hrs. 9 min.			7	1	3
C-11-RC-15	11″	11″ square columns; slag aggregate concrete (3690 psi); Reinforcement: vertical, four $1^1/_4$″ rebars; horizontal, $^3/_8$″ ties at $7^1/_2$″ pitch; Cover: $1^1/_2$″.	91 tons	2 hrs.			7	2, 3, 4, 5	2
C-11-RC-16	11″	11″ square columns; limestone aggregate concrete (5230 psi); Reinforcement: vertical, four $1^1/_4$″ rebars; horizontal, $^3/_8$″ ties at $7^1/_2$″ pitch; Cover: $1^1/_2$″.	91.5 tons	3 hrs. 41 min.			7	1	$3^1/_2$
C-11-RC-17	11″	11″ square columns; limestone aggregate concrete (5530 psi); Reinforcement: vertical, four $1^1/_4$″ rebars; horizontal, $^3/_8$″ ties at $7^1/_2$″ pitch; Cover: $1^1/_2$″.	91.5 tons	3 hrs. 47 min.			7	1	$3^1/_2$
C-11-RC-18	11″	11″ square columns; limestone aggregate concrete (5280 psi); Reinforcement: vertical, four $1^1/_4$″ rebars; horizontal, $^3/_8$″ ties at $7^1/_2$″ pitch; Cover: $1^1/_2$″.	91.5 tons	2 hrs.			7	2, 3, 4, 6	2
C-11-RC-19	11″	11″ square columns; limestone aggregate concrete (4180 psi); Reinforcement: vertical, four $^5/_8$″ rebars; horizontal, $^3/_8$″ ties at 7″ pitch; Cover: $1^1/_2$″.	71.4 tons	2 hrs.			7	2, 7	2
C-11-RC-20	11″	11″ square columns; gravel concrete (4530 psi); Reinforcement: vertical, four $^5/_8$″ rebars; horizontal, $^3/_8$″ ties at 7″ pitch; Cover: $1^1/_2$″ with $^1/_2$″ plaster.	58.8 tons	2 hrs.			7	2, 3, 9	$1^1/_4$
C-11-RC-21	11″	11″ square columns; gravel concrete (3520 psi); Reinforcement: vertical, four $^5/_8$″ rebars; horizontal, $^3/_8$″ ties at 7″ pitch; Cover: $1^1/_2$″.	Variable	1 hr. 24 min.			7	1, 8	2

(Continued)

TABLE 2.1.2—REINFORCED CONCRETE COLUMNS MINIMUM DIMENSION 10″ TO LESS THAN 12″—continued

ITEM CODE	MINIMUM DIMENSION	CONSTRUCTION DETAILS	PERFORMANCE		REFERENCE NUMBER			NOTES	REC. HOURS
			LOAD	TIME	PRE-BMS-92	BMS-92	POST-BMS-92		
C-11-RC-22	11″	11″ square columns; aggregate concrete (3710 psi); Reinforcement: vertical, four $^5/_8$″ rebars; horizontal, $^3/_8$″ ties at 7″ pitch; Cover: $1^1/_2$″.	58.8 tons	2 hrs.			7	2, 3, 10	2
C-11-RC-23	11″	11″ square columns; aggregate concrete (3190 psi); Reinforcement: vertical, four $^5/_8$″ rebars; horizontal, $^3/_8$″ ties at 7″ pitch; Cover: $1^1/_2$″.	58.8 tons	2 hrs.			7	2, 3, 10	2
C-11-RC-24	11″	11″ square columns; aggregate concrete (4860 psi); Reinforcement: vertical, four $^5/_8$″ rebars; horizontal, $^3/_8$″ ties at 7″ pitch; Cover: $1^1/_2$″.	86.1 tons	1 hr. 20 min.			7	1	$1^1/_3$
C-11-RC-25	11″	11″ square columns; aggregate concrete (4850 psi); Reinforcement: vertical, four $^5/_8$″ rebars; horizontal, $^3/_8$″ ties at 7″ pitch; Cover: $1^1/_2$″.	58.8 tons	1 hr. 59 min.			7	1	$1^3/_4$
C-11-RC-26	11″	11″ square columns; aggregate concrete (3834 psi); Reinforcement: vertical, four $^5/_8$″ rebars; horizontal, $^5/_{16}$″ ties at $4^1/_2$″ pitch; Cover: $1^1/_2$″.	71.4 tons	53 min.			7	1	$^3/_4$

For SI: 1 inch = 25.4 mm, 1 pound per square inch = 0.00689 MPa, 1 ton = 8.896 kN.

Notes:

1. Failure mode - collapse.
2. Passed 2 hour fire exposure.
3. Passed hose stream test.
4. Reloaded effectively after 48 hours but collapsed at load in excess of original test load.
5. Failing load was 150 tons.
6. Failing load was 112 tons.
7. Failed during hose stream test.
8. Range of load 58.8 tons (initial) to 92 tons (92 minutes) to 60 tons (80 minutes).
9. Collapsed at 44 tons in reload after 96 hours.
10. Withstood reload after 72 hours.
11. Collapsed on reload after 48 hours.

TABLE 2.1.3—REINFORCED CONCRETE COLUMNS MINIMUM DIMENSION 12″ TO LESS THAN 14″

ITEM CODE	MINIMUM DIMENSION	CONSTRUCTION DETAILS	PERFORMANCE		REFERENCE NUMBER			NOTES	REC. HOURS
			LOAD	TIME	PRE-BMS-92	BMS-92	POST-BMS-92		
C-12-RC-1	12″	12″ square columns; gravel aggregate concrete (2647 psi); Reinforcement: vertical, four $^{5}/_{8}$″ rebars; horizontal, $^{5}/_{16}$″ ties at $4^{1}/_{2}$″ pitch; Cover: 2″.	78.2 tons	38 min.		1	7	1	$^{1}/_{2}$
C-12-RC-2	12″	Reinforced columns with $1^{1}/_{2}$″ concrete outside of reinforced steel; Gross diameter or side of column: 12″; Group I, Column A.	—	6 hrs.		1		2, 3	6
C-12-RC-3	12″	Description as per C-12-RC-2; Group I, Column B.	—	4 hrs.		1		2, 3	4
C-12-RC-4	12″	Description as per C-12-RC-2; Group II, Column A.	—	4 hrs.		1		2, 3	4
C-12-RC-5	12″	Description as per C-12-RC-2; Group II, Column B.	—	2 hrs. 30 min.		1		2, 3	$2^{1}/_{2}$
C-12-RC-6	12″	Description as per C-12-RC-2; Group III, Column A.	—	3 hrs.		1		2, 3	3
C-12-RC-7	12″	Description as per C-12-RC-2; Group III, Column B.	—	2 hrs.		1		2, 3	2
C-12-RC-8	12″	Description as per C-12-RC-2; Group IV, Column A.	—	2 hrs.		1		2, 3	2
C-12-RC-9	12″	Description as per C-12-RC-2; Group IV, Column B.	—	1 hr. 30 min.		1		2, 3	$1^{1}/_{2}$

For SI: 1 inch = 25.4 mm, 1 pound per square inch = 0.00689 MPa, 1 pound per square yard = 5.3 N/m^2.

Notes:

1. Failure mode - unspecified structural.
2. Group I: includes concrete having calcareous aggregate containing a combined total of not more than 10 percent of quartz, chert and flint for the coarse aggregate.

   Group II: includes concrete having trap-rock aggregate applied without metal ties and also concrete having cinder, sandstone or granite aggregate, if held in place with wire mesh or expanded metal having not larger than 4-inch mesh, weighing not less than 1.7 lbs./yd.^{2}, placed not more than 1 inch from the surface of the concrete.

   Group III: includes concrete having cinder, sandstone or granite aggregate tied with No. 5 gage steel wire, wound spirally over the column section on a pitch of 8 inches, or equivalent ties, and concrete having siliceous aggregates containing a combined total of 60 percent or more of quartz, chert and flint, if held in place with wire mesh or expanded metal having not larger than 4-inch mesh, weighing not less than 1.7 lbs./yd.^{2}, placed not more than 1 inch from the surface of the concrete.

   Group IV: includes concrete having siliceous aggregates containing a combined total of 60 percent or more of quartz, chert and flint, and tied with No. 5 gage steel wire wound spirally over the column section on a pitch of 8 inches, or equivalent ties.
3. Groupings of aggregates and ties are the same as for structural steel columns protected solidly with concrete, the ties to be placed over the vertical reinforcing bars and the mesh where required, to be placed within 1 inch from the surface of the column.

   Column A: working loads are assumed as carried by the area of the column inside of the lines circumscribing the reinforcing steel.

   Column B: working loads are assumed as carried by the gross area of the column.

TABLE 2.1.4—REINFORCED CONCRETE COLUMNS MINIMUM DIMENSION 14″ TO LESS THAN 16″

ITEM CODE	MINIMUM DIMENSION	CONSTRUCTION DETAILS	PERFORMANCE		REFERENCE NUMBER			NOTES	REC. HOURS
			LOAD	TIME	PRE-BMS-92	BMS-92	POST-BMS-92		
C-14-RC-1	14″	14″ square columns; gravel aggregate concrete (4295 psi); Reinforcement: vertical four $^3/_4$″ rebars; horizontal: $^1/_4$″ ties at 9″ pitch; Cover: $1^1/_2$″.	86 tons	1 hr. 22 min.			7	1	$1^1/_4$
C-14-RC-2	14″	Reinforced concrete columns with $1^1/_2$″ concrete outside reinforcing steel; Gross diameter or side of column: 12″; Group I, Column A.	—	7 hrs.		1		2, 3	7
C-14-RC-3	14″	Description as per C-14-RC-2; Group II, Column B.	—	5 hrs.		1		2, 3	5
C-14-RC-4	14″	Description as per C-14-RC-2; Group III, Column A.	—	5 hrs.		1		2, 3	5
C-14-RC-5	14″	Description as per C-14-RC-2; Group IV, Column B.	—	3 hrs. 30 min.		1		2, 3	$3^1/_2$
C-14-RC-6	14″	Description as per C-14-RC-2; Group III, Column A.	—	4 hrs.		1		2, 3	4
C-14-RC-7	14″	Description as per C-14-RC-2; Group III, Column B.	—	2 hrs. 30 min.		1		2, 3	$2^1/_2$
C-14-RC-8	14″	Description as per C-14-RC-2; Group IV, Column A.	—	2 hrs. 30 min.		1		2, 3	$2^1/_2$
C-14-RC-9	14″	Description as per C-14-RC-2; Group IV, Column B.	—	1 hr. 30 min.		1		2, 3	$1^1/_2$

For SI: 1 inch = 25.4 mm, 1 pound per square inch = 0.00689 MPa, 1 pound per square yard = 5.3 N/m^2.

Notes:

1. Failure mode - main rebars buckled between links at various points.
2. Group I: includes concrete having calcareous aggregate containing a combined total of not more than 10 percent of quartz, chert and flint for the coarse aggregate.

   Group II: includes concrete having trap-rock aggregate applied without metal ties and also concrete having cinder, sandstone or granite aggregate, if held in place with wire mesh or expanded metal having not larger than 4-inch mesh, weighing not less than 1.7 lbs./yd.^{2}, placed not more than 1 inch from the surface of the concrete.

   Group III: includes concrete having cinder, sandstone or granite aggregate tied with No. 5 gage steel wire, wound spirally over the column section on a pitch of 8 inches, or equivalent ties, and concrete having siliceous aggregates containing a combined total of 60 percent or more of quartz, chert and flint, if held in place with wire mesh or expanded metal having not larger than 4-inch mesh, weighing not less than 1.7 lbs./yd.^{2}, placed not more than 1 inch from the surface of the concrete.

   Group IV: includes concrete having siliceous aggregates containing a combined total of 60 percent or more of quartz, chert and flint, and tied with No. 5 gage steel wire wound spirally over the column section on a pitch of 8 inches, or equivalent ties.
3. Groupings of aggregates and ties are the same as for structural steel columns protected solidly with concrete, the ties to be placed over the vertical reinforcing bars and the mesh where required, to be placed within 1 inch from the surface of the column.

   Column A: working loads are assumed as carried by the area of the column inside of the lines circumscribing the reinforcing steel.

   Column B: working loads are assumed as carried by the gross area of the column.

FIGURE 2.1.5—REINFORCED CONCRETE COLUMNS
MINIMUM DIMENSION 16″ TO LESS THAN 18″

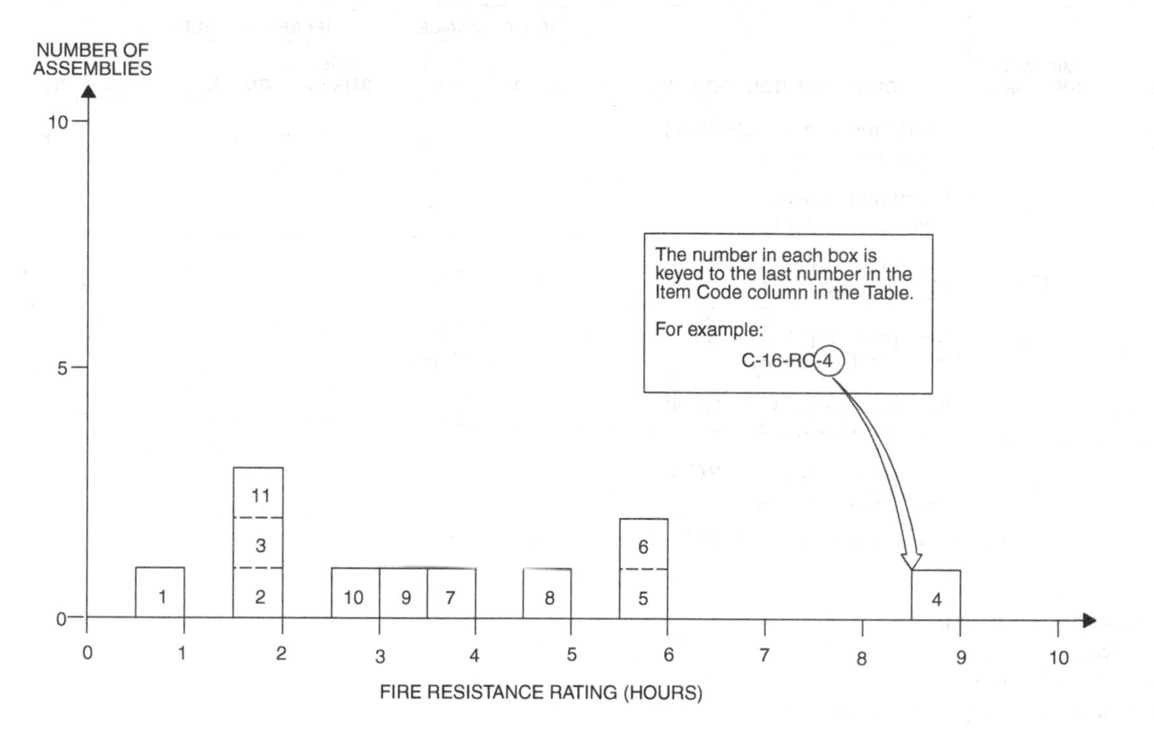

TABLE 2.1.5—REINFORCED CONCRETE COLUMNS
MINIMUM DIMENSION 16″ TO LESS THAN 18″

ITEM CODE	MINIMUM DIMENSION	CONSTRUCTION DETAILS	PERFORMANCE		REFERENCE NUMBER			NOTES	REC. HOURS
			LOAD	TIME	PRE-BMS-92	BMS-92	POST-BMS-92		
C-16-RC-1	16″	16″ square columns; gravel aggregate concrete (4550 psi); Reinforcement: vertical, eight $1^3/_8$″ rebars; horizontal, $^5/_{16}$″ ties at 6″ pitch $1^3/_8$″ below column surface and $^5/_{16}$″ ties at 6″ pitch linking center rebars of each face forming a smaller square in column cross section.	237 tons	1 hr			7	1, 2, 3	1
C-16-RC-2	16″	16″ square columns; gravel aggregate concrete (3360 psi); Reinforcement: vertical, eight $1^3/_8$″ rebars; horizontal, $^5/_{16}$″ ties at 6″ pitch; Cover: $1^3/_8$″.	210 tons	2 hrs.			7	2, 4, 5, 6	2
C-16-RC-3	16″	16″ square columns; gravel aggregate concrete (3980 psi); Reinforcement: vertical, four $^7/_8$″ rebars; horizontal, $^3/_8$″ ties at 6″ pitch; Cover: 1″.	123.5 tons	2 hrs.			7	2, 4, 7	2
C-16-RC-4	16″	Reinforced concrete columns with $1^1/_2$″ concrete outside reinforcing steel; Gross diameter or side of column: 16″; Group I, Column A.	—	9 hrs.		1		8, 9	9
C-16-RC-5	16″	Description as per C-16-RC-4; Group I, Column B.	—	6 hrs.		1		8, 9	6

(Continued)

TABLE 2.1.5—REINFORCED CONCRETE COLUMNS MINIMUM DIMENSION 16″ TO LESS THAN 18″—continued

ITEM CODE	MINIMUM DIMENSION	CONSTRUCTION DETAILS	PERFORMANCE		REFERENCE NUMBER			NOTES	REC. HOURS
			LOAD	TIME	PRE-BMS-92	BMS-92	POST-BMS-92		
C-16-RC-6	16″	Description as per C-16-RC-4; Group II, Column A.	—	6 hrs.		1		8, 9	6
C-16-RC-7	16″	Description as per C-16-RC-4; Group II, Column B.	—	4 hrs.		1		8, 9	4
C-16-RC-8	16″	Description as per C-16-RC-4; Group III, Column A.	—	5 hrs.		1		8, 9	5
C-16-RC-9	16″	Description as per C-16-RC-4; Group III, Column B.	—	3 hrs. 30 min.		1		8, 9	$3^1/_2$
C-16-RC-10	16″	Description as per C-16-RC-4; Group IV, Column A.	—	3 hrs.		1		8, 9	3
C-16-RC-11	16″	Description as per C-16-RC-4; Group IV, Column B.	—	2 hrs.		1		8, 9	2

For SI: 1 inch = 25.4 mm, 1 pound per square inch = 0.00689 MPa, 1 pound per square yard = 5.3 N/m^2.

Notes:

1. Column passed 1-hour fire test.
2. Column passed hose stream test.
3. No reload specified.
4. Column passed 2-hour fire test.
5. Column reloaded successfully after 24 hours.
6. Reinforcing details same as C-16-RC-1.
7. Column passed reload after 72 hours.
8. Group I: includes concrete having calcareous aggregate containing a combined total of not more than 10 percent of quartz, chert and flint for the coarse aggregate.

   Group II: includes concrete having trap-rock aggregate applied without metal ties and also concrete having cinder, sandstone or granite aggregate, if held in place with wire mesh or expanded metal having not larger than 4-inch mesh, weighing not less than 1.7 lbs./yd.^{2}, placed not more than 1 inch from the surface of the concrete.

   Group III: includes concrete having cinder, sandstone or granite aggregate tied with No. 5 gage steel wire, wound spirally over the column section on a pitch of 8 inches, or equivalent ties, and concrete having siliceous aggregates containing a combined total of 60 percent or more of quartz, chert and flint, if held in place with wire mesh or expanded metal having not larger than 4-inch mesh, weighing not less than 1.7 lbs./yd.^{2}, placed not more than 1 inch from the surface of the concrete.

   Group IV: includes concrete having siliceous aggregates containing a combined total of 60 percent or more of quartz, chert and flint, and tied with No. 5 gage steel wire wound spirally over the column section on a pitch of 8 inches, or equivalent ties.
9. Groupings of aggregates and ties are the same as for structural steel columns protected solidly with concrete, the ties to be placed over the vertical reinforcing bars and the mesh where required, to be placed within 1 inch from the surface of the column.

   Column A: working loads are assumed as carried by the area of the column inside of the lines circumscribing the reinforcing steel.

   Column B: working loads are assumed as carried by the gross area of the column.

TABLE 2.1.6—REINFORCED CONCRETE COLUMNS MINIMUM DIMENSION 18″ TO LESS THAN 20″

ITEM CODE	MINIMUM DIMENSION	CONSTRUCTION DETAILS	PERFORMANCE		REFERENCE NUMBER			NOTES	REC. HOURS
			LOAD	TIME	PRE-BMS-92	BMS-92	POST-BMS-92		
C-18-RC-1	18″	Reinforced concrete columns with $1^1/_2$″ concrete outside reinforced steel; Gross diameter or side of column: 18″; Group I, Column A.	—	11 hrs.		1		1, 2	11
C-18-RC-2	18″	Description as per C-18-RC-1; Group I, Column B.	—	8 hrs.		1		1, 2	8
C-18-RC-3	18″	Description as per C-18-RC-1; Group II, Column A.	—	7 hrs.		1		1, 2	7
C-18-RC-4	18″	Description as per C-18-RC-1; Group II, Column B.	—	5 hrs.		1		1, 2	5
C-18-RC-5	18″	Description as per C-18-RC-1; Group III, Column A.	—	6 hrs.		1		1, 2	6
C-18-RC-6	18″	Description as per C-18-RC-1; Group III, Column B.	—	4 hrs.		1		1, 2	4
C-18-RC-7	18″	Description as per C-18-RC-1; Group IV, Column A.	—	3 hrs. 30 min.		1		1, 2	$3^1/_2$
C-18-RC-8	18″	Description as per C-18-RC-1; Group IV, Column B.	—	2 hrs. 30 min.		1		1, 2	$2^1/_2$

For SI: 1 inch = 25.4 mm, 1 pound per square yard = 5.3 N/m^2.

Notes:

1. Group I: includes concrete having calcareous aggregate containing a combined total of not more than 10 percent of quartz, chert and flint for the coarse aggregate.

   Group II: includes concrete having trap-rock aggregate applied without metal ties and also concrete having cinder, sandstone or granite aggregate, if held in place with wire mesh or expanded metal having not larger than 4-inch mesh, weighing not less than 1.7 lbs./yd.^{2}, placed not more than 1 inch from the surface of the concrete.

   Group III: includes concrete having cinder, sandstone or granite aggregate tied with No. 5 gage steel wire, wound spirally over the column section on a pitch of 8 inches, or equivalent ties, and concrete having siliceous aggregates containing a combined total of 60 percent or more of quartz, chert and flint, if held in place with wire mesh or expanded metal having not larger than 4-inch mesh, weighing not less than 1.7 lbs./yd.^{2}, placed not more than 1 inch from the surface of the concrete.

   Group IV: includes concrete having siliceous aggregates containing a combined total of 60 percent or more of quartz, chert and flint and, tied with No. 5 gage steel wire wound spirally over the column section on a pitch of 8 inches, or equivalent ties.

2. Groupings of aggregates and ties are the same as for structural steel columns protected solidly with concrete, the ties to be placed over the vertical reinforcing bars and the mesh where required, to be placed within 1 inch from the surface of the column.

   Column A: working loads are assumed as carried by the area of the column inside of the lines circumscribing the reinforcing steel.

   Column B: working loads are assumed as carried by the gross area of the column.

FIGURE 2.1.7—REINFORCED CONCRETE COLUMNS MINIMUM DIMENSION 20″ TO LESS THAN 22″

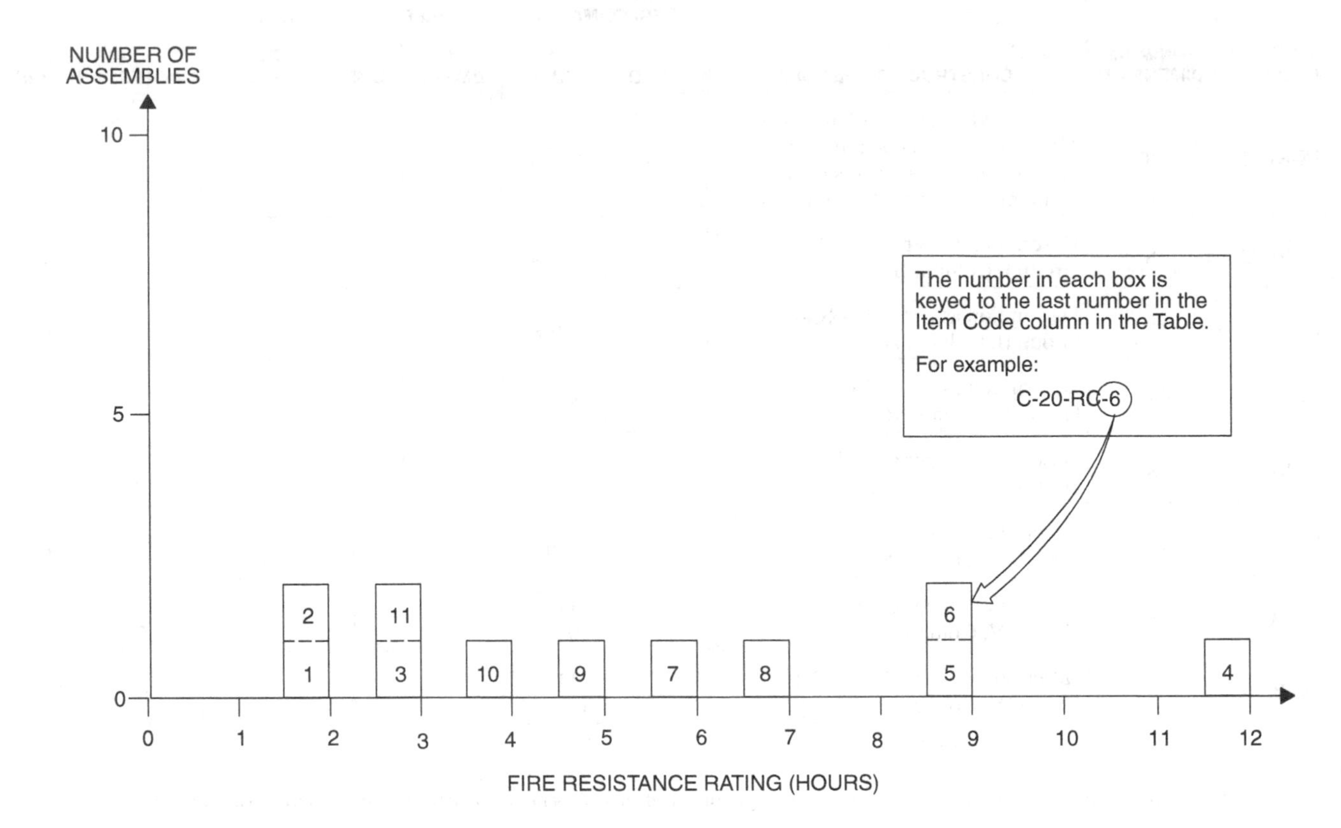

TABLE 2.1.7—REINFORCED CONCRETE COLUMNS MINIMUM DIMENSION 20″ TO LESS THAN 22″

ITEM CODE	MINIMUM DIMENSION	CONSTRUCTION DETAILS	PERFORMANCE		REFERENCE NUMBER			NOTES	REC. HOURS
			LOAD	TIME	PRE-BMS-92	BMS-92	POST-BMS-92		
C-20-RC-1	20″	20″ square columns; gravel aggregate concrete (6690 psi); Reinforcement: vertical, four $1^3/_4$″ rebars; horizontal, $^3/_8$″ wire at 6″ pitch; Cover $1^3/_4$″.	367 tons	2 hrs.			7	1, 2, 3	2
C-20-RC-2	20″	20″ square columns; gravel aggregate concrete (4330 psi); Reinforcement: vertical, four $1^3/_4$″ rebars; horizontal, $^3/_8$″ ties at 6″ pitch; Cover $1^3/_4$″.	327 tons	2 hrs.			7	1, 2, 4	2
C-20-RC-3	$20^1/_4$″	20″ square columns; gravel aggregate concrete (4230 psi); Reinforcement: vertical, four $1^1/_8$″ rebars; horizontal, $^3/_8$″ wire at 5″ pitch; Cover $1^1/_8$″.	199 tons	2 hrs. 56 min.			7	5	$2^3/_4$
C-20-RC-4	20″	Reinforced concrete columns with $1^1/_2$″ concrete outside of reinforcing steel; Gross diameter or side of column: 20″; Group I, Column A.	—	12 hrs.		1		6, 7	12
C-20-RC-5	20″	Description as per C-20-RC-4; Group I, Column B.	—	9 hrs.		1		6, 7	9
C-20-RC-6	20″	Description as per C-20-RC-4; Group II, Column A.	—	9 hrs.		1		6, 7	9

(Continued)

TABLE 2.1.7—REINFORCED CONCRETE COLUMNS MINIMUM DIMENSION 20″ TO LESS THAN 22″—continued

ITEM CODE	MINIMUM DIMENSION	CONSTRUCTION DETAILS	PERFORMANCE		REFERENCE NUMBER			NOTES	REC. HOURS
			LOAD	TIME	PRE-BMS-92	BMS-92	POST-BMS-92		
C-20-RC-7	20″	Description as per C-20-RC-4; Group II, Column B.	—	6 hrs		1		6, 7	6
C-20-RC-8	20″	Description as per C-20-RC-4; Group III, Column A.	—	7 hrs.		1		6, 7	7
C-20-RC-9	20″	Description as per C-20-RC-4; Group III, Column B.	—	5 hrs.		1		6, 7	5
C-20-RC-10	20″	Description as per C-20-RC-4; Group IV, Column A.	—	4 hrs.		1		6, 7	4
C-20-RC-11	20″	Description as per C-20-RC-4; Group IV, Column B.	—	3 hrs.		1		6, 7	3

For SI: 1 inch = 25.4 mm, 1 pound per square yard = 5.3 N/m^2, 1 ton = 8.896 kN.

Notes:

1. Passed 2-hour fire test.
2. Passed hose stream test.
3. Failed during reload at 300 tons.
4. Passed reload after 72 hours.
5. Failure mode - collapse.
6. Group I: includes concrete having calcareous aggregate containing a combined total of not more than 10 percent of quartz, chert and flint for the coarse aggregate.

   Group II: includes concrete having trap-rock aggregate applied without metal ties and also concrete having cinder, sandstone or granite aggregate, if held in place with wire mesh or expanded metal having not larger than 4-inch mesh, weighing not less than 1.7 lbs./yd.^{2}, placed not more than 1 inch from the surface of the concrete.

   Group III: includes concrete having cinder, sandstone or granite aggregate tied with No. 5 gage steel wire, wound spirally over the column section on a pitch of 8 inches, or equivalent ties, and concrete having siliceous aggregates containing a combined total of 60 percent or more of quartz, chert and flint, if held in place with wire mesh or expanded metal having not larger than 4-inch mesh, weighing not less than 1.7 lbs./yd.^{2}, placed not more than 1 inch from the surface of the concrete.

   Group IV: includes concrete having siliceous aggregates containing a combined total of 60 percent or more of quartz, chert and flint, and tied with No. 5 gage steel wire wound spirally over the column section on a pitch of 8 inches, or equivalent ties.
7. Groupings of aggregates and ties are the same as for structural steel columns protected solidly with concrete, the ties to be placed over the vertical reinforcing bars and the mesh where required, to be placed within 1 inch from the surface of the column.

   Column A: working loads are assumed as carried by the area of the column inside of the lines circumscribing the reinforcing steel.

   Column B: working loads are assumed as carried by the gross area of the column.

TABLE 2.1.8—HEXAGONAL REINFORCED CONCRETE COLUMNS MINIMUM DIMENSION 12″ TO LESS THAN 14″

ITEM CODE	MINIMUM DIMENSION	CONSTRUCTION DETAILS	PERFORMANCE		REFERENCE NUMBER			NOTES	REC. HOURS
			LOAD	TIME	PRE-BMS-92	BMS-92	POST-BMS-92		
C-12-HRC-1	12″	12″ hexagonal columns; gravel aggregate concrete (4420 psi); Reinforcement: vertical, eight $^1/_2$″ rebars; horizontal, $^5/_{16}$″ helical winding at 1$^1/_2$″ pitch; Cover: $^1/_2$″.	88 tons	58 min.			7	1	$^3/_4$
C-12-HRC-2	12″	12″ hexagonal columns; gravel aggregate concrete (3460 psi); Reinforcement: vertical, eight $^1/_2$″ rebars; horizontal, $^5/_{16}$″ helical winding at 1$^1/_2$″ pitch; Cover: $^1/_2$″.	78.7 tons	1 hr.			7	2	1

For SI: 1 inch = 25.4 mm, 1 pound per square inch = 0.00689 MPa, 1 ton = 8.896 kN.

Notes:

1. Failure mode - collapse.
2. Test stopped at 1 hour.

TABLE 2.1.9—HEXAGONAL REINFORCED CONCRETE COLUMNS MINIMUM DIMENSION 14″ TO LESS THAN 16″

ITEM CODE	MINIMUM DIMENSION	CONSTRUCTION DETAILS	PERFORMANCE		REFERENCE NUMBER			NOTES	REC. HOURS
			LOAD	TIME	PRE-BMS-92	BMS-92	POST-BMS-92		
C-14-HRC-1	14″	14″ hexagonal columns; gravel aggregate concrete (4970 psi); Reinforcement: vertical, eight $^{1}/_{2}$″ rebars; horizontal, $^{5}/_{16}$″ helical winding on 2″ pitch; Cover: $^{1}/_{2}$″.	90 tons	2 hrs.			7	1, 2, 3	2

For SI: 1 inch = 25.4 mm, 1 pound per square inch = 0.00689 MPa, 1 ton = 8.896 kN.

Notes:

1. Withstood 2-hour fire test.
2. Withstood hose stream test.
3. Withstood reload after 48 hours.

TABLE 2.1.10—HEXAGONAL REINFORCED CONCRETE COLUMNS DIAMETER — 16″ TO LESS THAN 18″

ITEM CODE	MINIMUM DIMENSION	CONSTRUCTION DETAILS	PERFORMANCE		REFERENCE NUMBER			NOTES	REC. HOURS
			LOAD	TIME	PRE-BMS-92	BMS-92	POST-BMS-92		
C-16-HRC-1	16″	16″ hexagonal columns; gravel concrete (6320 psi); Reinforcement: vertical, eight $^{5}/_{8}$″ rebars; horizontal, $^{5}/_{16}$″ helical winding on $^{3}/_{4}$″ pitch; Cover: $^{1}/_{2}$″.	140 tons	1 hr. 55 min.			7	1	$1^{3}/_{4}$
C-16-HRC-2	16″	16″ hexagonal columns; gravel aggregate concrete (5580 psi); Reinforcement: vertical, eight $^{5}/_{8}$″ rebars; horizontal, $^{5}/_{16}$″ helical winding on $1^{3}/_{4}$″ pitch; Cover: $^{1}/_{2}$″	124 tons	2 hrs.			7	2	2

For SI: 1 inch = 25.4 mm, 1 pound per square inch = 0.00689 MPa, 1 ton = 8.896 kN.

Notes:

1. Failure mode - collapse.
2. Failed on furnace removal.

TABLE 2.1.11—HEXAGONAL REINFORCED CONCRETE COLUMNS DIAMETER — 20″ TO LESS THAN 22″

ITEM CODE	MINIMUM DIMENSION	CONSTRUCTION DETAILS	PERFORMANCE		REFERENCE NUMBER			NOTES	REC. HOURS
			LOAD	TIME	PRE-BMS-92	BMS-92	POST-BMS-92		
C-20-HRC-1	20″	20″ hexagonal columns; gravel concrete (6080 psi); Reinforcement: vertical, $^{3}/_{4}$″ rebars; horizontal, $^{5}/_{6}$″ helical winding on $1^{3}/_{4}$″ pitch; Cover: $^{1}/_{2}$″.	211 tons	2 hrs.			7	1	2
C-20-HRC-2	20″	20″ hexagonal columns; gravel concrete (5080 psi); Reinforcement: vertical, $^{3}/_{4}$″ rebars; horizontal, $^{5}/_{16}$″ wire on $1^{3}/_{4}$″ pitch; Cover: $^{1}/_{2}$″.	184 tons	2 hrs. 15 min.			7	2, 3, 4	$2^{1}/_{4}$

For SI: 1 inch = 25.4 mm, 1 pound per square inch = 0.00689 MPa, 1 ton = 8.896 kN.

Notes:

1. Column collapsed on furnace removal.
2. Passed $2^{1}/_{4}$-hour fire test.
3. Passed hose stream test.
4. Withstood reload after 48 hours.

TABLE 2.2—ROUND CAST IRON COLUMNS

ITEM CODE	MINIMUM DIMENSION	CONSTRUCTION DETAILS	PERFORMANCE		REFERENCE NUMBER			NOTES	REC. HOURS
			LOAD	TIME	PRE-BMS-92	BMS-92	POST-BMS-92		
C-7-CI-1	7″ O.D.	Column: .6″ minimum metal thickness; unprotected.	—	30 min.		1			$^{1}/_{2}$
C-7-CI-2	7″ O.D.	Column: .6″ minimum metal thickness concrete filled, outside unprotected.	—	45 min.		1			$^{3}/_{4}$
C-11-CI-3	11″ O.D.	Column: .6″ minimum metal thickness; Protection: 1$^{1}/_{2}$″ portland cement plaster on high ribbed metal lath, $^{1}/_{2}$″ broken air space.	—	3 hrs.		1			3
C-11-CI-4	11″ O.D.	Column: .6″ minimum metal thickness; Protection: 2″ concrete other than siliceous aggregate.	—	2 hrs. 30 min.		1			2$^{1}/_{2}$
C-12-CI-5	12.5″ O.D.	Column: 7″ O.D. .6″ minimum metal thickness; Protection: 2″ porous hollow tile, $^{3}/_{4}$″ mortar between tile and column, outside wire ties.	—	3 hrs.		1			3
C-7-CI-6	7.6″ O.D.	Column: 7″ I.D., $^{3}/_{10}$″ minimum metal thickness, concrete filled unprotected.	—	30 min.		1			$^{1}/_{2}$
C-8-CI-7	8.6″ O.D.	Column: 8″ I.D., $^{3}/_{10}$″ minimum metal thickness; concrete filled reinforced with four 3$^{1}/_{2}$″ × $^{3}/_{8}$″ angles, in fill; unprotected outside.	—	1 hr.		1			1

For SI: 1 inch = 25.4 mm.

FIGURE 2.3—STEEL COLUMNS—GYPSUM ENCASEMENTS

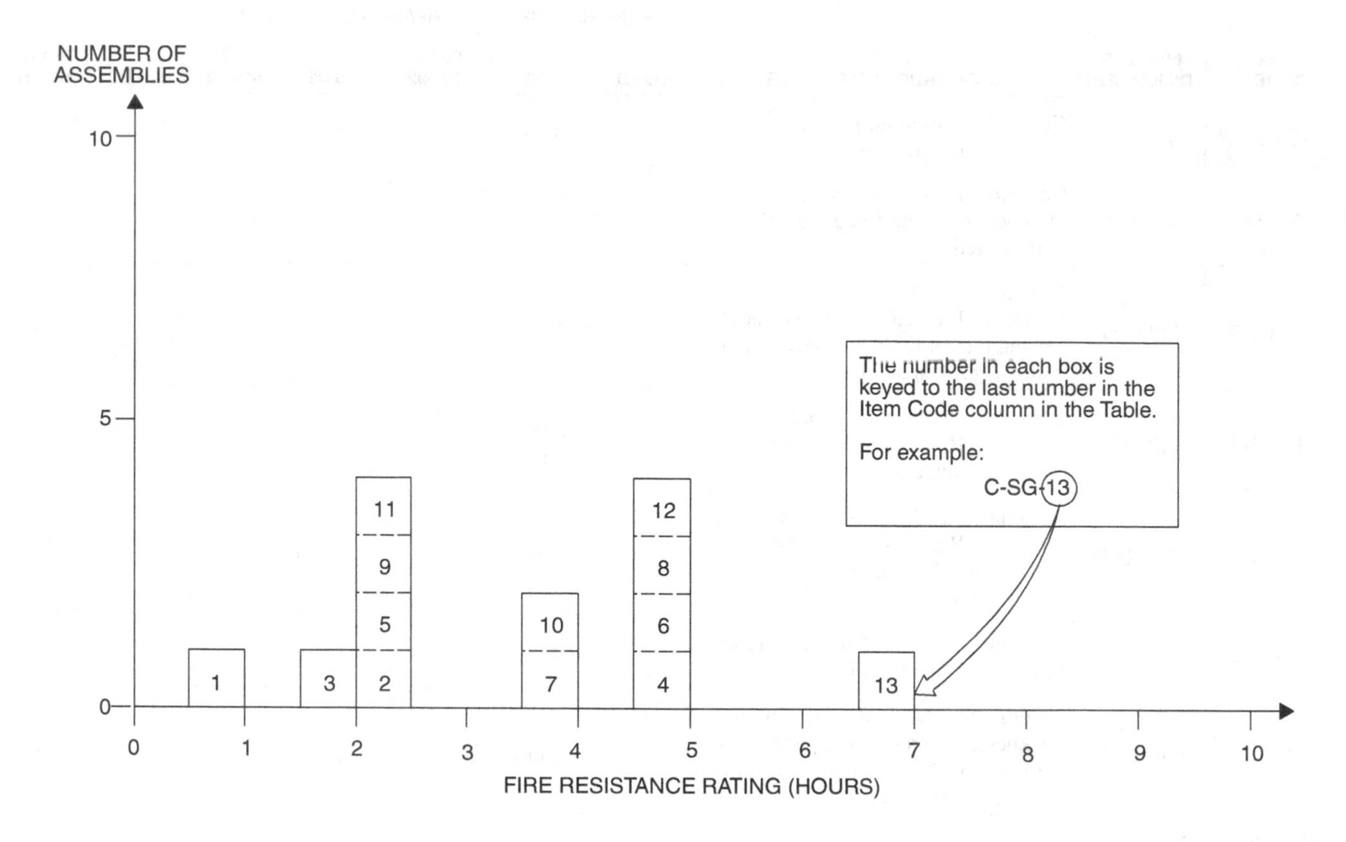

TABLE 2.3—STEEL COLUMNS—GYPSUM ENCASEMENTS

ITEM CODE	MINIMUM AREA OF SOLID MATERIAL	CONSTRUCTION DETAILS	PERFORMANCE		REFERENCE NUMBER			NOTES	REC. HOURS
			LOAD	TIME	PRE-BMS-92	BMS-92	POST-BMS-92		
C-SG-1	—	Steel protected with $^3/_4''$ 1:3 sanded gypsum or 1″ 1:2$^1/_2$ portland cement plaster on wire or lath; one layer.	—	1 hr.		1			1
C-SG-2	—	Same as C-SG-1; two layers.	—	2 hrs. 30 min.		1			2$^1/_2$
C-SG-3	130 in.^{2}	2″ solid blocks with wire mesh in horizontal joints; 1″ mortar on flange; reentrant space filled with block and mortar.	—	2 hrs.		1			2
C-SG-4	150 in.^{2}	Same as C-130-SG-3 with $^1/_2''$ sanded gypsum plaster.	—	5 hrs.		1			5
C-SG-5	130 in.^{2}	2″ solid blocks with wire mesh in horizontal joints; 1″ mortar on flange; reentrant space filled with gypsum concrete.	—	2 hrs. 30 min.		1			2$^1/_2$
C-SG-6	150 in.^{2}	Same as C-130-SG-5 with $^1/_2''$ sanded gypsum plaster.	—	5 hrs.		1			5
C-SG-7	300 in.^{2}	4″ solid blocks with wire mesh in horizontal joints; 1″ mortar on flange; reentrant space filled with block and mortar.	—	4 hrs.		1			4
C-SG-8	300 in.^{2}	Same as C-300-SG-7 with reentrant space filled with gypsum concrete.	—	5 hrs.		1			5

TABLE 2.3—STEEL COLUMNS—GYPSUM ENCASEMENTS—continued

ITEM CODE	MINIMUM AREA OF SOLID MATERIAL	CONSTRUCTION DETAILS	PERFORMANCE		REFERENCE NUMBER			NOTES	REC. HOURS
			LOAD	TIME	PRE-BMS-92	BMS-92	POST-BMS-92		
C-SG-9	85 in.^{2}	2″ solid blocks with cramps at horizontal joints; mortar on flange only at horizontal joints; reentrant space not filled.	—	2 hrs. 30 min.		1			$2^1/_2$
C-SG-10	105 in.^{2}	Same as C-85-SG-9 with $^1/_2$″ sanded gypsum plaster.	—	4 hrs.		1			4
C-SG-11	95 in.^{2}	3″ hollow blocks with cramps at horizontal joints; mortar on flange only at horizontal joints; reentrant space not filled.	—	2 hrs. 30 min.		1			$2^1/_2$
C-SG-12	120 in.^{2}	Same as C-95-SG-11 with $^1/_2$″ sanded gypsum plaster.	—	5 hrs.		1			5
C-SG-13	130 in.^{2}	2″ neat fibered gypsum reentrant space filled poured solid and reinforced with 4″ × 4″ wire mesh $^1/_2$″ sanded gypsum plaster.	—	7 hrs.		1			7

For SI: 1 inch = 25.4 mm, 1 square inch = 645 mm^2.

TABLE 2.4—TIMBER COLUMNS MINIMUM DIMENSION

ITEM CODE	MINIMUM DIMENSION	CONSTRUCTION DETAILS	PERFORMANCE		REFERENCE NUMBER			NOTES	REC. HOURS
			LOAD	TIME	PRE-BMS-92	BMS-92	POST-BMS-92		
C-11-TC-1	11″	With unprotected steel plate cap.	—	30 min.		1		1, 2	$^1/_2$
C-11-TC-2	11″	With unprotected cast iron cap and pintle.	—	45 min.		1		1, 2	$^3/_4$
C-11-TC-3	11″	With concrete or protected steel or cast iron cap.	—	1 hr. 15 min.		1		1, 2	$1^1/_4$
C-11-TC-4	11″	With $^3/_8$″ gypsum wallboard over column and over cast iron or steel cap.	—	1 hr. 15 min.		1		1, 2	$1^1/_4$
C-11-TC-5	11″	With 1″ portland cement plaster on wire lath over column and over cast iron or steel cap; $^3/_4$″ air space.	—	2 hrs.		1		1, 2	2

For SI: 1 inch = 25.4 mm, 1 square inch = 645 mm^2.

Notes:

1. Minimum area: 120 square inches.
2. Type of wood: long leaf pine or Douglas fir.

TABLE 2.5.1.1—STEEL COLUMNS—CONCRETE ENCASEMENTS MINIMUM DIMENSION LESS THAN 6″

ITEM CODE	MINIMUM DIMENSION	CONSTRUCTION DETAILS	PERFORMANCE		REFERENCE NUMBER			NOTES	REC. HOURS
			LOAD	TIME	PRE-BMS-92	BMS-92	POST-BMS-92		
C-5-SC-1	5″	5″ × 6″ outer dimensions; 4″ × 3″ × 10 lbs. "H" beam; Protection: gravel concrete (4900 psi) 6″ × 4″ - 13 SWG mesh.	12 tons	1 hr. 29 min.			7	1	$1^1/_4$

For SI: 1 inch = 25.4 mm, 1 pound per square inch = 0.00689 MPa, 1 ton = 8.896 kN.

Notes:

1. Failure mode - collapse.

TABLE 2.5.1.2—STEEL COLUMNS—CONCRETE ENCASEMENTS 6″ TO LESS THAN 8″ THICK

ITEM CODE	MINIMUM DIMENSION	CONSTRUCTION DETAILS	PERFORMANCE		REFERENCE NUMBER			NOTES	REC. HOURS
			LOAD	TIME	PRE-BMS-92	BMS-92	POST-BMS-92		
C-7-SC-1	7″	7″ × 8″ column; 4″ × 3″ × 10 lbs. "H" beam; Protection: brick filled concrete (6220 psi); 6″ × 4″ mesh - 13 SWG; 1″ below column surface.	12 tons	2 hrs. 46 min.			7	1	$2^3/_4$
C-7-SC-2	7″	7″ × 8″ column; 4″ × 3″ × 10 lbs. "H" beam; Protection: gravel concrete (5140 psi); 6″ × 4″ 13 SWG mesh 1″ below surface.	12 tons	3 hrs. 1 min.			7	1	3
C-7-SC-3	7″	7″ × 8″ column; 4″ × 3″ × 10 lbs. "H" beam; Protection: concrete (4540 psi); 6″ × 4″ - 13 SWG mesh; 1″ below column surface.	12 tons	3 hrs. 9 min.			7	1	3
C-7-SC-4	7″	7″ × 8″ column; 4″ × 3″ × 10 lbs. "H" beam; Protection: gravel concrete (5520 psi); 4″ × 4″ mesh; 16 SWG.	12 tons	2 hrs. 50 min.			7	1	$2^3/_4$

For SI: 1 inch = 25.4 mm, 1 pound per square inch = 0.00689 MPa, 1 ton = 8.896 kN.

Notes:

1. Failure mode - collapse.

FIGURE 2.5.1.3—STEEL COLUMNS—CONCRETE ENCASEMENTS MINIMUM DIMENSION 8″ TO LESS THAN 10″

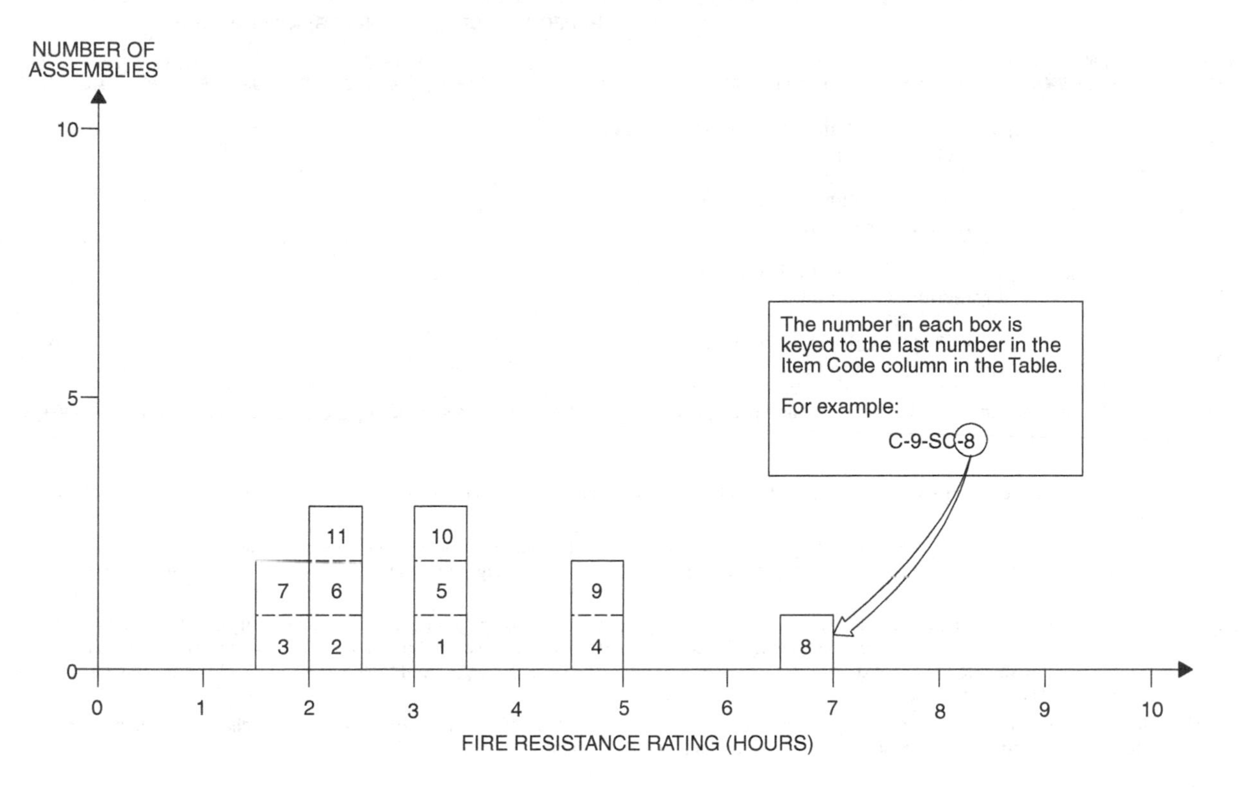

TABLE 2.5.1.3—STEEL COLUMNS—CONCRETE ENCASEMENTS MINIMUM DIMENSION 8″ TO LESS THAN 10″

ITEM CODE	MINIMUM DIMENSION	CONSTRUCTION DETAILS	PERFORMANCE		REFERENCE NUMBER			NOTES	REC. HOURS
			LOAD	TIME	PRE-BMS-92	BMS-92	POST-BMS-92		
C-8-SC-1	$8^1/_2$″	$8^1/_2$″ × 10″ column; 6″ × $4^1/_2$″ × 20 lbs. "H" beam; Protection: gravel concrete (5140 psi); 6″ × 4″ - 13 SWG mesh.	39 tons	3 hrs. 8 min.			7	1	3
C-8-SC-2	8″	8″ × 10″ column; 8″ × 6″ × 35 lbs. "I" beam; Protection: gravel concrete (4240 psi); 6″ × 4″ - 13 SWG mesh; $^1/_2$″ cover.	90 tons	2 hrs. 1 min.			7	1	2
C-8-SC-3	8″	8″ × 10″ concrete encased column; 8″ × 6″ × 35 lbs. "H" beam; protection: aggregate concrete (3750 psi); 4″ mesh - 16 SWG reinforcing $^1/_2$″ below column surface.	90 tons	1 hr. 58 min.			7	1	$1^3/_4$
C-8-SC-4	8″	6″ × 6″ steel column; 2″ outside protection; Group I.	—	5 hrs.		1		2	5
C-8-SC-5	8″	6″ × 6″ steel column; 2″ outside protection; Group II.	—	3 hrs. 30 min.		1		2	$3^1/_2$
C-8-SC-6	8″	6″ × 6″ steel column; 2″ outside protection; Group III.	—	2 hrs. 30 min.		1		2	$2^1/_2$
C-8-SC-7	8″	6″ × 6″ steel column; 2″ outside protection; Group IV.	—	1 hr. 45 min.		1		2	$1^3/_4$

(Continued)

TABLE 2.5.1.3—STEEL COLUMNS—CONCRETE ENCASEMENTS MINIMUM DIMENSION 8″ TO LESS THAN 10″—continued

ITEM CODE	MINIMUM DIMENSION	CONSTRUCTION DETAILS	PERFORMANCE		REFERENCE NUMBER			NOTES	REC. HOURS
			LOAD	TIME	PRE-BMS-92	BMS-92	POST-BMS-92		
C-9-SC-8	9″	6″ × 6″ steel column; 3″ outside protection; Group I.	—	7 hrs.		1		2	7
C-9-SC-9	9″	6″ × 6″ steel column; 3″ outside protection; Group II.	—	5 hrs.		1		2	5
C-9-SC-10	9″	6″ × 6″ steel column; 3″ outside protection; Group III.	—	3 hrs. 30 min.		1		2	$3^1/_2$
C-9-SC-11	9″	6″ × 6″ steel column; 3″ outside protection; Group IV.	—	2 hrs. 30 min.		1		2	$2^1/_2$

For SI: 1 inch = 25.4 mm, 1 pound = 0.004448 kN, 1 pound per square inch = 0.00689 MPa, 1 pound per square yard = 5.3 N/m^2, 1 ton = 8.896 kN.

Notes:

1. Failure mode - collapse.
2. Group I: includes concrete having calcareous aggregate containing a combined total of not more than 10 percent of quartz, chert and flint for the coarse aggregate.

   Group II: includes concrete having trap-rock aggregate applied without metal ties and also concrete having cinder, sandstone or granite aggregate, if held in place with wire mesh or expanded metal having not larger than 4-inch mesh, weighing not less than 1.7 lbs./yd.^{2}, placed not more than 1 inch from the surface of the concrete.

   Group III: includes concrete having cinder, sandstone or granite aggregate tied with No. 5 gage steel wire, wound spirally over the column section on a pitch of 8 inches, or equivalent ties, and concrete having siliceous aggregates containing a combined total of 60 percent or more of quartz, chert and flint, if held in place with wire mesh or expanded metal having not larger than 4-inch mesh, weighing not less than 1.7 lbs./yd.^{2}, placed not more than 1 inch from the surface of the concrete.

   Group IV: includes concrete having siliceous aggregates containing a combined total of 60 percent or more of quartz, chert and flint, and tied with No. 5 gage steel wire wound spirally over the column section on a pitch of 8 inches, or equivalent ties.

FIGURE 2.5.1.4—STEEL COLUMNS—CONCRETE ENCASEMENTS MINIMUM DIMENSION 10″ TO LESS THAN 12″

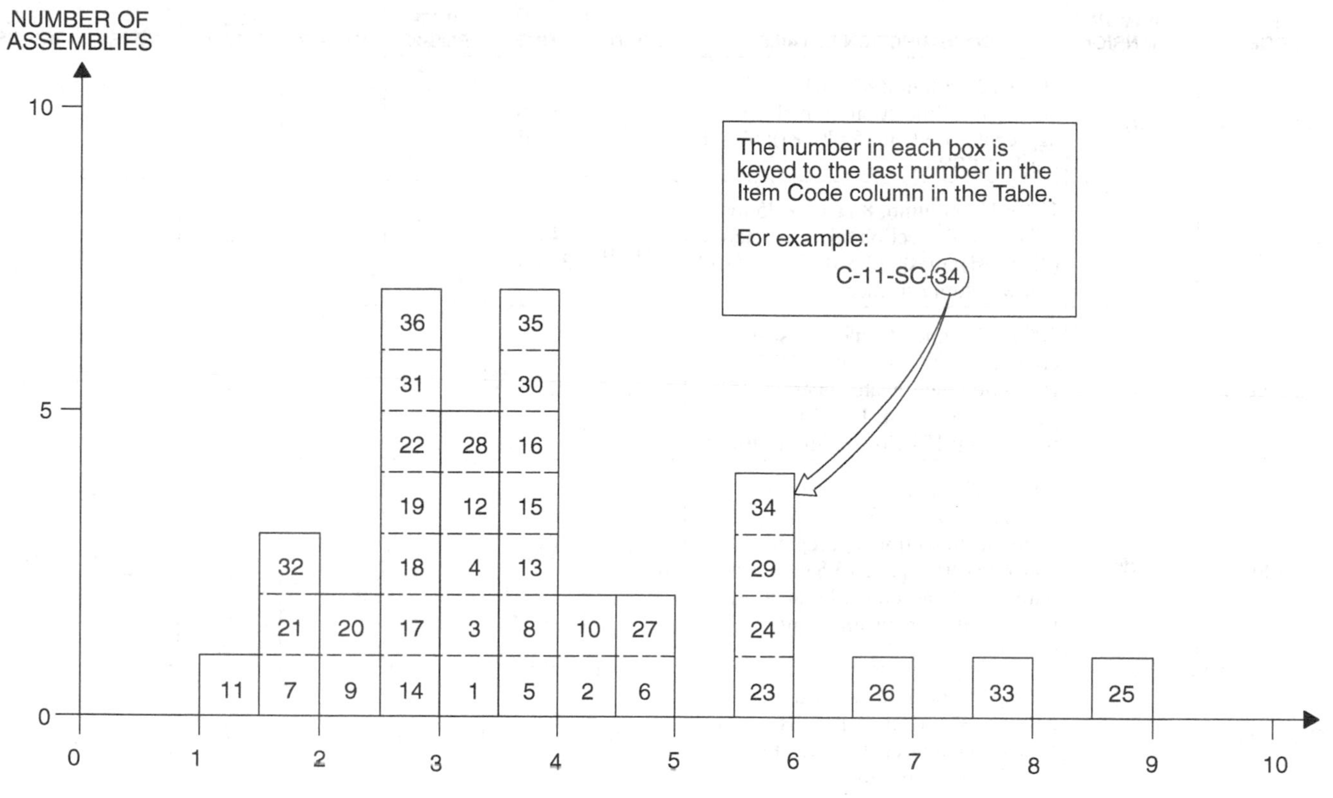

TABLE 2.5.1.4—STEEL COLUMNS—CONCRETE ENCASEMENTS MINIMUM DIMENSION 10″ TO LESS THAN 12″

ITEM CODE	MINIMUM DIMENSION	CONSTRUCTION DETAILS	PERFORMANCE		REFERENCE NUMBER			NOTES	REC. HOURS
			LOAD	TIME	PRE-BMS-92	BMS-92	POST-BMS-92		
C-10-SC-1	10″	10″ × 12″ concrete encased steel column; 8″ × 6″ × 35 lbs. "H" beam; Protection: gravel aggregate concrete (3640 psi); Mesh 6″ × 4″ 13 SWG, 1″ below column surface.	90 tons	3 hrs. 7 min.			7	1,2	3
C-10-SC-2	10″	10″ × 16″ column; 8″ × 6″ × 35 lbs. "H" beam; Protection: clay brick concrete (3630 psi); 6″ × 4″ mesh; 13 SWG, 1″ below column surface.	90 tons	4 hrs. 6 min.			7	2	4
C-10-SC-3	10″	10″ × 12″ column; 8″ × 6″ × 35 lbs. "H" beam; Protection: crushed stone and sand concrete (3930 psi); 6″ × 4″ - 13 SWG mesh; 1″ below column surface.	90 tons	3 hrs. 17 min.			7	2	$3^1/_4$
C-10-SC-4	10″	10″ × 12″ column; 8″ × 6″ × 35 lbs. "H" beam; Protection: crushed basalt and sand concrete (4350 psi); 6″ × 4″ - 13 SWG mesh; 1″ below column surface.	90 tons	3 hrs. 22 min.			7	2	$3^1/_3$

(Continued)

TABLE 2.5.1.4—STEEL COLUMNS—CONCRETE ENCASEMENTS MINIMUM DIMENSION 10″ TO LESS THAN 12″—continued

ITEM CODE	MINIMUM DIMENSION	CONSTRUCTION DETAILS	PERFORMANCE		REFERENCE NUMBER			NOTES	REC. HOURS
			LOAD	TIME	PRE-BMS-92	BMS-92	POST-BMS-92		
C-10-SC-5	10″	10″ × 12″ column; 8″ × 6″ × 35 lbs. "H" beam; Protection: gravel aggregate concrete (5570 psi); 6″ × 4″ mesh; 13 SWG.	90 tons	3 hrs. 39 min.			7	2	$3^1/_2$
C-10-SC-6	10″	10″ × 16″ column; 8″ × 6″ × 35 lbs. "I" beam; Protection: gravel concrete (4950 psi); mesh; 6″ × 4″ 13 SWG 1″ below column surface.	90 tons	4 hrs. 32 min.			7	2	$4^1/_2$
C-10-SC-7	10″	10″ × 12″ concrete encased steel column; 8″ × 6″ × 35 lbs. "H" beam; Protection: aggregate concrete (1370 psi); 6″ × 4″ mesh; 13 SWG reinforcing 1″ below column surface.	90 tons	2 hrs.			7	3, 4	2
C-10-SC-8	10″	10″ × 12″ concrete encased steel column; 8″ × 6″ × 35 lbs. "H" column; Protection: aggregate concrete (4000 psi); 13 SWG iron wire loosely around column at 6″ pitch about 2″ beneath column surface.	86 tons	3 hrs. 36 min.			7	2	$3^1/_2$
C-10-SC-9	10″	10″ × 12″ concrete encased steel column; 8″ × 6″ × 35 lbs. "H" beam; Protection: aggregate concrete (3290 psi); 2″ cover minimum.	86 tons	2 hrs. 8 min.			7	2	2
C-10-SC-10	10″	10″ × 14″ concrete encased steel column; 8″ × 6″ × 35 lbs. "H" column; Protection: crushed brick filled concrete (5310 psi); 6″ × 4″ mesh; 13 SWG reinforcement 1″ below column surface.	90 tons	4 hrs. 28 min.			7	2	$4^1/_3$
C-10-SC-11	10″	10″ × 14″ concrete encased column; 8″ × 6″ 35 lbs. "H" beam; Protection: aggregate concrete (342 psi); 6″ × 4″ mesh; 13 SWG reinforcement 1″ below surface.	90 tons	1 hr. 2 min.			7	2	1
C-10-SC-12	10″	10″ × 12″ concrete encased steel column; 8″ × 6″ × 35 lbs. "H" beam; Protection: aggregate concrete (4480 psi); four $^3/_8$″ vertical bars at "H" beam edges with $^3/_{16}$″ spacers at beam surface at 3′ pitch and $^3/_{16}$″ binders at 10″ pitch; 2″ concrete cover.	90 tons	3 hrs. 2 min.			7	2	3

(Continued)

TABLE 2.5.1.4—STEEL COLUMNS—CONCRETE ENCASEMENTS MINIMUM DIMENSION 10″ TO LESS THAN 12″—continued

ITEM CODE	MINIMUM DIMENSION	CONSTRUCTION DETAILS	PERFORMANCE		REFERENCE NUMBER			NOTES	REC. HOURS
			LOAD	TIME	PRE-BMS-92	BMS-92	POST-BMS-92		
C-10-SC-13	10″	10″ × 12″ concrete encased steel column; 8″ × 6″ × 35 lbs. "H" beam; Protection: aggregate concrete (5070 psi); 6″ × 4″ mesh; 13 SWG reinforcing at 6″ beam sides wrapped and held by wire ties across (open) 8″ beam face; reinforcements wrapped in 6″ × 4″ mesh; 13 SWG throughout; $^1/_2$″ cover to column surface.	90 tons	3 hrs. 59 min.			7	2	$3^3/_4$
C-10-SC-14	10″	10″ × 12″ concrete encased steel column; 8″ × 6″ × 35 lbs. "H" beam; Protection: aggregate concrete (4410 psi); 6″ × 4″ mesh; 13 SWG reinforcement $1^1/_4$″ below column surface; $^1/_2$″ limestone cement plaster with $^3/_8$″ gypsum plaster finish.	90 tons	2 hrs. 50 min.			7	2	$2^3/_4$
C-10-SC-15	10″	10″ × 12″ concrete encased steel column; 8″ × 6″ × 35 lbs. "H" beam; Protection: crushed clay brick filled concrete (4260 psi); 6″ × 4″ mesh; 13 SWG reinforcing 1″ below column surface.	90 tons	3 hrs. 54 min.			7	2	$3^3/_4$
C-10-SC-16	10″	10″ × 12″ concrete encased steel column; 8″ × 6″ × 35 lbs. "H" beam; Protection: limestone aggregate concrete (4350 psi); 6″ × 4″ mesh; 13 SWG reinforcing 1″ below column surface.	90 tons	3 hrs. 54 min.			7	2	$3^3/_4$
C-10-SC-17	10″	10″ × 12″ concrete encased steel column; 8″ × 6″ × 35 lbs. "H" beam; Protection: limestone aggregate concrete (5300 psi); 6″ × 4″; 13 SWG wire mesh 1″ below column surface.	90 tons	3 hrs.			7	4, 5	3
C-10-SC-18	10″	10″ × 12″ concrete encased steel column; 8″ × 6″ × 35 lbs. "H" beam; Protection: limestone aggregate concrete (4800 psi) with 6″ × 4″; 13 SWG mesh reinforcement 1″ below surface.	90 tons	3 hrs.			7	4, 5	3
C-10-SC-19	10″	10″ × 14″ concrete encased steel column; 12″ × 8″ × 65 lbs. "H" beam; Protection: aggregate concrete (3900 psi); 4″ mesh; 16 SWG reinforcing $^1/_2$″ below column surface.	118 tons	2 hrs. 42 min.			7	2	2
C-10-SC-20	10″	10″ × 14″ concrete encased steel column; 12″ × 8″ × 65 lbs. "H" beam; Protection: aggregate concrete (4930 psi); 4″ mesh; 16 SWG reinforcing $^1/_2$″ below column surface.	177 tons	2 hrs. 8 min.			7	2	2

(Continued)

TABLE 2.5.1.4—STEEL COLUMNS—CONCRETE ENCASEMENTS MINIMUM DIMENSION 10″ TO LESS THAN 12″—continued

ITEM CODE	MINIMUM DIMENSION	CONSTRUCTION DETAILS	PERFORMANCE		REFERENCE NUMBER			NOTES	REC. HOURS
			LOAD	TIME	PRE-BMS-92	BMS-92	POST-BMS-92		
C-10-SC-21	$10^3/_8$″	$10^3/_8$″ × $12^3/_8$″ concrete encased steel column; 8″ × 6″ × 35 lbs. "H" beam; Protection: aggregate concrete (835 psi) with 6″ × 4″ mesh; 13 SWG reinforcing $1^3/_{16}$″ below column surface; $^3/_{16}$″ gypsum plaster finish.	90 tons	2 hrs.			7	3, 4	2
C-11-SC-22	11″	11″ × 13″ concrete encased steel column; 8″ × 6″ × 35 lbs. "H" beam; Protection: "open texture" brick filled concrete (890 psi) with 6″ × 4″ mesh; 13 SWG reinforcing $1^1/_2$″ below column surface; $^3/_8$″ lime cement plaster; $^1/_8$″ gypsum plaster finish.	90 tons	3 hrs.			7	6, 7	3
C-11-SC-23	11″	11″ × 12″ column; 4″ × 3″ × 10 lbs. "H" beam; gravel concrete (4550 psi); 6″ × 4″ - 13 SWG mesh reinforcing; 1″ below column surface.	12 tons	6 hrs.			7	7, 8	6
C-11-SC-24	11″	11″ × 12″ column; 4″ × 3″ × 10 lbs. "H" beam; Protection: gravel aggregate concrete (3830 psi); with 4″ × 4″ mesh; 16 SWG, 1″ below column surface.	16 tons	5 hrs. 32 min.			7	2	$5^1/_2$
C-10-SC-25	10″	6″ × 6″ steel column with 4″ outside protection; Group I.	—	9 hrs.		1		9	9
C-10-SC-26	10″	Description as per C-SC-25; Group II.	—	7 hrs.		1		9	7
C-10-SC-27	10″	Description as per C-10-SC-25; Group III.	—	5 hrs.		1		9	5
C-10-SC-28	10″	Description as per C-10-SC-25; Group IV.	—	3 hrs. 30 min.		1		9	$3^1/_2$
C-10-SC-29	10″	8″ × 8″ steel column with 2″ outside protection; Group I.	—	6 hrs.		1		9	6
C-10-SC-30	10″	Description as per C-10-SC-29; Group II.	—	4 hrs.		1		9	4
C-10-SC-31	10″	Description as per C-10-SC-29; Group III.	—	3 hrs.		1		9	3
C-10-SC-32	10″	Description as per C-10-SC-29; Group IV.	—	2 hrs.		1		9	2
C-11-SC-33	11″	8″ × 8″ steel column with 3″ outside protection; Group I.	—	8 hrs.		1		9	8
C-11-SC-34	11″	Description as per C-10-SC-33; Group II.	—	6 hrs.		1		9	6
C-11-SC-35	11″	Description as per C-10-SC-33; Group III.	—	4 hrs.		1		9	4

(Continued)

**TABLE 2.5.1.4—STEEL COLUMNS—CONCRETE ENCASEMENTS
MINIMUM DIMENSION 10″ TO LESS THAN 12″—continued**

ITEM CODE	MINIMUM DIMENSION	CONSTRUCTION DETAILS	PERFORMANCE		REFERENCE NUMBER			NOTES	REC. HOURS
			LOAD	TIME	PRE-BMS-92	BMS-92	POST-BMS-92		
C-11-SC-36	11″	Description as per C-10-SC-33; Group IV.	—	3 hrs.		1		9	3

For SI: 1 inch = 25.4 mm, 1 pound = 0.004448 kN, 1 pound per square inch = 0.00689 MPa, 1 pound per square yard = 5.3 N/m^2, 1 ton = 8.896 kN.

Notes:

1. Tested under total restraint load to prevent expansion - minimum load 90 tons.
2. Failure mode - collapse.
3. Passed 2-hour fire test (Grade “C,” British).
4. Passed hose stream test.
5. Column tested and passed 3-hour grade fire resistance (British).
6. Column passed 3-hour fire test.
7. Column collapsed during hose stream testing.
8. Column passed 6-hour fire test.
9. Group I: includes concrete having calcareous aggregate containing a combined total of not more than 10 percent of quartz, chert and flint for the coarse aggregate.

   Group II: includes concrete having trap-rock aggregate applied without metal ties and also concrete having cinder, sandstone or granite aggregate, if held in place with wire mesh or expanded metal having not larger than 4-inch mesh, weighing not less than 1.7 lbs./yd.^{2}, placed not more than 1 inch from the surface of the concrete.

   Group III: includes concrete having cinder, sandstone or granite aggregate tied with No. 5 gage steel wire, wound spirally over the column section on a pitch of 8 inches, or equivalent ties, and concrete having siliceous aggregates containing a combined total of 60 percent or more of quartz, chert and flint, if held in place with wire mesh or expanded metal having not larger than 4-inch mesh, weighing not less than 1.7 lbs./yd.^{2}, placed not more than 1 inch from the surface of the concrete.

   Group IV: includes concrete having siliceous aggregates containing a combined total of 60 percent or more of quartz, chert and flint, and tied with No. 5 gage steel wire wound spirally over the column section on a pitch of 8 inches, or equivalent ties.

FIGURE 2.5.1.5—STEEL COLUMNS—CONCRETE ENCASEMENTS
MINIMUM DIMENSION 12″ TO LESS THAN 14″

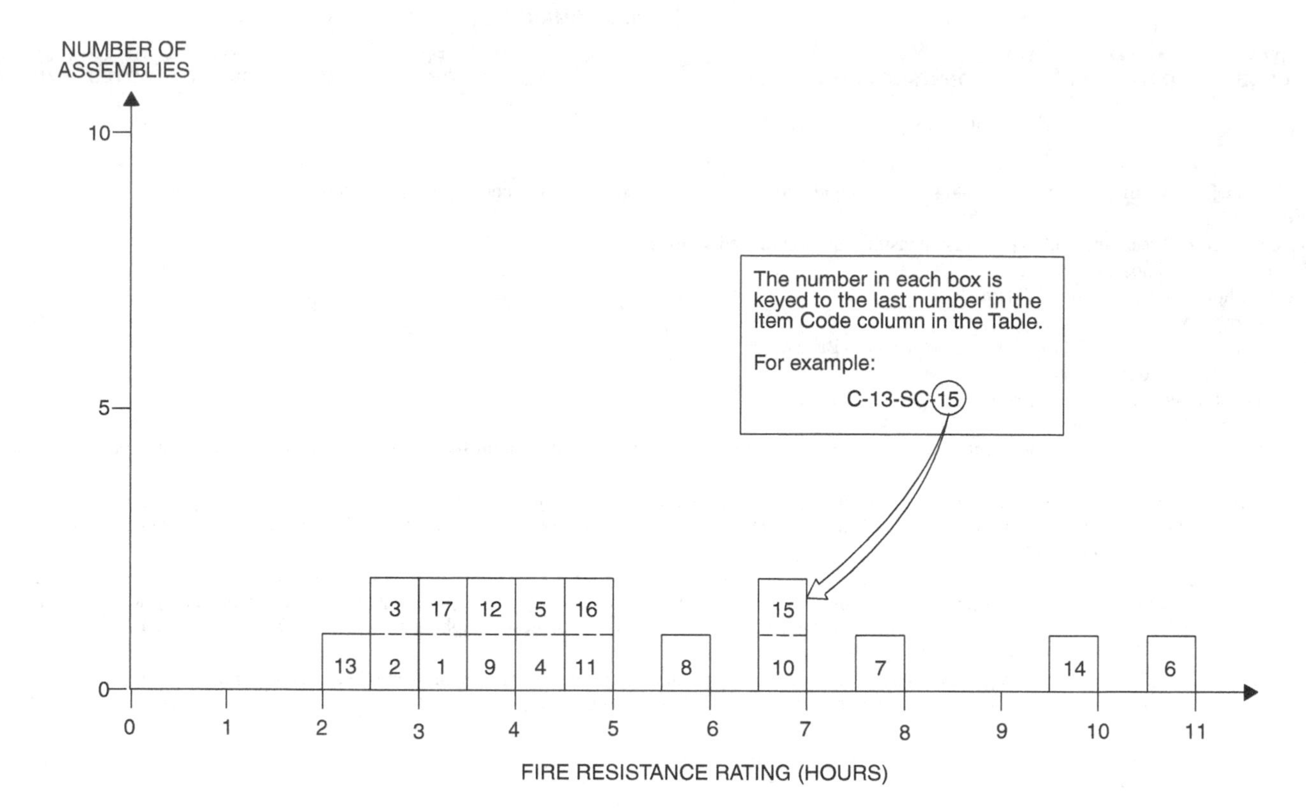

TABLE 2.5.1.5—STEEL COLUMNS—CONCRETE ENCASEMENTS
MINIMUM DIMENSION 12″ TO LESS THAN 14″

ITEM CODE	MINIMUM DIMENSION	CONSTRUCTION DETAILS	PERFORMANCE		REFERENCE NUMBER			NOTES	REC. HOURS
			LOAD	TIME	PRE-BMS-92	BMS-92	POST-BMS-92		
C-12-SC-1	12″	12″ × 14″ concrete encased steel column; 8″ × 6″ × 35 lbs. "H" beam; Protection: aggregate concrete (4150 psi) with 4″ mesh; 16 SWG reinforcing 1″ below column surface.	120 tons	3 hrs. 24 min.			7	1	$3^1/_3$
C-12-SC-2	12″	12″ × 16″ concrete encased column; 8″ × 6″ × 35 lbs. "H" beam; Protection: aggregate concrete (4300 psi) with 4″ mesh; 16 SWG reinforcing 1″ below column surface.	90 tons	2 hrs. 52 min.			7	1	$2^3/_4$
C-12-SC-3	12″	12″ × 16″ concrete encased steel column; 12″ × 8″ × 65 lbs. "H" column; Protection: gravel aggregate concrete (3550 psi) with 4″ mesh; 16 SWG reinforcement 1″ below column surface.	177 tons	2 hrs. 31 min.			7	1	$2^1/_2$
C-12-SC-4	12″	12″ × 16″ concrete encased column; 12″ × 8″ × 65 lbs. "H" beam; Protection: aggregate concrete (3450 psi) with 4″ mesh; 16 SWG reinforcement 1″ below column surface.	118 tons	4 hrs. 4 min.			7	1	4

(Continued)

TABLE 2.5.1.5—STEEL COLUMNS—CONCRETE ENCASEMENTS MINIMUM DIMENSION 12″ TO LESS THAN 14″—continued

ITEM CODE	MINIMUM DIMENSION	CONSTRUCTION DETAILS	PERFORMANCE		REFERENCE NUMBER			NOTES	REC. HOURS
			LOAD	TIME	PRE-BMS-92	BMS-92	POST-BMS-92		
C-12-SC-5	$12^1/_2$″	$12^1/_2$″ × 14″ column; 6″ × $4^1/_2$″ × 20 lbs. “H” beam; Protection: gravel aggregate concrete (3750 psi) with 4″ × 4″ mesh; 16 SWG reinforcing 1″ below column surface.	52 tons	4 hrs. 29 min.			7	1	$4^1/_3$
C-12-SC-6	12″	8″ × 8″ steel column; 2″ outside protection; Group I.	—	11 hrs.			1	2	11
C-12-SC-7	12″	Description as per C-12-SC-6; Group II.	—	8 hrs.		1		2	8
C-12-SC-8	12″	Description as per C-12-SC-6; Group III.	—	6 hrs.		1		2	6
C-12-SC-9	12″	Description as per C-12-SC-6; Group IV.	—	4 hrs.		1		2	4
C-12-SC-10	12″	10″ × 10″ steel column; 2″ outside protection; Group I.	—	7 hrs.		1		2	7
C-12-SC-11	12″	Description as per C-12-SC-10; Group II.	—	5 hrs.		1		2	5
C-12-SC-12	12″	Description as per C-12-SC-10; Group III.	—	4 hrs.		1		2	4
C-12-SC-13	12″	Description as per C-12-SC-10; Group IV.	—	2 hrs. 30 min.		1		2	$2^1/_2$
C-13-SC-14	13″	10″ × 10″ steel column; 3″ outside protection; Group I.	—	10 hrs.		1		2	10
C-13-SC-15	13″	Description as per C-12-SC-14; Group II.	—	7 hrs.		1		2	7
C-13-SC-16	13″	Description as per C-12-SC-14; Group III.	—	5 hrs.		1		2	5
C-13-SC-17	13″	Description as per C-12-SC-14; Group IV.	—	3 hrs. 30 min.		1		2	$3^1/_2$

For SI: 1 inch = 25.4 mm, 1 pound = 0.004448 kN, 1 pound per square inch = 0.00689 MPa, 1 pound per square yard = 5.3 N/m^2, 1 ton = 8.896 kN.

Notes:

1. Failure mode - collapse.
2. Group I: includes concrete having calcareous aggregate containing a combined total of not more than 10 percent of quartz, chert and flint for the coarse aggregate.

   Group II: includes concrete having trap-rock aggregate applied without metal ties and also concrete having cinder, sandstone or granite aggregate, if held in place with wire mesh or expanded metal having not larger than 4-inch mesh, weighing not less than 1.7 lbs./yd.^{2}, placed not more than 1 inch from the surface of the concrete.

   Group III: includes concrete having cinder, sandstone or granite aggregate tied with No. 5 gage steel wire, wound spirally over the column section on a pitch of 8 inches, or equivalent ties, and concrete having siliceous aggregates containing a combined total of 60 percent or more of quartz, chert and flint, if held in place with wire mesh or expanded metal having not larger than 4-inch mesh, weighing not less than 1.7 lbs./yd.^{2}, placed not more than 1 inch from the surface of the concrete.

   Group IV: includes concrete having siliceous aggregates containing a combined total of 60 percent or more of quartz, chert and flint, and tied with No. 5 gage steel wire wound spirally over the column section on a pitch of 8 inches, or equivalent ties.

FIGURE 2.5.1.6—STEEL COLUMNS—CONCRETE ENCASEMENTS MINIMUM DIMENSION 14″ TO LESS THAN 16″

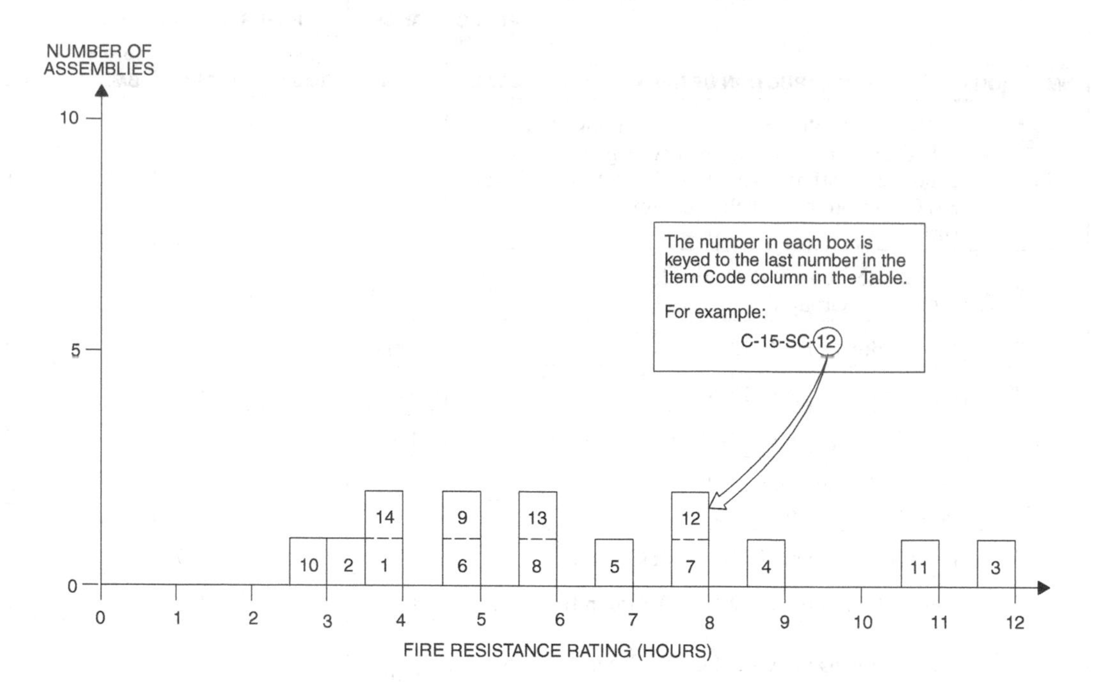

TABLE 2.5.1.6—STEEL COLUMNS—CONCRETE ENCASEMENTS MINIMUM DIMENSION 14″ TO LESS THAN 16″

ITEM CODE	MINIMUM DIMENSION	CONSTRUCTION DETAILS	PERFORMANCE		REFERENCE NUMBER			NOTES	REC. HOURS
			LOAD	TIME	PRE-BMS-92	BMS-92	POST-BMS-92		
C-14-SC-1	14″	24″ × 16″ concrete encased steel column; 8″ × 6″ × 35 lbs. "H" column; Protection: aggregate concrete (4240 psi); 4″ mesh - 16 SWG reinforcing 1″ below column surface.	90 tons	3 hrs. 40 min.			7	1	3
C-14-SC-2	14″	14″ × 18″ concrete encased steel column; 12″ × 8″ × 65 lbs. "H" beam; Protection: gravel aggregate concrete (4000 psi) with 4″ - 16 SWG wire mesh reinforcement 1″ below column surface.	177 tons	3 hrs. 20 min.			7	1	3
C-14-SC-3	14″	10″ × 10″ steel column; 4″ outside protection; Group I.	—	12 hrs.		1		2	12
C-14-SC-4	14″	Description as per C-14-SC-3; Group II.	—	9 hrs.		1		2	9
C-14-SC-5	14″	Description as per C-14-SC-3; Group III.	—	7 hrs.		1		2	7
C-14-SC-6	14″	Description as per C-14-SC-3; Group IV.	—	5 hrs.		1		2	5
C-14-SC-7	14″	12″ × 12″ steel column; 2″ outside protection; Group I.	—	8 hrs.		1		2	8
C-14-SC-8	14″	Description as per C-14-SC-7; Group II.	—	6 hrs.		1		2	6

(Continued)

TABLE 2.5.1.6—STEEL COLUMNS—CONCRETE ENCASEMENTS MINIMUM DIMENSION 14″ TO LESS THAN 16″—continued

ITEM CODE	MINIMUM DIMENSION	CONSTRUCTION DETAILS	PERFORMANCE		REFERENCE NUMBER			NOTES	REC. HOURS
			LOAD	TIME	PRE-BMS-92	BMS-92	POST-BMS-92		
C-14-SC-9	14″	Description as per C-14-SC-7; Group III.	—	5 hrs.		1		2	5
C-14-SC-10	14″	Description as per C-14-SC-7; Group IV	—	3 hrs.		1		2	3
C-15-SC-11	15″	12″ × 12″ steel column; 3″ outside protection; Group I.	—	11 hrs.		1		2	11
C-15-SC-12	15″	Description as per C-15-SC-11; Group II.	—	8 hrs.		1		2	8
C-15-SC-13	15″	Description as per C-15-SC-11; Group III.	—	6 hrs.		1		2	6
C-15-SC-14	15″	Description as per C-15-SC-11; Group IV.	—	4 hrs.		1		2	4

For SI: 1 inch = 25.4 mm, 1 pound = 0.004448 kN, 1 pound per square inch = 0.00689 MPa, 1 pound per square yard = 5.3 N/m^2, 1 ton = 8.896 kN.

Notes:

1. Collapse.
2. Group I: includes concrete having calcareous aggregate containing a combined total of not more than 10 percent of quartz, chert and flint for the coarse aggregate.

   Group II: includes concrete having trap-rock aggregate applied without metal ties and also concrete having cinder, sandstone or granite aggregate, if held in place with wire mesh or expanded metal having not larger than 4-inch mesh, weighing not less than 1.7 lbs./yd.^{2}, placed not more than 1 inch from the surface of the concrete.

   Group III: includes concrete having cinder, sandstone or granite aggregate tied with No. 5 gage steel wire, wound spirally over the column section on a pitch of 8 inches, or equivalent ties, and concrete having siliceous aggregates containing a combined total of 60 percent or more of quartz, chert and flint, if held in place with wire mesh or expanded metal having not larger than 4-inch mesh, weighing not less than 1.7 lbs./yd.^{2}, placed not more than 1 inch from the surface of the concrete.

   Group IV: includes concrete having siliceous aggregates containing a combined total of 60 percent or more of quartz, chert and flint, and tied with No. 5 gage steel wire wound spirally over the column section on a pitch of 8 inches, or equivalent ties.

TABLE 2.5.1.7—STEEL COLUMNS—CONCRETE ENCASEMENTS MINIMUM DIMENSION 16″ TO LESS THAN 18″

ITEM CODE	MINIMUM DIMENSION	CONSTRUCTION DETAILS	PERFORMANCE		REFERENCE NUMBER			NOTES	REC. HOURS
			LOAD	TIME	PRE-BMS-92	BMS-92	POST-BMS-92		
C-16-SC-13	16″	12″ × 12″ steel column; 4″ outside protection; Group I.	—	14 hrs.		1		1	14
C-16-SC-2	16″	Description as per C-16-SC-1; Group II.	—	10 hrs.		1		1	10
C-16-SC-3	16″	Description as per C-16-SC-1; Group III.	—	8 hrs.		1		1	8
C-16-SC-4	16″	Description as per C-16-SC-1; Group IV.	—	5 hrs.		1		1	5

For SI: 1 inch = 25.4 mm.

Notes:

1. Group I: includes concrete having calcareous aggregate containing a combined total of not more than 10 percent of quartz, chert and flint for the coarse aggregate.

   Group II: includes concrete having trap-rock aggregate applied without metal ties and also concrete having cinder, sandstone or granite aggregate, if held in place with wire mesh or expanded metal having not larger than 4-inch mesh, weighing not less than 1.7 lbs./yd.^{2}, placed not more than 1 inch from the surface of the concrete.

   Group III: includes concrete having cinder, sandstone or granite aggregate tied with No. 5 gage steel wire, wound spirally over the column section on a pitch of 8 inches, or equivalent ties, and concrete having siliceous aggregates containing a combined total of 60 percent or more of quartz, chert and flint, if held in place with wire mesh or expanded metal having not larger than 4-inch mesh, weighing not less than 1.7 lbs./yd.^{2}, placed not more than 1 inch from the surface of the concrete.

   Group IV: includes concrete having siliceous aggregates containing a combined total of 60 percent or more of quartz, chert and flint, and tied with No. 5 gage steel wire wound spirally over the column section on a pitch of 8 inches, or equivalent ties.

TABLE 2.5.2.1—STEEL COLUMNS—BRICK AND BLOCK ENCASEMENTS MINIMUM DIMENSION 10″ TO LESS THAN 12″

ITEM CODE	MINIMUM DIMENSION	CONSTRUCTION DETAILS	PERFORMANCE		REFERENCE NUMBER			NOTES	REC. HOURS
			LOAD	TIME	PRE-BMS-92	BMS-92	POST-BMS-92		
C-10-SB-1	$10^1/_2$″	$10^1/_2$″ × 13″ brick encased steel columns; 8″ × 6″ × 35 lbs. "H" beam; Protection. Fill of broken brick and mortar; 2″ brick on edge; joints broken in alternate courses; cement-sand grout; 13 SWG wire reinforcement in every third horizontal joint.	90 tons	3 hrs. 6 min.			7	1	3
C-10-SB-2	$10^1/_2$″	$10^1/_2$″ × 13″ brick encased steel columns; 8″ × 6″ × 35 lbs. "H" beam; Protection: 2″ brick; joints broken in alternate courses; cement-sand grout; 13 SWG iron wire reinforcement in alternate horizontal joints.	90 tons	2 hrs.			7	2, 3, 4	2
C-10-SB-3	10″	10″ × 12″ block encased columns; 8″ × 6″ × 35 lbs. "H" beam; Protection: 2″ foamed slag concrete blocks; 13 SWG wire at each horizontal joint; mortar at each joint.	90 tons	2 hrs.			7	5	2
C-10-SB-4	$10^1/_2$″	$10^1/_2$″ × 12″ block encased steel columns; 8″ × 6″ × 35 lbs. "H" beam; Protection: gravel aggregate concrete fill (unconsolidated) 2″ thick hollow clay tiles with mortar at edges.	86 tons	56 min.			7	1	$^3/_4$
C-10-SB-5	$10^1/_2$″	$10^1/_2$″ × 12″ block encased steel columns; 8″ × 6″ × 35 lbs. "H" beam; Protection: 2″ hollow clay tiles with mortar at edges.	86 tons	22 min.			7	1	$^1/_4$

For SI: 1 inch = 25.4 mm, 1 pound = 0.004448 kN, 1 ton = 8.896 kN.

Notes:

1. Failure mode - collapse.
2. Passed 2-hour fire test (Grade "C" - British).
3. Passed hose stream test.
4. Passed reload test.
5. Passed 2-hour fire exposure but collapsed immediately following hose stream test.

TABLE 2.5.2.2—STEEL COLUMNS—BRICK AND BLOCK ENCASEMENTS MINIMUM DIMENSION 12″ TO LESS THAN 14″

ITEM CODE	MINIMUM DIMENSION	CONSTRUCTION DETAILS	PERFORMANCE		REFERENCE NUMBER			NOTES	REC. HOURS
			LOAD	TIME	PRE-BMS-92	BMS-92	POST-BMS-92		
C-12-SB-1	12″	12″ × 15″ brick encased steel columns; 8″ × 6″ × 35 lbs. "H" beam; Protection: $2^5/_8$″ thick brick; joints broken in alternate courses; cement-sand grout; fill of broken brick and mortar.	90 tons	1 hr. 49 min.			7	1	$1^3/_4$

For SI: 1 inch = 25.4 mm, 1 pound = 0.004448 kN, 1 ton = 8.896 kN.

Notes:

1. Failure mode – collapse.

TABLE 2.5.2.3—STEEL COLUMNS—BRICK AND BLOCK ENCASEMENTS MINIMUM DIMENSION 14″ TO LESS THAN 16″

ITEM CODE	MINIMUM DIMENSION	CONSTRUCTION DETAILS	PERFORMANCE		REFERENCE NUMBER			NOTES	REC. HOURS
			LOAD	TIME	PRE-BMS-92	BMS-92	POST-BMS-92		
C-15-SB-1	15″	15″ × 17″ brick encased steel columns; 8″ × 6″ × 35 lbs. "H" beam; Protection: $4^1/_2$″ thick brick; joints broken in alternate courses; cement-sand grout; fill of broken brick and mortar.	45 tons	6 hrs.			7	1	6
C-15-SB-2	15″	15″ × 17″ brick encased steel columns; 8″ × 6″ × 35 lbs. "H" beam; Protection. Fill of broken brick and mortar; $4^1/_2$″ brick; joints broken in alternate courses; cement-sand grout.	86 tons	6 hrs.			7	2, 3, 4	6
C-15-SB-3	15″	15″ × 18″ brick encased steel columns; 8″ × 6″ × 35 lbs. "H" beam; Protection: $4^1/_2$″ brick work; joints alternating; cement-sand grout.	90 tons	4 hrs.			7	5, 6	4
C-15-SB-4	14″	14″ × 16″ block encased steel columns; 8″ × 6″ × 35 lbs. "H" beam; Protection: 4″ thick foam slag concrete blocks; 13 SWG wire reinforcement in each horizontal joint; mortar in joints.	90 tons	5 hrs. 52 min.			7	7	$4^3/_4$

For SI: 1 inch = 25.4 mm, 1 pound = 0.004448 kN, 1 ton = 8.896 kN.

Notes:

1. Only a nominal load was applied to specimen.
2. Passed 6-hour fire test (Grade "A" - British).
3. Passed (6 minute) hose stream test.
4. Reload not specified.
5. Passed 4-hour fire exposure.
6. Failed by collapse between first and second minute of hose stream exposure.
7. Mode of failure-collapse.

TABLE 2.5.3.1—STEEL COLUMNS—PLASTER ENCASEMENTS MINIMUM DIMENSION 6″ TO LESS THAN 8″

ITEM CODE	MINIMUM DIMENSION	CONSTRUCTION DETAILS	PERFORMANCE		REFERENCE NUMBER			NOTES	REC. HOURS
			LOAD	TIME	PRE-BMS-92	BMS-92	POST-BMS-92		
C-7-SP-1	$7^1/_2$″	$7^1/_2$″ × $9^1/_2$″ plaster protected steel columns; 8″ × 6″ × 35 lbs. "H" beam; Protection: 24 SWG wire metal lath; $1^1/_4$″ lime plaster.	90 tons	57 min.			7	1	$^3/_4$
C-7-SP-2	$7^7/_8$″	$7^7/_8$″ × 10″ plaster protected steel columns; 8″ × 6″ × 35 lbs. "H" beam; Protection: $^3/_8$″ gypsum bal wire wound with 16 SWG wire helically wound at 4″ pitch; $^1/_2$″ gypsum plaster.	90 tons	1 hr. 13 min.			7	1	1
C-7-SP-3	$7^1/_4$″	$7^1/_4$″ × $9^3/_8$″ plaster protected steel columns; 8″ × 6″ × 35 lbs. "H" beam; Protection: $^3/_8$″ gypsum board; wire helically wound 16 SWG at 4″ pitch; $^1/_4$″ gypsum plaster finish.	90 tons	1 hr. 14 min.			7	1	1

Notes:

1. Failure mode – collapse.

TABLE 2.5.3.2—STEEL COLUMNS—PLASTER ENCASEMENTS MINIMUM DIMENSION 8″ TO LESS THAN 10″

ITEM CODE	MINIMUM DIMENSION	CONSTRUCTION DETAILS	PERFORMANCE		REFERENCE NUMBER			NOTES	REC. HOURS
			LOAD	TIME	PRE-BMS-92	BMS-92	POST-BMS-92		
C-8-SP-1	8″	8″ × 10″ plaster protected steel columns; 8″ × 6″ × 35 lbs. "H" beam; Protection: 24 SWG wire lath; 1″ gypsum plaster.	86 tons	1 hr. 23 min.			7	1	$1^{1}/_{4}$
C-8-SP-2	$8^{1}/_{2}$″	$8^{1}/_{2}$″ × $10^{1}/_{2}$″ plaster protected steel columns; 8″ × 6″ × 35 lbs. "H" beam; Protection: 24 SWG metal lath wrap; $1^{1}/_{4}$″ gypsum plaster.	90 tons	1 hr. 36 min.			7	1	$1^{1}/_{2}$
C-9-SP-3	9″	9″ × 11″ plaster protected steel columns; 8″ × 6″ × 35 lbs. "H" beam; Protection: 24 SWG metal lath wrap; $^{1}/_{8}$″ M.S. ties at 12″ pitch wire netting $1^{1}/_{2}$″ × 22 SWG between first and second plaster coats; $1^{1}/_{2}$″ gypsum plaster.	90 tons	1 hr. 33 min.			7	1	$1^{1}/_{2}$
C-8-SP-4	$8^{3}/_{4}$″	$8^{3}/_{4}$″ × $10^{3}/_{4}$″ plaster protected steel columns; 8″ × 6″ × 35 lbs. "H" beam; Protection: $^{3}/_{4}$″ gypsum board; wire wound spirally (#16 SWG) at $1^{1}/_{2}$″ pitch; $^{1}/_{2}$″ gypsum plaster.	90 tons	2 hrs.			7	2, 3, 4	2

For SI: 1 inch = 25.4 mm, 1 pound = 0.004448 kN, 1 ton = 8.896 kN.

Notes:

1. Failure mode - collapse.
2. Passed 2 hour fire exposure test (Grade "C" - British).
3. Passed hose stream test.

TABLE 2.5.4.1—STEEL COLUMNS—MISCELLANEOUS ENCASEMENTS MINIMUM DIMENSION 6″ TO LESS THAN 8″

ITEM CODE	MINIMUM DIMENSION	CONSTRUCTION DETAILS	PERFORMANCE		REFERENCE NUMBER			NOTES	REC. HOURS
			LOAD	TIME	PRE-BMS-92	BMS-92	POST-BMS-92		
C-7-SM-1	$7^{5}/_{8}$″	$7^{5}/_{8}$″ × $9^{1}/_{2}$″ (asbestos plaster) protected steel columns; 8″ × 6″ × 35 lbs. "H" beam; Protection: 20 gage $^{1}/_{2}$″ metal lath; $^{9}/_{16}$″ asbestos plaster (minimum).	90 tons	1 hr. 52 min.			7	1	$1^{3}/_{4}$

For SI: 1 inch = 25.4 mm, 1 pound = 0.004448 kN, 1 ton = 8.896 kN.

Notes:

1. Failure mode - collapse.

TABLE 2.5.4.2—STEEL COLUMNS—MISCELLANEOUS ENCASEMENTS MINIMUM DIMENSION 8″ TO LESS THAN 10″

ITEM CODE	MINIMUM DIMENSION	CONSTRUCTION DETAILS	PERFORMANCE		REFERENCE NUMBER			NOTES	REC. HOURS
			LOAD	TIME	PRE-BMS-92	BMS-92	POST-BMS-92		
C-9-SM-1	$9^{5}/_{8}$″	$9^{5}/_{8}$″ × $11^{3}/_{8}$″ asbestos slab and cement plaster protected columns; 8″ × 6″ × 35 lbs. "H" beam; Protection: 1″ asbestos slab; wire wound; $^{5}/_{8}$″ plaster.	90 tons	2 hrs.			7	1, 2	2

For SI: 1 inch = 25.4 mm, 1 pound = 0.004448 kN, 1 ton = 8.896 kN.

Notes:

1. Passed 2 hour fire exposure test.
2. Collapsed during hose stream test.

TABLE 2.5.4.3—STEEL COLUMNS—MISCELLANEOUS ENCASEMENTS MINIMUM DIMENSION 10″ TO LESS THAN 12″

ITEM CODE	MINIMUM DIMENSION	CONSTRUCTION DETAILS	PERFORMANCE		REFERENCE NUMBER			NOTES	REC. HOURS
			LOAD	TIME	PRE-BMS-92	BMS-92	POST-BMS-92		
C-11-SM-1	$11^1/_2$″	$11^1/_2$″ × $13^1/_2$″ wood wool and plaster protected steel columns; 8″ × 6″ × 35 lbs. "H" beam; Protection: wood-wool-cement paste as fill and to 2″ cover over beam; $^3/_4$″ gypsum plaster finish.	90 tons	2 hrs.			7	1, 2, 3	2
C-10-SM-1	10″	10″ × 12″ asbestos protected steel columns; 8″ × 6″ × 35 lbs. "H" beam; Protection: sprayed on asbestos paste to 2″ cover over column.	90 tons	4 hrs.			7	2, 3, 4	4

For SI: 1 inch = 25.4 mm, 1 pound = 0.004448 kN, 1 ton = 8.896 kN.

Notes:

1. Passed 2 hour fire exposure (Grade "C" - British).
2. Passed hose stream test.
3. Passed reload test.
4. Passed 4 hour fire exposure test.

TABLE 2.5.4.4—STEEL COLUMNS—MISCELLANEOUS ENCASEMENTS MINIMUM DIMENSION 12″ TO LESS THAN 14″

ITEM CODE	MINIMUM DIMENSION	CONSTRUCTION DETAILS	PERFORMANCE		REFERENCE NUMBER			NOTES	REC. HOURS
			LOAD	TIME	PRE-BMS-92	BMS-92	POST-BMS-92		
C-12-SM-1	12″	12″ × $14^1/_4$″ cement and asbestos protected columns; 8″ × 6″ × 35 lbs. "H" beam; Protection: fill of asbestos packing pieces 1″ thick 1′ 3″ o.c.; cover of 2″ molded asbestos inner layer; 1″ molded asbestos outer layer; held in position by 16 SWG nichrome wire ties; wash of refractory cement on outer surface.	86 tons	4 hrs. 43 min.			7	1, 2, 3	$4^2/_3$

For SI: 1 inch = 25.4 mm, 1 pound = 0.004448 kN, 1 ton = 8.896 kN.

Notes:

1. Passed 4 hour fire exposure (Grade "B" - British).
2. Passed hose stream test.
3. Passed reload test.

SECTION III—FLOOR/CEILING ASSEMBLIES

FIGURE 3.1—FLOOR/CEILING ASSEMBLIES—REINFORCED CONCRETE

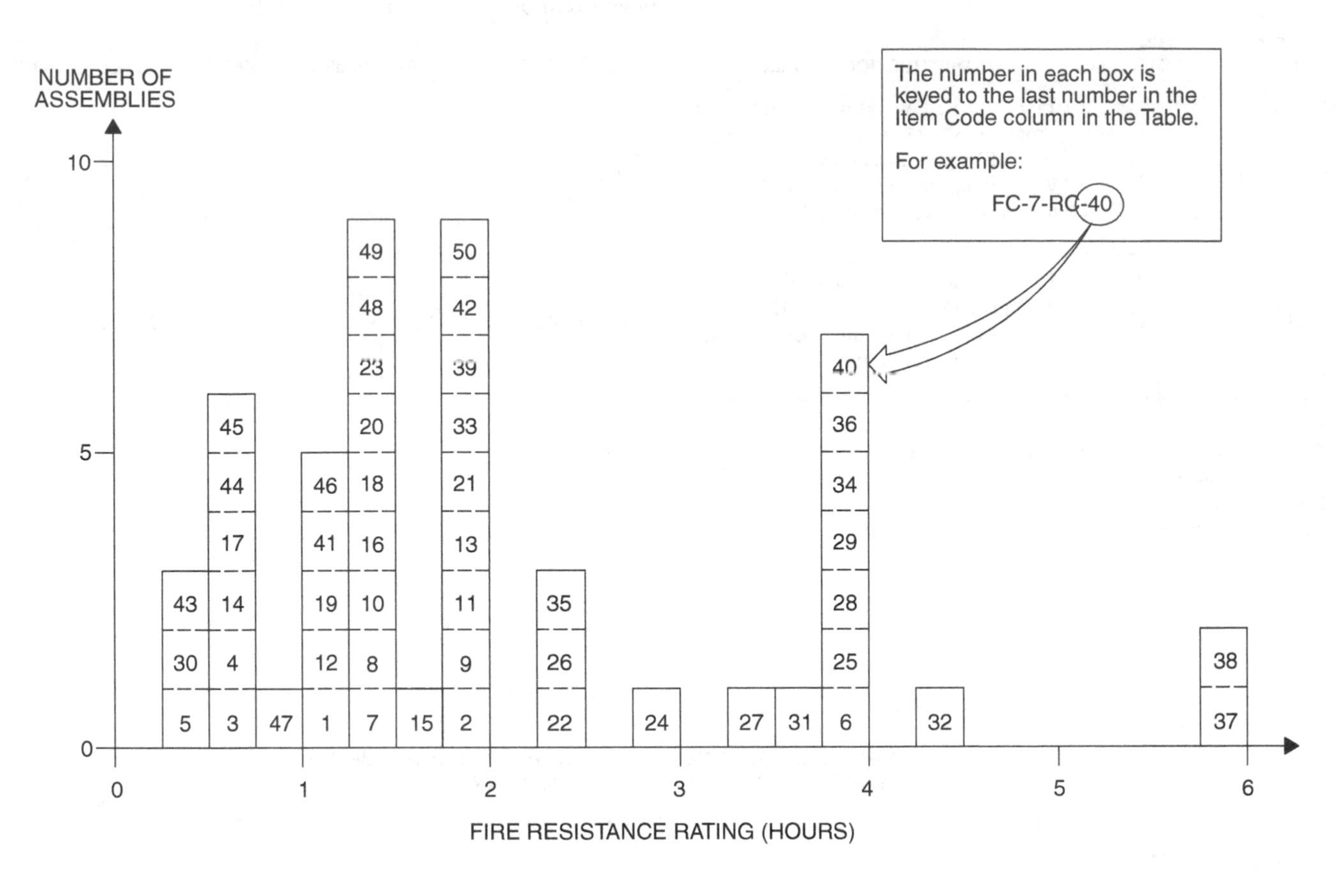

TABLE 3.1—FLOOR/CEILING ASSEMBLIES—REINFORCED CONCRETE

ITEM CODE	ASSEMBLY THICKNESS	CONSTRUCTION DETAILS	PERFORMANCE		REFERENCE NUMBER			NOTES	REC. HOURS
			LOAD	TIME	PRE-BMS-92	BMS-92	POST-BMS-92		
F/C-3-RC-1	$3^3/_4''$	$3^3/_4''$ thick floor; $3^1/_4''$ (5475 psi) concrete deck; $^1/_2''$ plaster under deck; $^3/_8''$ main reinforcement bars at $5^1/_2''$ pitch with $^7/_8''$ concrete cover; $^3/_8''$ main reinforcement bars at $4^1/_2''$ pitch perpendicular with $^1/_2''$ concrete cover; 13′1″ span restrained.	195 psf	24 min.			7	1, 2	$^1/_3$
F/C-3-RC-2	$3^1/_4''$	$3^1/_4''$ deep (3540 psi) concrete deck; $^3/_8''$ main reinforcement bars at $5^1/_2''$ pitch with $^7/_8''$ cover; $^3/_8''$ main reinforcement bars at $4^1/_2''$ pitch perpendicular with $^1/_2''$ cover; 13′1″ span restrained.	195 psf	2 hrs.			7	1, 3, 4	2
F/C-3-RC-3	$3^1/_4''$	$3^1/_4''$ deep (4175 psi) concrete deck; $^3/_8''$ main reinforcement bars at $5^1/_2''$ pitch with $^7/_8''$ cover; $^3/_8''$ main reinforcement bars at $4^1/_2''$ pitch perpendicular with $^1/_2''$ cover; 13′1″ span restrained.	195 psf	31 min.			7	1, 5	$^1/_2$

(Continued)

TABLE 3.1—FLOOR/CEILING ASSEMBLIES—REINFORCED CONCRETE—continued

ITEM CODE	ASSEMBLY THICKNESS	CONSTRUCTION DETAILS	PERFORMANCE		REFERENCE NUMBER			NOTES	REC. HOURS
			LOAD	TIME	PRE-BMS-92	BMS-92	POST-BMS-92		
F/C-3-RC-4	3 1/4″	3 1/4″ deep (4355 psi) concrete deck; 3/8″ main reinforcement bars at 5 1/2″ pitch with 7/8″ cover; 3/8″ main reinforcement bars at 4 1/2″ pitch perpendicular with 1/2″ cover; 13′1″ span restrained.	195 psf	41 min.			7	1, 5, 6	1/2
F/C-3-RC-5	3 1/4″	3 1/4″ thick (3800 psi) concrete deck; 3/8″ main reinforcement bars at 5 1/2″ pitch with 7/8″ cover; 3/8″ main reinforcement bars at 4 1/2″ pitch perpendicular with 1/2″ cover; 13′1″ span restrained.	195 psf	1 hr. 5 min.			7	1, 5	1
F/C-4-RC-6	4 1/4″	4 1/4″ thick; 3 1/4″ (4000 psi) concrete deck; 1″ sprayed asbestos lower surface; 3/8″ main reinforcement bars at 5 7/8″ pitch with 7/8″ concrete cover; 3/8″ main reinforcement bars at 4 1/2″ pitch perpendicular with 1/2″ concrete cover; 13′1″ span restrained.	195 psf	4 hrs.			7	1, 7	4
F/C-4-RC-7	4″	4″ (5025 psi) concrete deck; 1/4″ reinforcement bars at 7 1/2″ pitch with 3/4″ cover; 3/8″ main reinforcement bars at 3 3/4″ pitch perpendicular with 1/2″ cover; 13′1″ span restrained.	140 psf	1 hr. 16 min.			7	1, 2	1 1/4
F/C-4-RC-8	4″	4″ thick (4905 psi) deck; 1/4″ reinforcement bars at 7 1/2″ pitch with 7/8″ cover; 3/8″ main reinforcement bars at 3 3/4″ pitch perpendicular with 1/2″ cover; 13′1″ span restrained.	100 psf	1 hr. 23 min.			7	1, 2	1 1/3
F/C-4-RC-9	4″	4″ deep (4370 psi); 1/4″ reinforcement bars at 6″ pitch with 3/4″ cover; 1/4″ main reinforcement bars at 4″ pitch perpendicular with 1/2″ cover; 13′1″ span restrained.	150 psf	2 hrs.			7	1, 3	2
F/C-4-RC-10	4″	4″ thick (5140 psi) deck; 1/4″ reinforcement bars at 7 1/2″ pitch with 7/8″ cover; 3/8″ main reinforcement bars at 3 3/4″ pitch perpendicular with 1/2″ cover; 13′1″ span restrained.	140 psf	1 hr. 16 min.			7	1, 5	1 1/4
F/C-4-RC-11	4″	4″ thick (4000 psi) concrete deck; 3″ × 1 1/2″ × 4 lbs. R.S.J.; 2′ 6″ C.R.S.; flush with top surface; 4″ × 6″ x 13 SWG mesh reinforcement 1″ from bottom of slab; 6′6″ span restrained.	150 psf	2 hrs.			7	1, 3	2
F/C-4-RC-12	4″	4″ deep (2380 psi) concrete deck; 3″ × 1 1/2″ × 4 lbs. R.S.J.; 2′ 6″ C.R.S.; flush with top surface; 4″ × 6″ x 13 SWG mesh reinforcement 1″ from bottom surface; 6′6″ span restrained.	150 psf	1 hr. 3 min.			7	1, 2	1

(Continued)

TABLE 3.1—FLOOR/CEILING ASSEMBLIES—REINFORCED CONCRETE—continued

ITEM CODE	ASSEMBLY THICKNESS	CONSTRUCTION DETAILS	PERFORMANCE		REFERENCE NUMBER			NOTES	REC. HOURS
			LOAD	TIME	PRE-BMS-92	BMS-92	POST-BMS-92		
F/C-4-RC-13	4 1/2″	4 1/2″ thick (5200 psi) deck; 1/4″ reinforcement bars at 7 1/4″ pitch with 7/8″ cover; 3/8″ main reinforcement bars at 3 3/4″ pitch perpendicular with 1/2″ cover; 13′1″ span restrained.	140 psf	2 hrs.			7	1, 3	2
F/C-4-RC-14	4 1/2″	4 1/2″ deep (2525 psi) concrete deck; 1/4″ reinforcement bars at 7 1/2″ pitch with 7/8″ cover; 3/8″ main reinforcement bars at 3 3/8″ pitch perpendicular with 1/2″ cover; 13′1″ span restrained.	150 psf	42 min.			7	1, 5	2/3
F/C-4-RC-15	4 1/2″	4 1/2″ deep (4830 psi) concrete deck; 1 1/2″ × No. 15 gauge wire mesh; 3/8″ reinforcement bars at 15″ pitch with 1″ cover; 1/2″ main reinforcement bars at 6″ pitch perpendicular with 1/2″ cover; 12′ span simply supported.	75 psf	1 hr. 32 min.			7	1, 8	1 1/2
F/C-4-RC-16	4 1/2″	4 1/2″ deep (4595 psi) concrete deck; 1/4″ reinforcement bars at 7 1/2″ pitch with 7/8″ cover; 3/8″ main reinforcement bars at 3 1/2″ pitch perpendicular with 1/2″ cover; 12′ span simply supported.	75 psf	1 hr. 20 min.			7	1, 8	1 1/3
F/C-4-RC-17	4 1/2″	4 1/2″ deep (3625 psi) concrete deck; 1/4″ reinforcement bars at 7 1/2″ pitch with 7/8″ cover; 3/8″ main reinforcement bars at 3 1/2″ pitch perpendicular with 1/2″ cover; 12′ span simply supported.	75 psf	35 min.			7	1, 8	1/2
F/C-4-RC-18	4 1/2″	4 1/2″ deep (4410 psi) concrete deck; 1/4″ reinforcement bars at 7 1/2″ pitch with 7/8″ cover; 3/8″ main reinforcement bars at 3 1/2″ pitch perpendicular with 1/2″ cover; 12′ span simply supported.	85 psf	1 hr. 27 min.			7	1, 8	1 1/3
F/C-4-RC-19	4 1/2″	4 1/2″ deep (4850 psi) deck; 3/8″ reinforcement bars at 15″ pitch with 1″ cover; 1/2″ main reinforcement bars at 6″ pitch perpendicular with 1/2″ cover; 12′ span simply supported.	75 psf	2 hrs. 15 min.			7	1, 9	1 1/4
F/C-4-RC-20	4 1/2″	4 1/2″ deep (3610 psi) deck; 1/4″ reinforcement bars at 7 1/2″ pitch with 7/8″ cover; 3/8″ main reinforcement bars at 3 1/2″ pitch perpendicular with 1/2″ cover; 12′ span simply supported.	75 psf	1 hr. 22 min.			7	1, 8	1 1/3
F/C-5-RC-21	5″	5″ deep; 4 1/2″ (5830 psi) concrete deck; 1/2″ plaster finish bottom of slab; 1/4″ reinforcement bars at 7 1/2″ pitch with 7/8″ cover; 3/8″ main reinforcement bars at 3 1/2″ pitch perpendicular with 1/2″ cover; 12′ span simply supported.	69 psf	2 hrs.			7	1, 3	2

(Continued)

TABLE 3.1—FLOOR/CEILING ASSEMBLIES—REINFORCED CONCRETE—continued

ITEM CODE	ASSEMBLY THICKNESS	CONSTRUCTION DETAILS	PERFORMANCE		REFERENCE NUMBER			NOTES	REC. HOURS
			LOAD	TIME	PRE-BMS-92	BMS-92	POST-BMS-92		
F/C-5-RC-22	5″	$4^1/_2$″ (5290 psi) concrete deck; $^1/_2$″ plaster finish bottom of slab; $^1/_4$″ reinforcement bars at $7^1/_2$″ pitch with $^7/_8$″ cover; $^3/_8$″ main reinforcement bars at $3^1/_2$″ pitch perpendicular with $^1/_2$″ cover; 12′ span simply supported.	No load	2 hrs. 28 min.			7	1, 10, 11	$2^1/_4$
F/C-5-RC-23	5″	5″ (3020 psi) concrete deck; 3″ × $1^1/_2$″ × 4 lbs. R.S.J.; 2′ C.R.S. with 1″ cover on bottom and top flanges; 8′ span restrained.	172 psf	1 hr. 24 min.			7	1, 2, 12	$1^1/_2$
F/C-5-RC-24	$5^1/_2$″	5″ (5180 psi) concrete deck; $^1/_2$″ retarded plaster underneath slab; $^1/_4$″ reinforcement bars at $7^1/_2$″ pitch with $1^3/_8$″ cover; $^3/_8$″ main reinforcement bars at $3^1/_2$″ pitch perpendicular with 1″ cover; 12′ span simply supported.	60 psf	2 hrs. 48 min.			7	1, 10	$2^3/_4$
F/C-6-RC-25	6″	6″ deep (4800 psi) concrete deck; $^1/_4$″ reinforcement bars at $7^1/_2$″ pitch with $^7/_8$″ cover; $^3/_8$″ main reinforcement bars at $3^1/_2$″ pitch perpendicular with $^7/_8$″ cover; 13′1″ span restrained.	195 psf	4 hrs.			7	1, 7	4
F/C-6-RC-26	6″	6″ (4650 psi) concrete deck; $^1/_4$″ reinforcement bars at $7^1/_2$″ pitch with $^7/_8$″ cover; $^3/_8$″ main reinforcement bars at $3^1/_2$″ pitch perpendicular with $^1/_2$″ cover; 13′1″ span restrained.	195 psf	2 hrs. 23 min.			7	1, 2	$2^1/_4$
F/C-6-RC-27	6″	6″ deep (6050 psi) concrete deck; $^1/_4$″ reinforcement bars at $7^1/_2$″ pitch $^7/_8$″ cover; $^3/_8$″ reinforcement bars at $3^1/_2$″ pitch perpendicular with $^1/_2$″ cover; 13′1″ span restrained.	195 psf	3 hrs. 30 min.			7	1, 10	$3^1/_2$
F/C-6-RC-28	6″	6″ deep (5180 psi) concrete deck; $^1/_4$″ reinforcement bars at 8″ pitch $^3/_4$″ cover; $^1/_4$″ reinforcement bars at $5^1/_2$″ pitch perpendicular with $^1/_2$″ cover; 13′1″ span restrained.	150 psf	4 hrs.			7	1, 7	4
F/C-6-RC-29	6″	6″ thick (4180 psi) concrete deck; 4″ × 3″ × 10 lbs. R.S.J.; 2′6″ C.R.S. with 1″ cover on both top and bottom flanges; 13′1″ span restrained.	160 psf	3 hrs. 48 min.			7	1, 10	$3^3/_4$
F/C-6-RC-30	6″	6″ thick (3720 psi) concrete deck; 4″ × 3″ × 10 lbs. R.S.J.; 2′6″ C.R.S. with 1″ cover on both top and bottom flanges; 12′ span simply supported.	115 psf	29 min.			7	1, 5, 13	$^1/_4$
F/C-6-RC-31	6″	6″ deep (3450 psi) concrete deck; 4″ × $1^3/_4$″ × 5 lbs. R.S.J.; 2′6″ C.R.S. with 1″ cover on both top and bottom flanges; 12′ span simply supported.	25 psf	3 hrs. 35 min.			7	1, 2	$3^1/_2$

(Continued)

TABLE 3.1—FLOOR/CEILING ASSEMBLIES—REINFORCED CONCRETE—continued

ITEM CODE	ASSEMBLY THICKNESS	CONSTRUCTION DETAILS	PERFORMANCE		REFERENCE NUMBER			NOTES	REC. HOURS
			LOAD	TIME	PRE-BMS-92	BMS-92	POST-BMS-92		
F/C-6-RC-32	6″	6″ deep (4460 psi) concrete deck; 4″ × 1$^3/_4$″ × 5 lbs. R.S.J.; 2′ C.R.S.; with 1″ cover on both top and bottom flanges; 12′ span simply supported.	60 psf	4 hrs. 30 min.			7	1, 10	4$^1/_2$
F/C-6-RC-33	6″	6″ deep (4360 psi) concrete deck; 4″ × 1$^3/_4$″ × 5 lbs. R.S.J.; 2′ C.R.S.; with 1″ cover on both top and bottom flanges; 13′1″ span restrained.	60 psf	2 hrs.			7	1, 3	2
F/C-6-RC-34	6$^1/_4$″	6$^1/_4$″ thick; 4$^3/_4$″ (5120 psi) concrete core; 1″ T&G board flooring; $^1/_2$″ plaster undercoat; 4″ × 3″ × 10 lbs. R.S.J.; 3′ C.R.S. flush with top surface concrete; 12′ span simply supported; 2″ × 1′3″ clinker concrete insert.	100 psf	4 hrs.			7	1, 7	4
F/C-6-RC-35	6$^1/_4$″	4$^3/_4$″ (3600 psi) concrete core; 1″ T&G board flooring; $^1/_2$″ plaster undercoat; 4″ × 3″ × 10 lbs. R.S.J.; 3′ C.R.S.; flush with top surface concrete; 12′ span simply supported; 2″ × 1′3″ clinker concrete insert.	100 psf	2 hrs. 30 min.			7	1, 5	2$^1/_2$
F/C-6-RC-36	6$^1/_4$″	4$^3/_4$″ (2800 psi) concrete core; 1″ T&G board flooring; $^1/_2$″ plaster undercoat; 4″ × 3″ × 10 lbs. R.S.J.; 3′ C.R.S.; flush with top surface concrete; 12″ span simply supported; 2″ × 1′3″ clinker concrete insert.	80 psf	4 hrs.			7	1, 7	4
F/C-7-RC-37	7″	(3640 psi) concrete deck; $^1/_4$″ reinforcement bars at 6″ pitch with 1$^1/_2$″ cover; $^1/_4$″ reinforcement bars at 5″ pitch perpendicular with 1$^1/_2$″ cover; 13′1″ span restrained.	169 psf	6 hrs.			7	1, 14	6
F/C-7-RC-38	7″	(4060 psi) concrete deck; 4″ × 3″ × 10 lbs. R.S.J.; 2′6″ C.R.S. with 1$^1/_2$″ cover on both top and bottom flanges; 4″ × 6″ × 13 SWG mesh reinforcement 1$^1/_2$″ from bottom of slab; 13′1″ span restrained.	175 psf	6 hrs.			7	1, 14	6
F/C-7-RC-39	7$^1/_4$″	5$^3/_4$″ (4010 psi) concrete core; 1″ T&G board flooring; $^1/_2$″ plaster undercoat; 4″ × 3″ × 10 lbs. R.S.J.; 2′6″ C.R.S.; 1″ down from top surface of concrete; 12′ simply supported span; 2″ × 1′3″ clinker concrete insert.	95 psf	2 hrs.			7	1, 3	2
F/C-7-RC-40	7$^1/_4$″	5$^3/_4$″ (3220 psi) concrete core; 1″ T&G flooring; $^1/_2$″ plaster undercoat; 4″ × 3″ × 10 lbs. R.S.J.; 2′6″ C.R.S.; 1″ down from top surface of concrete; 12′ simply supported span; 2″ × 1′3″ clinker concrete insert.	95 psf	4 hrs.			7	1, 7	4

(Continued)

TABLE 3.1—FLOOR/CEILING ASSEMBLIES—REINFORCED CONCRETE—continued

ITEM CODE	ASSEMBLY THICKNESS	CONSTRUCTION DETAILS	PERFORMANCE		REFERENCE NUMBER			NOTES	REC. HOURS
			LOAD	TIME	PRE-BMS-92	BMS-92	POST-BMS-92		
F/C-7-RC-41	10″ ($2^1/_4$″ Slab)	Ribbed floor, see Note 15 for details; slab $2^1/_2$″ deep (3020 psi); $^1/_4$″ reinforcement bars at 6″ pitch with $^3/_4$″ cover; beams $7^1/_2$″ deep × 5″ wide; 24″ C.R.S.; $^5/_8$″ reinforcement bars two rows $^1/_2$″ vertically apart with 1″ cover; 13′1″ span restricted.	195 psf	1 hr. 4 min.			7	1, 2, 15	1
F/C-5-RC-42	$5^1/_2$″	Composite ribbed concrete slab assembly; see Note 17 for details.	See Note 16	2 hrs.			43	16, 17	2
F/C-3-RC-43	3″	2500 psi concrete; $^5/_8$″ cover; fully restrained at test.	See Note 16	30 min.			43	16	$^1/_2$
F/C-3-RC-44	3″	2000 psi concrete; $^5/_8$″ cover; free or partial restraint at test.	See Note 16	45 min.			43	16	$^3/_4$
F/C-4-RC-45	4″	2500 psi concrete; $^5/_8$″ cover; fully restrained at test.	See Note 16	40 min.			43	16	$^2/_3$
F/C-4-RC-46	4″	2000 psi concrete; $^3/_4$″ cover; free or partial restraint at test.	See Note 16	1 hr. 15 min.			43	16	$1^1/_4$
F/C-5-RC-47	5″	2500 psi concrete; $^3/_4$″ cover; fully restrained at test.	See Note 16	1 hr.			43	16	1
F/C-5-RC-48	5″	2000 psi concrete; $^3/_4$″ cover; free or partial restraint at test.	See Note 16	1 hr. 30 min.			43	16	$1^1/_2$
F/C-6-RC-49	6″	2500 psi concrete; 1″ cover; fully restrained at test.	See Note 16	1 hr. 30 min.			43	16	$1^1/_2$
F/C-6-RC-50	6″	2000 psi concrete; 1″ cover; free or partial restraint at test.	See Note 16	2 hrs.			43	16	2

For SI: 1 inch = 25.4 mm, 1 foot = 305 mm, 1 pound per square inch = 0.00689 MPa, 1 pound per square foot = 47.9 N/m^2.

Notes:

1. British test.
2. Failure mode - local back face temperature rise.
3. Tested for Grade "C" (2 hour) fire resistance.
4. Collapse imminent following hose stream.
5. Failure mode - flame thru.
6. Void formed with explosive force and report.
7. Achieved Grade "B" (4 hour) fire resistance (British).
8. Failure mode - collapse.
9. Test was run to 2 hours, but specimen was partially supported by the furnace at $1^1/_4$ hours.
10. Failure mode - average back face temperature.
11. Recommended endurance for nonload bearing performance only.
12. Floor maintained load bearing ability to 2 hours at which point test was terminated.
13. Test was run to 3 hours at which time failure mode 2 (above) was reached in spite of crack formation at 29 minutes.
14. Tested for Grade "A" (6 hour) fire resistance.
15.

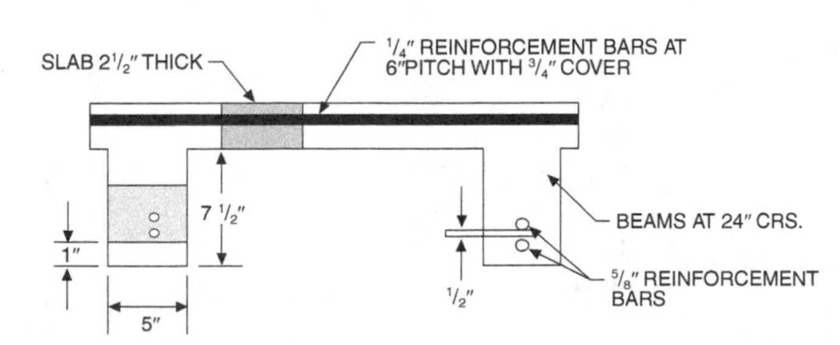

16. Load unspecified.
17. Total assembly thickness $5^1/_2$ inches. Three-inch thick blocks of molded excelsior bonded with portland cement used as inserts with $2^1/_2$-inch cover (concrete) above blocks and $^3/_4$-inch gypsum plaster below. Nine-inch wide ribs containing reinforcing steel of unspecified size interrupted 20-inch wide segments of slab composite (i.e., plaster, excelsior blocks, concrete cover).

FIGURE 3.2—FLOOR/CEILING ASSEMBLIES—STEEL STRUCTURAL ELEMENTS

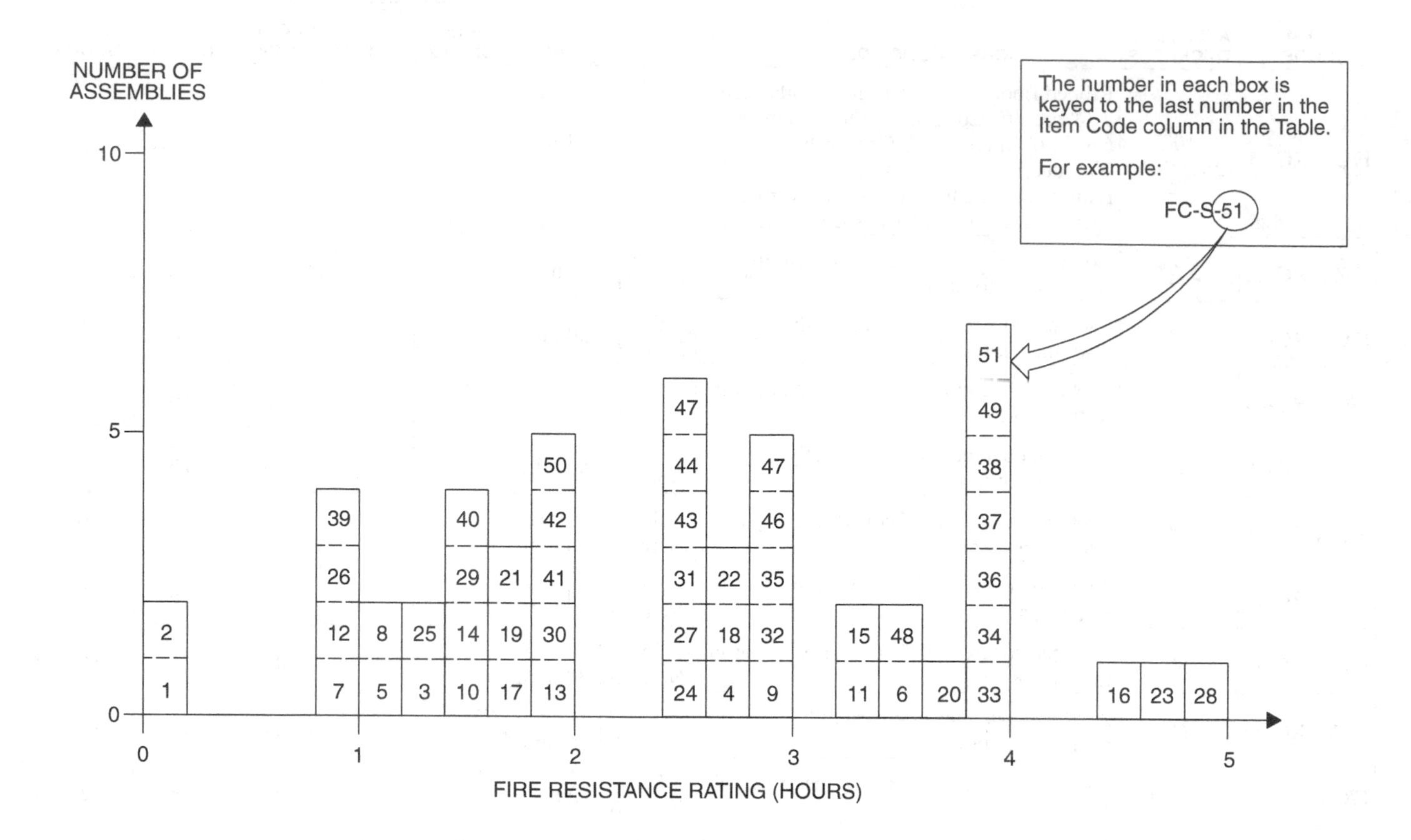

TABLE 3.2—FLOOR/CEILING ASSEMBLIES—STEEL STRUCTURAL ELEMENTS

ITEM CODE	MEMBRANE THICKNESS	CONSTRUCTION DETAILS	PERFORMANCE		REFERENCE NUMBER			NOTES	REC. HOURS
			LOAD	TIME	PRE-BMS-92	BMS-92	POST-BMS-92		
F/C-S-1	0″	- 10′ × 13′6″; S.J. 103 - 24″ o.c.; Deck: 2″ concrete; Membrane: none.	145 psf	7 min.			3	1, 2, 3, 8	0
F/C-S-2	0″	- 10′ × 13′6″; S.J. 103 - 24″ o.c.; Deck: 2″ concrete; Membrane: none	145 psf	7 min.			3	1, 2, 3, 8	0
F/C-S-3	$^{1}/_{2}$″	- 10′ × 13′6″; S.J. 103 - 24″ o.c.; Deck: 2″ concrete 1:2:4; Membrane: furring 12″ o.c.; Clips A, B, G; No extra reinforcement; $^{1}/_{2}$″ plaster - 1.5:2.5.	145 psf	1 hr. 15 min.			3	2, 3, 8	$1^{1}/_{4}$
F/C-S-4	$^{1}/_{2}$″	- 10′ × 13′6″; S.J. 103 - 24″ o.c.; Deck: 2″ concrete 1:2:4; Membrane: furring 16″ o.c.; Clips D, E, F, G; Diagonal wire reinforcement; $^{1}/_{2}$″ plaster - 1.5:2.5.	145 psf	2 hrs. 46 min.			3	3, 8	$2^{3}/_{4}$
F/C-S-5	$^{1}/_{2}$″	- 10′ × 13′6″; S.J. 103 - 24″ o.c.; Deck: 2″ concrete 1:2:4; Membrane: furring 16″ o.c.; Clips A, B, G; No extra reinforcement; $^{1}/_{2}$″ plaster - 1.5:2.5.	145 psf	1 hr. 4 min.			3	2, 3, 8	1

(Continued)

TABLE 3.2—FLOOR/CEILING ASSEMBLIES—STEEL STRUCTURAL ELEMENTS—continued

ITEM CODE	MEMBRANE THICKNESS	CONSTRUCTION DETAILS	PERFORMANCE		REFERENCE NUMBER			NOTES	REC. HOURS
			LOAD	TIME	PRE-BMS-92	BMS-92	POST-BMS-92		
F/C-S-6	$^1/_2$″	10′ × 13′6″; S.J. 103 - 24″ o.c.; Deck: 2″ concrete 1:2:4; Membrane: furring 16″ o.c.; Clips D, E, F, G; Hexagonal mesh reinforcement; $^1/_2$″ plaster.	145 psf	3 hrs. 28 min.			3	2, 3, 8	$2^1/_3$
F/C-S-7	$^1/_2$″	10′ × 13′6″; S.J. 103 - 24″ o.c.; Deck: 4 lbs. rib lath; 6″ × 6″ - 10 × 10 ga. reinforcement; 2″ deck gravel concrete; Membrane: furring 16″ o.c.; Clips C, E; Reinforcement: none; $^1/_2$″ plaster - 1.5:2.5 mill mix.	N/A	55 min.			3	5, 8	$^3/_4$
F/C-S-8	$^1/_2$″	Spec. 9′ × 4′4″; S.J. 103 bar joists - 18″ o.c.; Deck: 4 lbs. rib lath base; 6″ × 6″ - 10 × 10 ga. reinforcement; 2″ deck 1:2:4 gravel concrete; Membrane: furring, $^3/_4$″ C.R.S., 16″ o.c.; Clips C, E; Reinforcement: none; $^1/_2$″ plaster - 1.5:2.5 mill mix.	300 psf	1 hr. 10 min.			3	2, 3, 8	1
F/C-S-9	$^5/_8$″	10′ × 13′6″; S.J. 103 - 24″ o.c.; Deck: 2″ concrete 1:2:4; Membrane: furring 12″ o.c.; Clips A, B, G; Extra "A" clips reinforcement; $^5/_8$″ plaster - 1.5:2; 1.5:3.	145 psf	3 hrs.			3	6, 8	3
F/C-S-10	$^5/_8$″	18′ × 13′6″; Joists, S.J. 103 - 24″ o.c.; Deck: 4 lbs. rib lath; 6″ × 6″ - 10 × 10 ga. reinforcement; 2″ deck 1:2:3.5 gravel concrete; Membrane: furring, spacing 16″ o.c.; Clips C, E; Reinforcement: none; $^5/_8$″ plaster - 1.5:2.5 mill mix.	145 psf	1 hr. 25 min.			3	2, 3, 8	$1^1/_3$
F/C-S-11	$^5/_8$″	10′ × 13′6″; S.J. 103 - 24″ o.c.; Deck: 2″ concrete 1:2:4; Membrane: furring 12″ o.c.; Clips D, E, F, G; Diagonal wire reinforcement; $^5/_8$″ plaster - 1.5:2; 0.5:3.	145 psf	3 hrs. 15 min.			3	2, 4, 8	$3^1/_4$
F/C-S-12	$^5/_8$″	10′ × 13′6″; Joists, S.J. 103 - 24″ o.c.; Deck: 3.4 lbs. rib lath; 6″ × 6″ - 10 × 10 ga. reinforcement; 2″ deck 1:2:4 gravel concrete; Membrane: furring 16″ o.c.; Clips D, E, F, G; Reinforcement: none; $^5/_8$″ plaster - 1.5:2.5.	145 psf	1 hr.			3	7, 8	1
F/C-S-13	$^3/_4$″	Spec. 9′ × 4′4″; S.J. 103 - 18″ o.c.; Deck: 4 lbs. rib lath; 6″ × 6″ - 10 × 10 ga. reinforcement; 2″ deck 1:2:4 gravel concrete; Membrane: furring, $^3/_4$″ C.R.S., 16″ o.c.; Clips C, E; Reinforcement: none; $^3/_4$″ plaster - 1.5:2.5 mill mix.	300 psf	1 hr. 56 min.			3	3, 8	$1^3/_4$

(Continued)

TABLE 3.2—FLOOR/CEILING ASSEMBLIES—STEEL STRUCTURAL ELEMENTS—continued

ITEM CODE	MEMBRANE THICKNESS	CONSTRUCTION DETAILS	PERFORMANCE		REFERENCE NUMBER			NOTES	REC. HOURS
			LOAD	TIME	PRE-BMS-92	BMS-92	POST-BMS-92		
F/C-S-14	$^{7}/_{8}$″	Floor finish: 1″ concrete; plate cont. weld; 4″ - 7.7 lbs. "I" beams; Ceiling: $^{1}/_{4}$″ rods 12″ o.c.; $^{7}/_{8}$″ gypsum sand plaster.	105 psf	1 hr. 35 min.			6	2, 4, 9, 10	$1^{1}/_{2}$
F/C-S-15	1″	Floor finish: $1^{1}/_{2}$″ L.W. concrete; $^{1}/_{2}$″ limestone cement; plate cont. weld; 5″ - 10 lbs. "I" beams; Ceiling: $^{1}/_{4}$″ rods 12″ o.c. tack welded to beams metal lath; 1″ P. C. plaster.	165 psf	3 hrs. 20 min.			6	4, 9, 11	$3^{1}/_{3}$
F/C-S-16	1″	10′ × 13′6″; S.J. 103 - 24″ o.c.; Deck. 2″ concrete 1:2:4; Membrane: furring 12″ o.c.; Clips D, E, F, G; Hexagonal mesh reinforcement; 1″ thick plaster - 1.5:2; 1.5:3.	145 psf	4 hrs. 26 min.			3	2, 4, 8	$4^{1}/_{3}$
F/C-S-17	1″	10′ × 13′6″; Joists - S.J. 103 - 24″ o.c.; Deck: 3.4 lbs. rib lath; 6″ × 6″ - 10 × 10 ga. reinforcement; 2″ deck 1:2:4 gravel concrete; Membrane: furring 16″ o.c.; Clips D, E, F, G; 1″ plaster.	145 psf	1 hr. 42 min.			3	2, 4, 8	$1^{2}/_{3}$
F/C-S-18	$1^{1}/_{8}$″	10′ × 13′6″; S. J. 103 - 24″ o.c.; Deck: 2″ concrete 1:2:4; Membrane: furring 12″ o.c.; Clips C, E, F, G; Diagonal wire reinforcement; $1^{1}/_{8}$″ plaster.	145 psf	2 hrs. 44 min.			3	2, 4, 8	$2^{2}/_{3}$
F/C-S-19	$1^{1}/_{8}$″	10′ × 13′6″; Joists - S.J. 103 - 24″ o.c.; Deck: $1^{1}/_{2}$″ gypsum concrete over; $^{1}/_{2}$″ gypsum board; Membrane: furring 12″ o.c.; Clips D, E, F, G; $1^{1}/_{8}$″ plaster - 1.5:2; 1.5:3.	145 psf	1 hr. 40 min.			3	2, 3, 8	$1^{2}/_{3}$
F/C-S-20	$1^{1}/_{8}$″	$2^{1}/_{2}$″ cinder concrete; $^{1}/_{2}$″ topping; plate 6″ welds 12″ o.c.; 5″ - 18.9 lbs. "H" center; 5″ - 10 lbs. "I" ends; 1″ channels 18″ o.c.; $1^{1}/_{8}$″ gypsum sand plaster.	150 psf	3 hrs. 43 min.			6	2, 4, 9, 11	$3^{2}/_{3}$
F/C-S-21	$1^{1}/_{4}$″	10′ × 13′6″; Joists - S.J. 103 - 24″ o.c.; Deck: $1^{1}/_{2}$″ gypsum concrete over; $^{1}/_{2}$″ gypsum board base; Membrane: furring 12″ o.c.; Clips D, E, F, G; $1^{1}/_{4}$″ plaster - 1.5:2; 1.5:3.	145 psf	1 hr. 48 min.			3	2, 3, 8	$1^{2}/_{3}$
F/C-S-22	$1^{1}/_{4}$″	Floor finish: $1^{1}/_{2}$″ limestone concrete; $^{1}/_{2}$″ sand cement topping; plate to beams $3^{1}/_{2}$″; 12″ o.c. welded; 5″ - 10 lbs. "I" beams; 1″ channels 18″ o.c.; $1^{1}/_{4}$″ wood fiber gypsum sand plaster on metal lath.	292 psf	2 hrs. 45 min.			6	2, 4, 9, 10	$2^{3}/_{4}$
F/C-S-23	$1^{1}/_{2}$″	$2^{1}/_{2}$″ L.W. (gas exp.) concrete; Deck: $^{1}/_{2}$″ topping; plate $6^{1}/_{4}$″ welds 12″ o.c.; Beams: 5″ - 18.9 lbs. "H" center; 5″ - 10 lbs. "I" ends; Membrane: 1″ channels 18″ o.c.; $1^{1}/_{2}$″ gypsum sand plaster.	150 psf	4 hrs. 42 min.			6	2, 4, 9	$4^{2}/_{3}$

(Continued)

TABLE 3.2—FLOOR/CEILING ASSEMBLIES—STEEL STRUCTURAL ELEMENTS—continued

ITEM CODE	MEMBRANE THICKNESS	CONSTRUCTION DETAILS	PERFORMANCE		REFERENCE NUMBER			NOTES	REC. HOURS
			LOAD	TIME	PRE-BMS-92	BMS-92	POST-BMS-92		
F/C-S-24	$1^1/_2''$	Floor finish: $1^1/_2''$ limestone concrete; $^1/_2''$ cement topping; plate $3^1/_2''$ - 12″ o.c. welded; 5″ - 10 lbs. "I" beams; Ceiling: 1″ channels 18″ o.c.; $1^1/_2''$ gypsum plaster.	292 psf	2 hrs. 34 min.			6	2, 4, 9, 10	$2^1/_2$
F/C-S-25	$1^1/_2''$	Floor finish: $1^1/_2''$ gravel concrete on exp. metal; plate cont. weld; 4″ - 7.7 lbs. "I" beams; Ceiling: $^1/_4''$ rods 12″ o.c. welded to beams; $1^1/_2''$ fiber gypsum sand plaster.	70 psf	1 hr. 24 min.			6	2, 4, 9, 10	$1^1/_3$
F/C-S-26	$2^1/_2''$	Floor finish: bare plate; $6^1/_4''$ welding - 12″ o.c.; 5″ - 18.9 lbs. "H" girders (inner); 5″ - 10 lbs "I" girders (two outer); 1″ channels 18″ o.c.; 2″ reinforced gypsum tile; $^1/_2''$ gypsum sand plaster.	122 psf	1 hr.			6	7, 9, 11	1
F/C-S-27	$2^1/_2''$	Floor finish: 2″ gravel concrete; plate to beams $3^1/_2''$ - 12″ o.c. welded; 4″ - 7.7 lbs. "I" beams; 2″ gypsum ceiling tiles; $^1/_2''$ 1:3 gypsum sand plaster.	105 psf	2 hrs. 31 min.			6	2, 4, 9, 10	$2^1/_2$
F/C-S-28	$2^1/_2''$	Floor finish: $1^1/_2''$ gravel concrete; $^1/_2''$ gypsum asphalt; plate continuous weld; 4″ - 7.7 lbs. "I" beams; 12″ - 31.8 lbs. "I" beams - girder at 5′ from one end; 1″ channels 18″ o.c.; 2″ reinforcement gypsum tile; $^1/_2''$ 1:3 gypsum sand plaster.	200 psf	4 hrs. 55 min.			6	2, 4, 9, 11	$4^2/_3$
F/C-S-29	$^3/_4''$	Floor: 2″ reinforced concrete or 2″ precast reinforced gypsum tile; Ceiling: $^3/_4''$ portland cement-sand plaster 1:2 for scratch coat and 1:3 for brown coat with 15 lbs. hydrated lime and 3 lbs. of short asbestos fiber bag per cement or $^3/_4''$ sanded gypsum plaster 1:2 for scratch coat and 1:3 for brown coat.	See Note 12	1 hr. 30 min.		1		12, 13, 14	$1^1/_2$
F/C-S-30	$^3/_4''$	Floor: $2^1/_4''$ reinforced concrete or 2″ reinforced gypsum tile; the latter with $^1/_4''$ mortar finish; Ceiling: $^3/_4''$ sanded gypsum plaster; 1:2 for scratch coat and 1:3 for brown coat.	See Note 12	2 hrs.		1		12, 13, 14	2
F/C-S-31	$^3/_4''$	Floor: $2^1/_2''$ reinforced concrete or 2″ reinforced gypsum tile; the latter with $^1/_4''$ mortar finish; Ceiling: 1″ neat gypsum plaster or $^3/_4''$ gypsum-vermiculite plaster, ratio of gypsum to fine vermiculite 2:1 to 3:1.	See Note 12	2 hrs. 30 min.		1		12, 13, 14	$2^1/_2$

(Continued)

TABLE 3.2—FLOOR/CEILING ASSEMBLIES—STEEL STRUCTURAL ELEMENTS—continued

ITEM CODE	MEMBRANE THICKNESS	CONSTRUCTION DETAILS	PERFORMANCE		REFERENCE NUMBER			NOTES	REC. HOURS
			LOAD	TIME	PRE-BMS-92	BMS-92	POST-BMS-92		
F/C-S-32	$^3/_4''$	Floor: $2^1/_2''$ reinforced concrete or 2″ reinforced gypsum tile; the latter with $^1/_2''$ mortar finish; Ceiling: 1″ neat gypsum plaster or $^3/_4''$ gypsum-vermiculite plaster, ratio of gypsum to fine vermiculite 2:1 to 3:1.	See Note 12	3 hrs.		1		12, 13, 14	3
F/C-S-33	1″	Floor: $2^1/_2''$ reinforced concrete or 2″ reinforced gypsum slabs; the latter with $^1/_2''$ mortar finish; Ceiling: 1″ gypsum-vermiculite plaster applied on metal lath and ratio 2:1 to 3:1 gypsum to vermiculite by weight.	See Note 12	4 hrs.		1		12, 13, 14	4
F/C-S-34	$2^1/_2''$	Floor: 2″ reinforced concrete or 2″ precast reinforced portland cement concrete or gypsum slabs; precast slabs to be finished with $^1/_4''$ mortar top coat; Ceiling: 2″ precast reinforced gypsum tile, anchored into beams with metal ties or clips and covered with $^1/_2''$ 1:3 sanded gypsum plaster.	See Note 12	4 hrs.		1		12, 13, 14	4
F/C-S-35	1″	Floor: 1:3:6 portland cement, sand and gravel concrete applied directly to the top of steel units and $1^1/_2''$ thick at top of cells, plus $^1/_2''$ 1:$2^1/_2''$ cement-sand finish, total thickness at top of cells, 2″; Ceiling: 1″ neat gypsum plaster, back of lath 2″ or more from underside of cellular steel.	See Note 15	3 hrs.		1		15, 16, 17, 18	3
F/C-S-36	1″	Floor: same as F/C-S-35; Ceiling: 1″ gypsum-vermiculite plaster (ratio of gypsum to vermiculite 2:1 to 3:1), the back of lath 2″ or more from under-side of cellular steel.	See Note 15	4 hrs.		1		15, 16, 17, 18	4
F/C-S-37	1″	Floor: same as F/C-S-35; Ceiling: 1″ neat gypsum plaster; back of lath 9″ or more from underside of cellular steel.	See Note 15	4 hrs.		1		15, 16, 17, 18	4
F/C-S-38	1″	Floor: same as F/C-S-35; Ceiling: 1″ gypsum-vermiculite plaster (ratio of gypsum to vermiculite 2:1 to 3:1), the back of lath being 9″ or more from underside of cellular steel.	See Note 15	5 hrs.		1		15, 16, 17, 18	5
F/C-S-39	$^3/_4''$	Floor: asbestos paper 14 lbs./100 ft.^{2} cemented to steel deck with waterproof linoleum cement, wood screeds and $^7/_8''$ wood floor; Ceiling: $^3/_4''$ sanded gypsum plaster 1:2 for scratch coat and 1:3 for brown coat.	See Note 19	1 hr.		1		19, 20, 21, 22	1

(Continued)

TABLE 3.2—FLOOR/CEILING ASSEMBLIES—STEEL STRUCTURAL ELEMENTS—continued

ITEM CODE	MEMBRANE THICKNESS	CONSTRUCTION DETAILS	PERFORMANCE		REFERENCE NUMBER			NOTES	REC. HOURS
			LOAD	TIME	PRE-BMS-92	BMS-92	POST-BMS-92		
F/C-S-40	$^3/_4''$	Floor: $1^1/_2''$, 1:2:4 portland cement concrete; Ceiling: $^3/_4''$ sanded gypsum plaster 1:2 for scratch coat and 1:3 for brown coat.	See Note 19	1 hr. 30 min.		1		19, 20, 21, 22	$1^1/_2$
F/C-S-41	$^3/_4''$	Floor: 2″, 1:2:4 portland cement concrete; Ceiling: $^3/_4''$ sanded gypsum plaster, 1:2 for scratch coat and 1:3 for brown coat.	See Note 19	2 hrs.		1		19, 20, 21, 22	2
F/C-S-42	1″	Floor: 2″, 1:2:4 portland cement concrete; Ceiling: 1″ portland cement-sand plaster with 10 lbs. of hydrated lime for @ bag of cement 1:2 for scratch coat and $1:2^1/_2''$ for brown coat.	See Note 19	2 hrs.		1		19, 20, 21, 22	2
F/C-S-43	$1^1/_2''$	Floor: 2″, 1:2:4 portland cement concrete; Ceiling: $1^1/_2''$, 1:2 sanded gypsum plaster on ribbed metal lath.	See Note 19	2 hrs. 30 min.		1		19, 20, 21, 22	$2^1/_2$
F/C-S-44	$1^1/_8''$	Floor: 2″, 1:2:4 portland cement concrete; Ceiling: $1^1/_8''$, 1:1 sanded gypsum plaster.	See Note 19	2 hrs. 30 min.		1		19, 20, 21, 22	$2^1/_2$
F/C-S-45	1″	Floor: $2^1/_2''$, 1:2:4 portland cement concrete; Ceiling: 1″, 1:2 sanded gypsum plaster.	See Note 19	2 hrs. 30 min.		1		19, 20, 21, 22	$2^1/_2$
F/C-S-46	$^3/_4''$	Floor: $2^1/_2''$, 1:2:4 portland cement concrete; Ceiling: 1″ neat gypsum plaster or $^3/_4''$ gypsum-vermiculite plaster, ratio of gypsum to vermiculite 2:1 to 3:1.	See Note 19	3 hrs.		1		19, 20, 21, 22	3
F/C-S-47	$1^1/_8''$	Floor: $2^1/_2''$, 1:2:4 portland cement, sand and cinder concrete plus $^1/_2''$, $1:2^1/_2''$ cement-sand finish; total thickness 3″; Ceiling: $1^1/_8''$, 1:1 sanded gypsum plaster.	See Note 19	3 hrs.		1		19, 20, 21, 22	3
F/C-S-48	$1^1/_8''$	Floor: $2^1/_2''$, gas expanded portland cement-sand concrete plus $^1/_2''$, 1:2.5 cement-sand finish; total thickness 3″; Ceiling: $1^1/_8''$, 1:1 sanded gypsum plaster.	See Note 19	3 hrs. 30 min.		1		19, 20, 21, 22	$3^1/_2$
F/C-S-49	1″	Floor: $2^1/_2''$, 1:2:4 portland cement concrete; Ceiling: 1″ gypsum-vermiculite plaster; ratio of gypsum to vermiculite 2:1 to 3:1.	See Note 19	4 hrs.		1		19, 20, 21, 22	4
F/C-S-50	$2^1/_2''$	Floor: 2″, 1:2:4 portland cement concrete; Ceiling: 2″ interlocking gypsum tile supported on upper face of lower flanges of beams, $^1/_2''$ 1:3 sanded gypsum plaster.	See Note 19	2 hrs.		1		19, 20, 21, 22	2

(Continued)

TABLE 3.2—FLOOR/CEILING ASSEMBLIES—STEEL STRUCTURAL ELEMENTS—continued

ITEM CODE	MEMBRANE THICKNESS	CONSTRUCTION DETAILS	PERFORMANCE		REFERENCE NUMBER			NOTES	REC. HOURS
			LOAD	TIME	PRE-BMS-92	BMS-92	POST-BMS-92		
F/C-S-51	$2^1/_2''$	Floor: 2″, 1:2:4 portland cement concrete; Ceiling: 2″ precast metal reinforced gypsum tile, $^1/_2''$ 1:3 sanded gypsum plaster (tile clipped to channels which are clipped to lower flanges of beams).	See Note 19	4 hrs.		1		19, 20, 21, 22	4

For SI: 1 inch = 25.4 mm, 1 foot = 305 mm, 1 pound per square inch = 0.00689 MPa, 1 pound per square foot = 47.9 N/m^2.

Notes:

1. No protective membrane over structural steel.
2. Performance time indicates first endpoint reached only several tests were continued to points where other failures occurred.
3. Load failure.
4. Thermal failure.
5. This is an estimated time to load bearing failure. The same joist and deck specimen was used for a later test with different membrane protection.
6. Test stopped at 3 hours to reuse specimen; no endpoint reached.
7. Test stopped at 1 hour to reuse specimen; no endpoint reached.
8. All plaster used = gypsum.
9. Specimen size - 18 feet by $13^1/_2$ inches. Floor deck - base material - $^1/_4$-inch by 18-foot steel plate welded to "I" beams.
10. "I" beams - 24 inches o.c.
11. "I" beams - 48 inches o.c.
12. Apply to open web joists, pressed steel joists or rolled steel beams, which are not stressed beyond 18,000 lbs./in.^{2} in flexure for open-web pressed or light rolled joists, and 20,000 lbs./in.^{2} for American standard or heavier rolled beams.
13. Ratio of weight of portland cement to fine and coarse aggregates combined for floor slabs shall not be less than $1{:}6^1/_2$.
14. Plaster for ceiling shall be applied on metal lath which shall be tied to supports to give the equivalent of single No. 18 gage steel wires 5 inches o.c.
15. Load: maximum fiber stress in steel not to exceed 16,000 psi.
16. Prefabricated units 2 feet wide with length equal to the span, composed of two pieces of No. 18 gage formed steel welded together to give four longitudinal cells.
17. Depth not less than 3 inches and distance between cells no less than 2 inches.
18. Ceiling: metal lath tied to furring channels secured to runner channels hung from cellular steel.
19. Load: rolled steel supporting beams and steel plate base shall not be stressed beyond 20,000 psi in flexure.
    Formed steel (with wide upper flange) construction shall not be stressed beyond 16,000 psi.
20. Some type of expanded metal or woven wire shall be embedded to prevent cracking in concrete flooring.
21. Ceiling plaster shall be metal lath wired to rods or channels which are clipped or welded to steel construction. Lath shall be no smaller than 18 gage steel wire and not more than 7 inches o.c.
22. The securing rods or channels shall be at least as effective as single $^3/_{16}$-inch rods with 1-inch of their length bent over the lower flanges of beams with the rods or channels tied to this clip with 14 gage iron wire.

FIGURE 3.3—FLOOR/CEILING ASSEMBLIES—WOOD JOIST

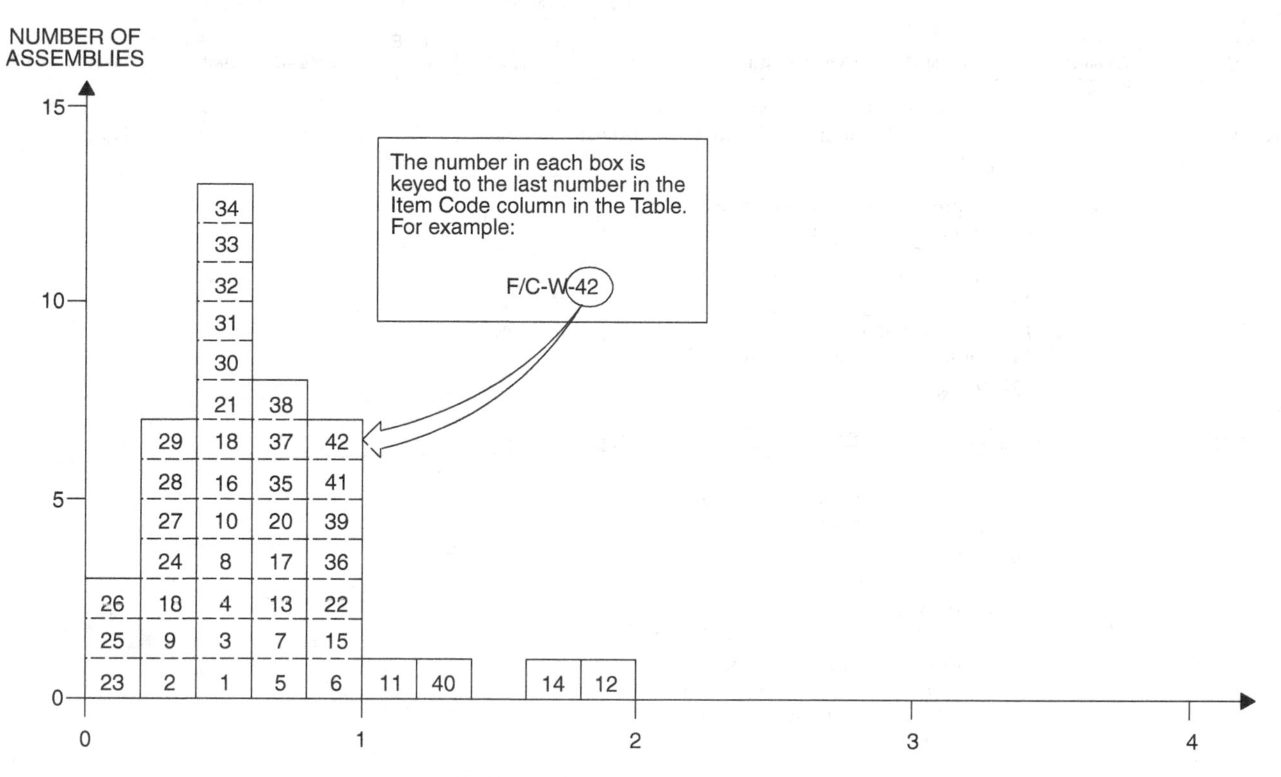

TABLE 3.3—FLOOR/CEILING ASSEMBLIES—WOOD JOIST

ITEM CODE	MEMBRANE THICKNESS	CONSTRUCTION DETAILS	PERFORMANCE		REFERENCE NUMBER			NOTES	REC. HOURS
			LOAD	TIME	PRE-BMS-92	BMS-92	POST-BMS-92		
F/C-W-1	$^{3}/_{8}''$	12′ clear span - 2″ × 9″ wood joists; 18″ o.c.; Deck: 1″ T&G; Filler: 3″ of ashes on $^{1}/_{2}''$ boards nailed to joist sides 2″ from bottom; 2″ air space; Membrane: $^{3}/_{8}''$ gypsum board.	60 psf	36 min.			7	1, 2	$^{1}/_{2}$
F/C-W-2	$^{1}/_{2}''$	12′ clear span - 2″ × 7″ joists; 15″ o.c.; Deck: 1″ nominal lumber; Membrane: $^{1}/_{2}''$ fiber board.	60 psf	22 min.			7	1, 2, 3	$^{1}/_{4}$
F/C-W-3	$^{1}/_{2}''$	12′ clear span - 2″ × 7″ wood joists; 16″ o.c.; 2″ × $1^{1}/_{2}''$ bridging at center; Deck: 1″ T&G; Membrane: $^{1}/_{2}''$ fiber board; 2 coats "distemper" paint.	30 psf	28 min.			7	1, 3, 15	$^{1}/_{3}$
F/C-W-4	$^{3}/_{16}''$	12′ clear span - 2″ × 7″ wood joists; 16″ o.c.; 2″ × $1^{1}/_{2}''$ bridging at center span; Deck: 1″ nominal lumber; Membrane: $^{1}/_{2}''$ fiber board under $^{3}/_{16}''$ gypsum plaster.	30 psf	32 min.			7	1, 2	$^{1}/_{2}$
F/C-W-5	$^{5}/_{8}''$	As per previous F/C-W-4 except membrane is $^{5}/_{8}''$ lime plaster.	70 psf	48 min.			7	1, 2	$^{3}/_{4}$
F/C-W-6	$^{5}/_{8}''$	As per previous F/C-W-5 except membrane is $^{5}/_{8}''$ gypsum plaster on 22 gage $^{3}/_{8}''$ metal lath.	70 psf	49 min.			7	1, 2	$^{3}/_{4}$

(Continued)

TABLE 3.3—FLOOR/CEILING ASSEMBLIES—WOOD JOIST—continued

ITEM CODE	MEMBRANE THICKNESS	CONSTRUCTION DETAILS	PERFORMANCE		REFERENCE NUMBER			NOTES	REC. HOURS
			LOAD	TIME	PRE-BMS-92	BMS-92	POST-BMS-92		
F/C-W-7	$^1/_2$″	As per previous F/C-W-6 except membrane is $^1/_2$″ fiber board under $^1/_2$″ gypsum plaster.	60 psf	43 min.			7	1, 2, 3	$^2/_3$
F/C-W-8	$^1/_2$″	As per previous F/C-W-7 except membrane is $^1/_2$″ gypsum board.	60 psf	33 min.			7	1, 2, 3	$^1/_2$
F/C-W-9	$^9/_{16}$″	12′ clear span - 2″ × 7″ wood joists; 15″ o.c.; 2″ × 1$^1/_2$″ bridging at center; Deck: 1″ nominal lumber; Membrane: $^3/_8$″ gypsum board; $^3/_{16}$″ gypsum plaster.	60 psf	24 min.			7	1, 2, 3	$^1/_3$
F/C-W-10	$^5/_8$″	As per F/C-W-9 except membrane is $^5/_8$″ gypsum plaster on wood lath.	60 psf	27 min.			7	1, 2, 3	$^1/_3$
F/C-W-11	$^7/_8$″	12′ clear span - 2″ × 9″ wood joists; 15″ o.c.; 2″ × 1$^1/_2$″ bridging at center span; Deck: 1″ T&G; Membrane: original ceiling joists have $^3/_8$″ plaster on wood lath; 4″ metal hangers attached below joists creating 15″ chases filled with mineral wool and closed with $^7/_8$″ plaster (gypsum) on $^3/_8$″ S.W.M. metal lath to form new ceiling surface.	75 psf	1 hr. 10 min.			7	1, 2	1
F/C-W-12	$^7/_8$″	12′ clear span - 2″ × 9″ wood joists; 15″ o.c.; 2″ × 1$^1/_2$″ bridging at center; Deck: 1″ T&G; Membrane: 3″ mineral wood below joists; 3″ hangers to channel below joists; $^7/_8$″ gypsum plaster on metal lath attached to channels.	75 psf	2 hrs.			7	1, 4	2
F/C-W-13	$^7/_8$″	12′ clear span - 2″ × 9″ wood joists; 16″ o.c.; 2″ × 1$^1/_2$″ bridging at center span; Deck: 1″ T&G on 1″ bottoms on $^3/_4$″ glass wool strips on $^3/_4$″ gypsum board nailed to joists; Membrane: $^3/_4$″ glass wool strips on joists; $^3/_8$″ perforated gypsum lath; $^1/_2$″ gypsum plaster.	60 psf	41 min.			7	1, 3	$^2/_3$
F/C-W-14	$^7/_8$″	12′ clear span - 2″ × 9″ wood joists; 15″ o.c.; Deck: 1″ T&G; Membrane: 3″ foam concrete in cavity on $^1/_2$″ boards nailed to joists; wood lath nailed to 1″ × 1$^1/_4$″ straps 14 o.c. across joists; $^7/_8$″ gypsum plaster.	60 psf	1 hr. 40 min.			7	1, 5	1$^2/_3$
F/C-W-15	$^7/_8$″	12′ clear span - 2″ × 9″ wood joists; 18″ o.c.; Deck: 1″ T&G; Membrane: 2″ foam concrete on $^1/_2$″ boards nailed to joist sides 2″ from joist bottom; 2″ air space; 1″ × 1$^1/_4$″ wood straps 14″ o.c. across joists; $^7/_8$″ lime plaster on wood lath.	60 psf	53 min.			7	1, 2	$^3/_4$

(Continued)

TABLE 3.3—FLOOR/CEILING ASSEMBLIES—WOOD JOIST—continued

ITEM CODE	MEMBRANE THICKNESS	CONSTRUCTION DETAILS	PERFORMANCE		REFERENCE NUMBER			NOTES	REC. HOURS
			LOAD	TIME	PRE-BMS-92	BMS-92	POST-BMS-92		
F/C-W-16	$^7/_8$″	12′ clear span - 2″ × 9″ wood joists; Deck: 1″ T&G; Membrane: 3″ ashes on $^1/_2$″ boards nailed to joist sides 2″ from joist bottom; 2″ air space; 1″ × 1$^1/_4$″ wood straps 14″ o.c.; $^7/_8$″ gypsum plaster on wood lath.	60 psf	28 min.			7	1, 2	$^1/_3$
F/C-W-17	$^7/_8$″	As per previous F/C-W-16 but with lime plaster mix.	60 psf	41 min.			7	1, 2	$^2/_3$
F/C-W-18	$^7/_8$″	12′ clear span - 2″ × 9″ wood joists; 18″ o.c.; 2″ × 1$^1/_2$″ bridging at center; Deck: 1″ T&G; Membrane: $^7/_8$″ gypsum plster on wood lath.	60 psf	36 min.			7	1, 2	$^1/_2$
F/C-W-19	$^7/_8$″	As per previous F/C-W-18 except with lime plaster membrane and deck is 1″ nominal boards (plain edge).	60 psf	19 min.			7	1, 2	$^1/_4$
F/C-W-20	$^7/_8$″	As per F/C-W-19, except deck is 1″ T&G boards.	60 psf	43 min.			7	1, 2	$^2/_3$
F/C-W-21	1″	12′ clear span - 2″ × 9″ wood joists; 16″ o.c.; 2″ × 1$^1/_2$″ bridging at center; Deck: 1″ T&G; Membrane: $^3/_8$″ gypsum base board; $^5/_8$″ gypsum plaster.	70 psf	29 min.			7	1, 2	$^1/_3$
F/C-W-22	1$^1/_8$″	12′ clear span - 2″ × 9″ wood joists; 16″ o.c.; 2″ × 2″ wood bridging at center; Deck: 1″ T&G; Membrane: hangers, channel with $^3/_8$″ gypsum baseboard affixed under $^3/_4$″ gypsum plaster.	60 psf	1 hr.			7	1, 2, 3	1
F/C-W-23	$^3/_8$″	Deck: 1″ nominal lumber; Joists: 2″ × 7″; 15″ o.c.; Membrane: $^3/_8$″ plasterboard with plaster skim coat.	60 psf	11$^1/_2$ min.			12	2, 6	$^1/_6$
F/C-W-24	$^1/_2$″	Deck: 1″ T&G lumber; Joists: 2″ × 9″; 16″ o.c.; Membrane: $^1/_2$″ plasterboard.	60 psf	18 min.			12	2, 7	$^1/_4$
F/C-W-25	$^1/_2$″	Deck: 1″ T&G lumber; Joists: 2″ × 7″; 16″ o.c.; Membrane: $^1/_2$″ fiber insulation board.	30 psf	8 min.			12	2, 8	$^2/_{15}$
F/C-W-26	$^1/_2$″	Deck: 1″ nominal lumber; Joists: 2″ × 7″; 15″ o.c.; Membrane: $^1/_2$″ fiber insulation board.	60 psf	8 min.			12	2, 9	$^2/_{15}$
F/C-W-27	$^5/_8$″	Deck: 1″ nominal lumber; Joists: 2″ × 7″; 15″ o.c.; Membrane: $^5/_8$″ gypsum plaster on wood lath.	60 psf	17 min.			12	2, 10	$^1/_4$
F/C-W-28	$^5/_8$″	Deck: 1″ T&G lumber; Joists: 2″ × 9″; 16″ o.c.; Membrane: $^1/_2$″ fiber insulation board; $^1/_2$″ plaster.	60 psf	20 min.			12	2, 11	$^1/_3$
F/C-W-29	No Membrane	Exposed wood joists.	See Note 13	15 min.		1		1, 12, 13, 14	$^1/_4$

(Continued)

TABLE 3.3—FLOOR/CEILING ASSEMBLIES—WOOD JOIST—continued

ITEM CODE	MEMBRANE THICKNESS	CONSTRUCTION DETAILS	PERFORMANCE		REFERENCE NUMBER			NOTES	REC. HOURS
			LOAD	TIME	PRE-BMS-92	BMS-92	POST-BMS-92		
F/C-W-30	$^3/_8''$	Gypsum wallboard: $^3/_8''$ or $^1/_2''$ with $1^1/_2''$ No. 15 gage nails with $^3/_{16}''$ heads spaced 6″ centers with asbestos paper applied with paperhangers' paste and finished with casein paint.	See Note 13	25 min.		1		1, 12, 13, 14	$^1/_2$
F/C-W-31	$^1/_2''$	Gypsum wallboard: $^1/_2''$ with $1^3/_4''$ No. 12 gage nails with $^1/_2''$ heads, 6″ o.c., and finished with casein paint.	See Note 13	25 min.		1		1, 12, 13, 14	$^1/_2$
F/C-W-32	$^1/_2''$	Gypsum wallboard: $^1/_2''$ with $1^1/_2''$ No. 12 gage nails with $^1/_2''$ heads, 18″ o.c., with asbestos paper applied with paperhangers' paste and secured with $1^1/_2''$ No. 15 gage nails with $^3/_{16}''$ heads and finished with casein paint; combined nail spacing 6″ o.c.	See Note 13	30 min.		1		1, 12, 13, 14	$^1/_2$
F/C-W-33	$^3/_8''$	Gypsum wallboard: two layers $^3/_8''$ secured with $1^1/_2''$ No. 15 gage nails with $^3/_8''$ heads, 6″ o.c.	See Note 13	30 min.		1		1, 12, 13, 14	$^1/_2$
F/C-W-34	$^1/_2''$	Perforated gypsum lath: $^3/_8''$, plastered with $1^1/_8''$ No. 13 gage nails with $^5/_{16}''$ heads, 4″ o.c.; $^1/_2''$ sanded gypsum plaster.	See Note 13	30 min.		1		1, 12, 13, 14	$^1/_2$
F/C-W-35	$^1/_2''$	Same as F/C-W-34, except with $1^1/_8''$ No. 13 gage nails with $^3/_8''$ heads, 4″ o.c.	See Note 13	45 min.		1		1, 12, 13, 14	$^3/_4$
F/C-W-36	$^1/_2''$	Perforated gypsum lath: $^3/_8''$, nailed with $1^1/_8''$ No. 13 gage nails with $^3/_8''$ heads, 4″ o.c.; joints covered with 3″ strips of metal lath with $1^3/_4''$ No. 12 nails with $^1/_2''$ heads, 5″ o.c.; $^1/_2''$ sanded gypsum plaster.	See Note 13	1 hr.		1		1, 12, 13, 14	1
F/C-W-37	$^1/_2''$	Gypsum lath: $^3/_8''$ and lower layer of $^3/_8''$ perforated gypsum lath nailed with $1^3/_4''$ No. 13 nails with $^5/_{16}''$ heads, 4″ o.c.; $^1/_2''$ sanded gypsum plaster or $^1/_2''$ portland cement plaster.	See Note 13	45 min.		1		1, 12, 13, 14	$^3/_4$
F/C-W-38	$^3/_4''$	Metal lath: nailed with $1^1/_4''$ No. 11 nails with $^3/_8''$ heads or 6d common driven 1″ and bent over, 6″ o.c.; $^3/_4''$ sanded gypsum plaster.	See Note 13	45 min.		1		1, 12, 13, 14	$^3/_4$
F/C-W-39	$^3/_4''$	Same as F/C-W-38, except nailed with $1^1/_2''$ No. 11 barbed roof nails with $^7/_{16}''$ heads, 6″ o.c.	See Note 13	1 hr.		1		1, 12, 13, 14	1

(Continued)

TABLE 3.3—FLOOR/CEILING ASSEMBLIES—WOOD JOIST—continued

ITEM CODE	MEMBRANE THICKNESS	CONSTRUCTION DETAILS	PERFORMANCE		REFERENCE NUMBER			NOTES	REC. HOURS
			LOAD	TIME	PRE-BMS-92	BMS-92	POST-BMS-92		
F/C-W-40	$^3/_4''$	Same as F/C-W-38, except with lath nailed to joists with additional supports for lath 27″ o.c.; attached to alternate joists and consisting of two nails driven $1^1/_4''$, 2″ above bottom on opposite sides of the joists, one loop of No. 18 wire slipped over each nail; the ends twisted together below lath.	See Note 13	1 hr. 15 min.		1		1, 12, 13, 14	$1^1/_4$
F/C-W-41	$^3/_4''$	Metal lath: nailed with $1^1/_2''$ No. 11 barbed roof nails with $^7/_{16}''$ heads, 6 o.c., with $^3/_4''$ portland cement plaster for scratch coat and 1:3 for brown coat, 3 lbs. of asbestos fiber and 15 lbs. of hydrated lime/94 lbs. bag of cement.	See Note 13	1 hr.		1		1, 12, 13, 14	1
F/C-W-42	$^3/_4''$	Metal lath: nailed with 8d, No. $11^1/_2$ gage barbed box nails, $2^1/_2''$ driven, $1^1/_4''$ on slant and bent over, 6″ o.c.; $^3/_4''$ sanded gypsum plaster, 1:2 for scratch coat and 1:3 for below coat.	See Note 13	1 hr.		1		1, 12, 13, 14	1

For SI: 1 inch = 25.4 mm, 1 foot = 305 mm, 1 pound per square inch = 0.00689 MPa, 1 pound per square foot = 47.9 N/m^2.

Notes:

1. Thickness indicates thickness of first membrane protection on ceiling surface.
2. Failure mode - flame thru.
3. Failure mode - collapse.
4. No endpoint reached at termination of test
5. Failure imminent - test terminated.
6. Joist failure - 11.5 minutes; flame thru - 13 minutes; collapse - 24 minutes.
7. Joist failure - 17 minutes; flame thru - 18 minutes; collapse - 33 minutes.
8. Joist failure - 18 minutes; flame thru - 8 minutes; collapse - 30 minutes.
9. Joist failure - 12 minutes; flame thru - 8 minutes; collapse - 22 minutes.
10. Joist failure - 11 minutes; flame thru - 17 minutes; collapse - 27 minutes.
11. Joist failure - 17 minutes; flame thru - 20 minutes; collapse - 43 minutes.
12. Joists: 2-inch by 10-inch southern pine or Douglas fir; No. 1 common or better. Subfloor: $^3/_4$-inch wood sheating diaphragm of asbestos paper, and finish of tongue-and-groove wood flooring.
13. Loadings: not more than 1,000 psi maximum fiber stress in joists.
14. Perforations in gypsum lath are to be not less than $^3/_4$-inch diameter with one perforation for not more than 16/in.^{2} diameter.
15. "Distemper" is a British term for a water-based paint such as white wash or calcimine.

FIGURE 3.4—FLOOR/CEILING ASSEMBLIES—HOLLOW CLAY TILE WITH REINFORCED CONCRETE

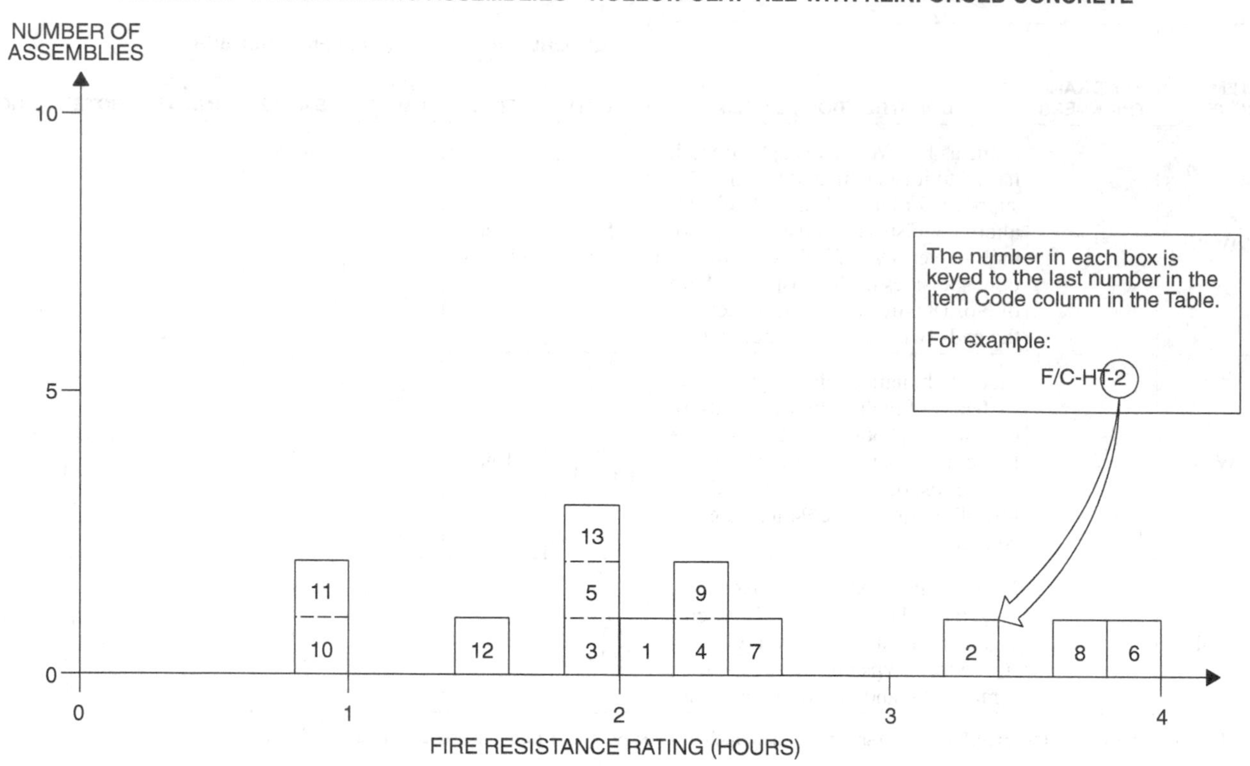

TABLE 3.4—FLOOR/CEILING ASSEMBLIES—HOLLOW CLAY TILE WITH REINFORCED CONCRETE

ITEM CODE	ASSEMBLY THICKNESS	CONSTRUCTION DETAILS	PERFORMANCE		REFERENCE NUMBER			NOTES	REC. HOURS
			LOAD	TIME	PRE-BMS-92	BMS-92	POST-BMS-92		
F/C-HT-1	6″	Cover: $1^1/_2$″ concrete (6080 psi); three cell hollow clay tiles, 12″ × 12″ × 4″; $3^1/_4$″ concrete between tiles including two $^1/_2$″ rebars with $^3/_4$″ concrete cover; $^1/_2$″ plaster cover, lower.	75 psf	2 hrs. 7 min.			7	1, 2, 3	2
F/C-HT-2	6″	Cover: $1^1/_2$″ concrete (5840 psi); three cell hollow clay tiles, 12″ × 12″ × 4″; $3^1/_4$″ concrete between tiles including two $^1/_2$″ rebars each with $^1/_2$″ concrete cover and $^5/_8$″ filler tiles between hollow tiles; $^1/_2$″ plaster cover, lower.	61 psf	3 hrs. 23 min.			7	3, 4, 6	$3^1/_3$
F/C-HT-3	6″	Cover: $1^1/_2$″ concrete (6280 psi); three cell hollow clay tiles, 12″ × 12″ × 4″; $3^1/_4$″ concrete between tiles including two $^1/_2$″ rebars with $^1/_2$″ cover; $^1/_2$″ plaster cover, lower.	122 psf	2 hrs.			7	1, 3, 5, 8	2
F/C-HT-4	6″	Cover: $1^1/_2$″ concrete (6280 psi); three cell hollow clay tiles, 12″ × 12″ × 4″; $3^1/_4$″ concrete between tiles including two $^1/_2$″ rebars with $^3/_4$″ cover; $^1/_2$″ plaster cover, lower.	115 psf	2 hrs. 23 min.			7	1, 3, 7	$2^1/_3$

(Continued)

TABLE 3.4—FLOOR/CEILING ASSEMBLIES—HOLLOW CLAY TILE WITH REINFORCED CONCRETE—continued

ITEM CODE	ASSEMBLY THICKNESS	CONSTRUCTION DETAILS	PERFORMANCE		REFERENCE NUMBER			NOTES	REC. HOURS
			LOAD	TIME	PRE-BMS-92	BMS-92	POST-BMS-92		
F/C-HT-5	6"	Cover: $1\frac{1}{2}$" concrete (6470 psi); three cell hollow clay tiles, 12" × 12" × 4"; $3\frac{1}{4}$" concrete between tiles including two $\frac{1}{2}$" rebars with $\frac{1}{2}$" cover; $\frac{1}{2}$" plaster cover, lower.	122 psf	2 hrs.			7	1, 3, 5, 8	2
F/C-HT-6	8"	Floor cover: $1\frac{1}{2}$" gravel cement (4300 psi); three cell, 12" × 12" × 6"; $3\frac{1}{2}$" space between tiles including two $\frac{1}{2}$" rebars with 1" cover from concrete bottom; $\frac{1}{2}$" plaster cover, lower.	165 psf	4 hrs.			7	1, 3, 9, 10	4
F/C-HT-7	9" (nom.)	Deck: $\frac{7}{8}$" T&G on 2" × $1\frac{1}{2}$" bottoms (18" o.c.) $1\frac{1}{2}$" concrete cover (4600 psi); three cell hollow clay tiles, 12" × 12" × 4"; 3" concrete between tiles including one $\frac{3}{4}$" rebar $\frac{3}{4}$" from tile bottom; $\frac{3}{4}$" plaster cover.	95 psf	2 hrs. 26 min.			7	4, 11, 12, 13	$2\frac{1}{3}$
F/C-HT-8	9" (nom.)	Deck: $\frac{7}{8}$" T&G on 2" × $1\frac{1}{2}$" bottoms (18" o.c.) $1\frac{1}{2}$" concrete cover (3850 psi); three cell hollow clay tiles, 12" × 12" × 4"; 3" concrete between tiles including one $\frac{3}{4}$" rebar $\frac{3}{4}$" from tile bottoms; $\frac{1}{2}$" plaster cover.	95 psf	3 hrs. 28 min.			7	4, 11, 12, 13	
F/C-HT-9	9" (nom.)	Deck: $\frac{7}{8}$" T&G on 2" × $1\frac{1}{2}$" bottoms (18" o.c.) $1\frac{1}{2}$" concrete cover (4200 psi); three cell hollow clay tiles, 12" × 12" × 4"; 3" concrete between tiles including one $\frac{3}{4}$" rebar $\frac{3}{4}$" from tile bottoms; $\frac{1}{2}$" plaster cover.	95 psf	2 hrs. 14 min.			7	3, 5, 8, 11	
F/C-HT-10	$5\frac{1}{2}$"	Fire clay tile (4" thick); $1\frac{1}{2}$" concrete cover; for general details, see Note 15.	See Note 14	1 hr.			43	15	1
F/C-HT-11	8"	Fire clay tile (6" thick); 2" cover.	See Note 14	1 hr.			43	15	1
F/C-HT-12	$5\frac{1}{2}$"	Fire clay tile (4" thick); $1\frac{1}{2}$" cover; $\frac{5}{8}$" gypsum plaster, lower.	See Note 14	1 hr. 30 min.			43	15	$1\frac{1}{2}$
F/C-HT-13	8"	Fire clay tile (6" thick); 2" cover; $\frac{5}{8}$" gypsum plaster, lower.	See Note 14	2 hrs.			43	15	$1\frac{1}{2}$

For SI: 1 inch = 25.4 mm, 1 foot = 305 mm, 1 pound per square inch = 0.00689MPa, 1 pound per square foot = 47.9 N/m^2.

Notes:

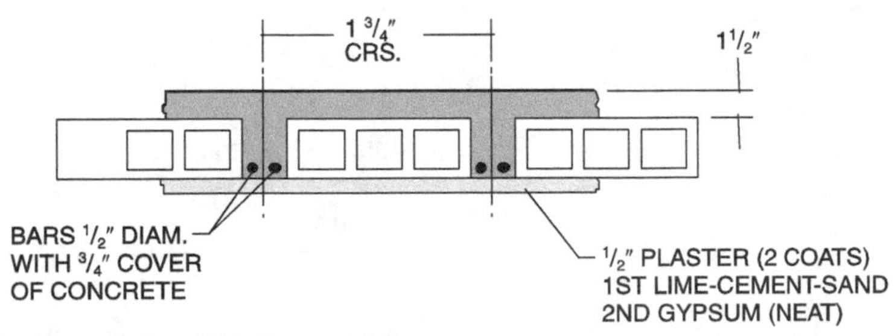

1. A generalized cross section of this floor type follows:
2. Failure mode - structural.
3. Plaster: base coat - lime-cement-sand; top coat - gypsum (neat).

TABLE 3.4—FLOOR/CEILING ASSEMBLIES—HOLLOW CLAY TILE WITH REINFORCED CONCRETE—continued

4. Failure mode - collapse.
5. Test stopped before any endpoints were reached.
6. A generalized cross section of this floor type follows:

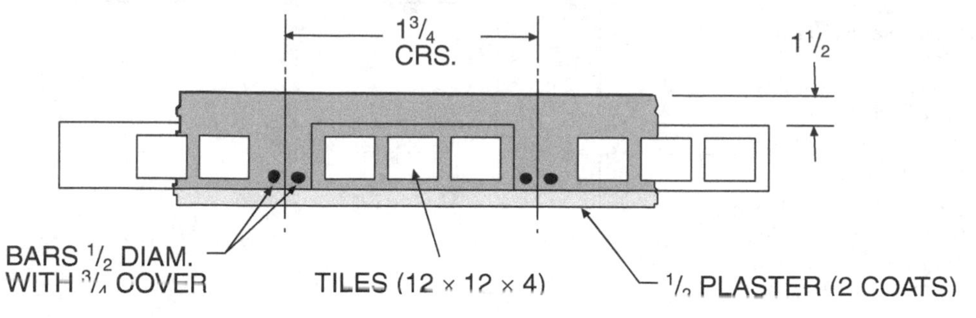

7. Failure mode - thermal - back face temperature rise.
8. Passed hose stream test.
9. Failed hose stream test.

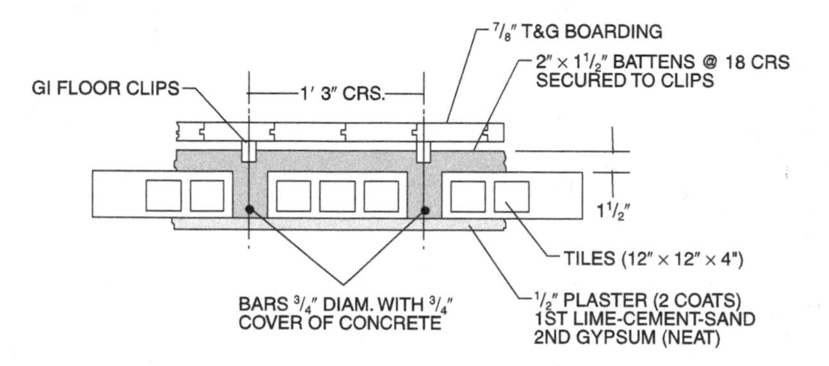

10. Test stopped at 4 hours before any endpoints were reached.
11. A generalized cross section of this floor type follows:
12. Plaster: base coat - retarded hemihydrate gypsum-sand; second coat - neat gypsum.
13. Concrete in Item 7 is P.C. based but with crushed brick aggregates while in Item 8 river sand and river gravels are used with the P.C.
14. Load - unspecified.
15. The 12-inch by 12-inch fire-clay tiles were laid end to end in rows spaced $2^1/_2$ inches or 4 inches apart. The reinforcing steel was placed between these rows and the concrete cast around them and over the tile to form the structural floor.

SECTION IV—BEAMS

TABLE 4.1.1—REINFORCED CONCRETE BEAMS DEPTH 10″ TO LESS THAN 12″

ITEM CODE	DEPTH	CONSTRUCTION DETAILS	PERFORMANCE		REFERENCE NUMBER			NOTES	REC. HOURS
			LOAD	TIME	PRE-BMS-92	BMS-92	POST-BMS-92		
B-11-RC-1	11″	24″ wide × 11″ deep reinforced concrete "T" beam (3290 psi); Details: see Note 5 figure.	8.8 tons	4 hrs. 2 min.			7	1, 2, 14	4
B-10-RC-2	10″	24″ wide × 10″ deep reinforced concrete "T" beam (4370 psi); Details: see Note 6 figure.	8.8 tons	1 hr. 53 min.			7	1, 3	$1^3/_4$
B-10-RC-3	$10^1/_2$″	24″ wide × $10^1/_2$″ deep reinforced concrete "T" beam (4450 psi); Details: see Note 7 figure.	8.8 tons	2 hrs. 40 min.			7	1, 3	$2^2/_3$
B-11-RC-4	11″	24″ wide × 11″ deep reinforced concrete "T" beam (2400 psi); Details: see Note 8 figure.	8.8 tons	3 hrs. 32 min.			7	1, 3, 14	$3^1/_2$
B-11-RC-5	11″	24″ wide × 11″ deep reinforced concrete "T" beam (4250 psi); Details: see Note 9 figure.	8.8 tons	3 hrs. 3 min.			7	1, 3, 14	3
B-11-RC-6	11″	Concrete flange: 4″ deep × 2′ wide (4895 psi) concrete; Concrete beam: 7″ deep × $6^1/_2$″ wide beam; "I" beam reinforcement; 10″ × $4^1/_2$″ × 25 lbs. R.S.J.; 1″ cover on flanges; Flange reinforcement: $^3/_8$″ diameter bars at 6″ pitch parallel to "T"; $^1/_4$″ diameter bars perpendicular to "T"; Beam reinforcement: 4″ × 6″ wire mesh No. 13 SWG; Span: 11′ restrained; Details: see Note 10 figure.	10 tons	6 hrs.			7	1, 4	6
B-11-RC-7	11″	Concrete flange: 6″ deep × 1′$6^1/_2$″ wide (3525 psi) concrete; Concrete beam: 5″ deep × 8″ wide precast concrete blocks $8^3/_4$″ long; "I" beam reinforcement; 7″ × 4″ × 16 lbs. R.S.J.; 2″ cover on bottom; $1^1/_2$″ cover on top; Flange reinforcement: two rows $^1/_2$″ diameter rods parallel to "T"; Beam reinforcement: $^1/_8$″ wire mesh perpendicular to 1″; Span: 1′3″ simply supported; Details: see Note 11 figure.	3.9 tons	4 hrs.			7	1, 2	4
B-11-RC-8	11″	Concrete flange: 4″ deep × 2′ wide (3525 psi) concrete; Concrete beam 7″ deep × $4^1/_2$″ wide; (scaled from drawing); "I" beam reinforcement; 10″ × $4^1/_2$″ × 25 lbs. R.S.J.; no concrete cover on bottom; Flange reinforcement: $^3/_8$″ diameter bars at 6 pitch parallel to "T"; $^1/_4$″ diameter bars perpendicular to "T"; Span: 11′ restricted.	10 tons	4 hrs.			7	1, 2, 12	4

(Continued)

TABLE 4.1.1—REINFORCED CONCRETE BEAMS DEPTH 10″ TO LESS THAN 12″—continued

ITEM CODE	DEPTH	CONSTRUCTION DETAILS	PERFORMANCE		REFERENCE NUMBER			NOTES	REC. HOURS
			LOAD	TIME	PRE-BMS-92	BMS-92	POST-BMS-92		
B-11-RC-9	$11^1/_2$″	24″ wide × $11^1/_2$″ deep reinforced concrete "T" beam (4390 psi); Details: see Note 12 figure.	8.8 tons	3 hrs. 24 min.			7	1, 3	$3^1/_3$

For SI: 1 inch = 25.4 mm, 1 foot = 305 mm, 1 pound = 0.004448 kN, 1 pound per square inch = 0.00689 MPa, 1 ton = 8.896 kN.

Notes:

1. Load concentrated at mid span.
2. Achieved 4 hour performance (Class "B," British).
3. Failure mode – collapse.
4. Achieved 6 hour performance (Class "A," British).
5.

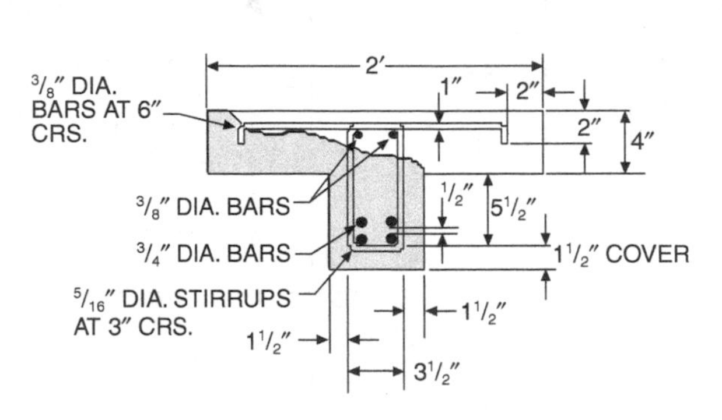

6.

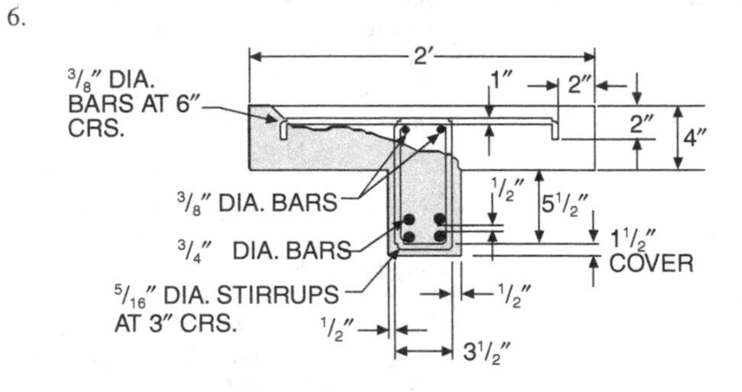

7.

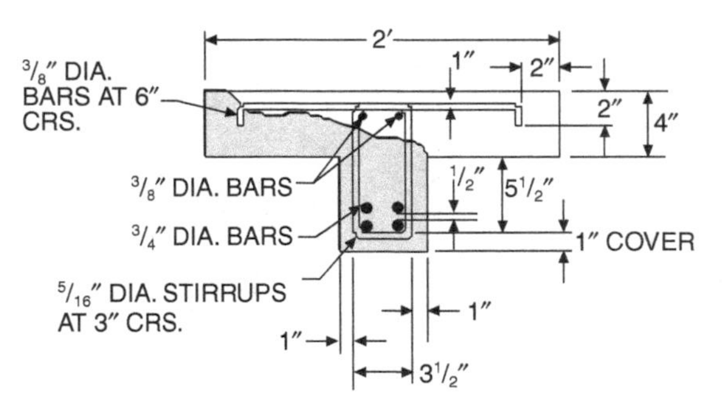

8.

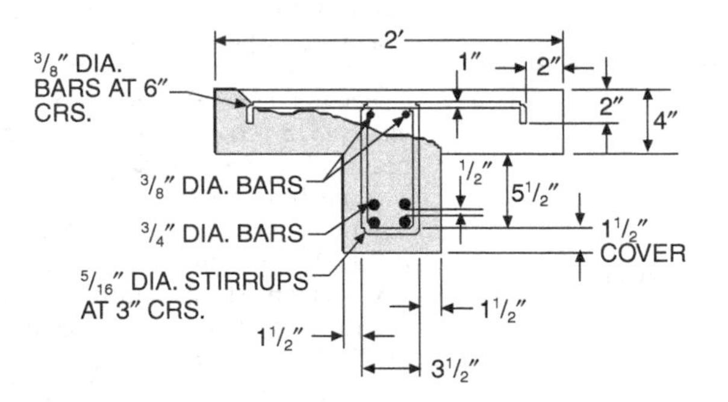

9.

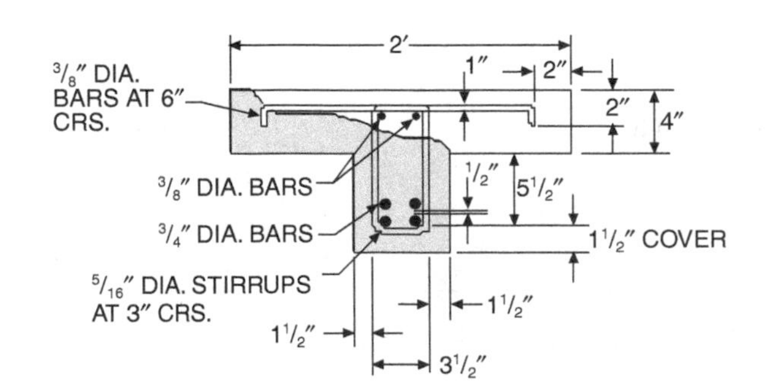

10.

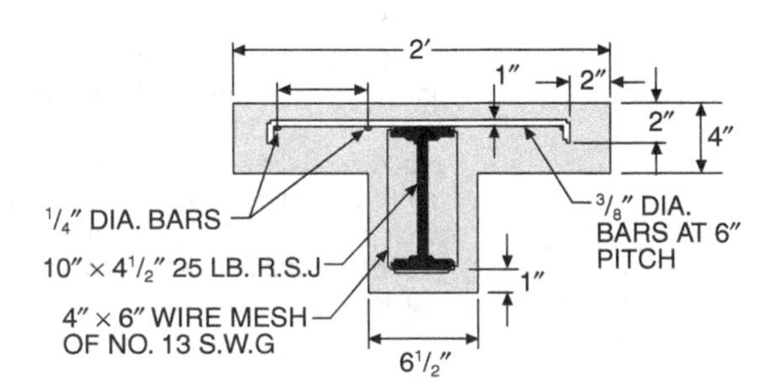

(Continued)

TABLE 4.1.1—REINFORCED CONCRETE BEAMS
DEPTH 10″ TO LESS THAN 12″—continued

11.

12.

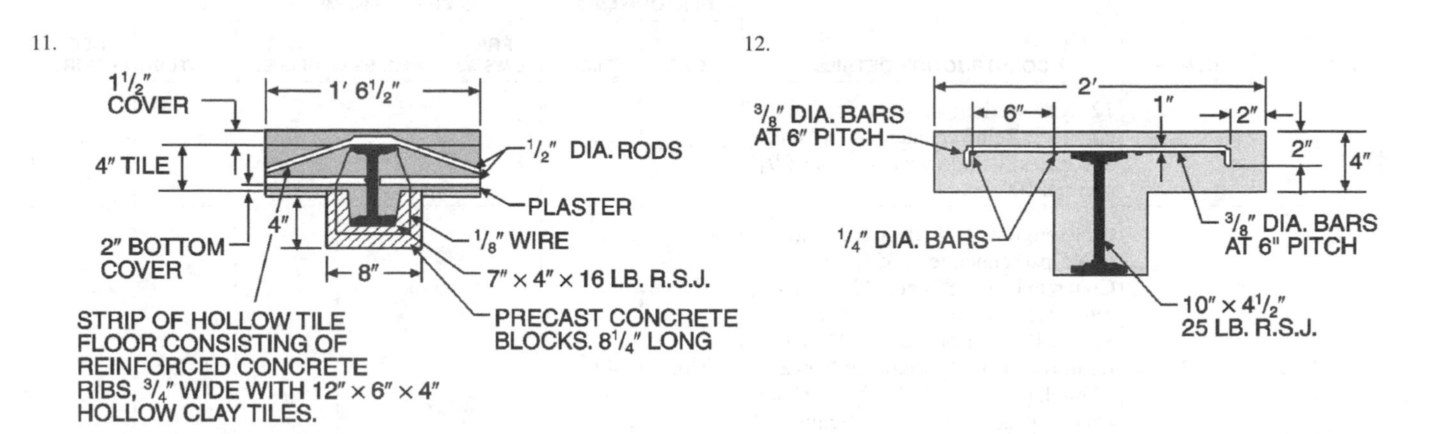

13.

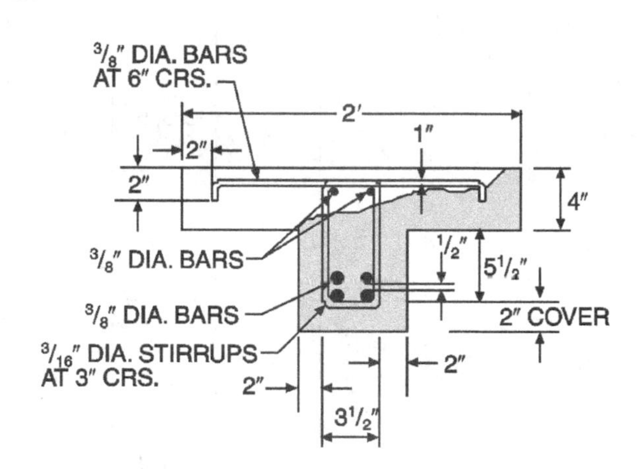

14. The different performances achieved by B-11-RC-1, B-11-RC-4 and B-11-RC-5 are attributable to differences in concrete aggregate compositions reported in the source document but unreported in this table. This demonstrates the significance of material composition in addition to other details.

TABLE 4.1.2—REINFORCED CONCRETE BEAMS DEPTH 12″ TO LESS THAN 14″

ITEM CODE	DEPTH	CONSTRUCTION DETAILS	PERFORMANCE		REFERENCE NUMBER			NOTES	REC. HOURS
			LOAD	TIME	PRE-BMS-92	BMS-92	POST-BMS-92		
B-12-RC-1	12″	12″ × 8″ section; 4160 psi aggregate concrete; Reinforcement: 4-$^{7}/_{8}$″ rebars at corners; 1″ below each surface; $^{1}/_{4}$″ stirrups 10″ o.c.	5.5 tons	2 hrs.			7	1	2
B-12-RC-2	12″	Concrete flange: 4″ deep × 2′ wide (3045 psi) concrete at 35 days; Concrete beam: 8″ deep; "I" beam reinforcement: 10″ × 4$^{1}/_{2}$″ × 25 lbs. R.S.J.; 1″ cover on flanges; Flange reinforcement: $^{3}/_{8}$″ diameter bars at 6″ pitch parallel to "T"; $^{1}/_{4}$″ diameter bars perpendicular to "T"; Beam reinforcement: 4″ × 6″ wire mesh No. 13 SWG; Span: 10′ 3″ simply supported.	10 tons	4 hrs.			7	2, 3, 5	4
B-13-RC-3	13″	Concrete flange: 4″ deep × 2′ wide (3825 psi) concrete at 46 days; Concrete beam: 9″ deep × 8$^{1}/_{2}$″ wide; (scaled from drawing); "I" beam reinforcement: 10″ × 4$^{1}/_{2}$″ × 25 lbs. R.S.J.; 3″ cover on bottom flange; 1″ cover on top flange; Flange reinforcement: $^{3}/_{8}$″ diameter bars at 6″ pitch parallel to "T"; $^{1}/_{4}$″ diameter bars perpendicular to "T"; Beam reinforcement: 4″ × 6″ wire mesh No. 13 SWG; Span: 11′ restrained.	10 tons	6 hrs.			7	2, 3, 6, 8, 9	4
B-12-RC-4	12″	Concrete flange: 4″ deep × 2′ wide (3720 psi) concrete at 42 days; Concrete beam: 8″ deep × 8$^{1}/_{2}$″ wide; (scaled from drawing); "I" beam reinforcement: 10″ × 4$^{1}/_{2}$″ × 25 lbs. R.S.J.; 2″ cover bottom flange; 1″ cover top flange; Flange reinforcement: $^{3}/_{8}$″ diameter bars at 6″ pitch parallel to "T"; $^{1}/_{4}$″ diameter bars perpendicular to "T"; Beam reinforcement: 4″ × 6″ wire mesh No. 13 SWG; Span: 11′ restrained.	10 tons	6 hrs.			7	1, 3, 4, 7, 8, 9	4

For SI: 1 inch = 25.4 mm, 1 foot = 305 mm, 1 pound = 0.004448 kN, 1 pound per square inch = 0.00689 MPa, 1 ton = 8.896 kN.

Notes:

1. Qualified for 2 hour use. (Grade "C," British) Test included hose stream and reload at 48 hours.
2. Load concentrated at mid span.
3. British test.
4. British test - qualified for 6 hour use (Grade "A").

(Continued)

TABLE 4.1.2—REINFORCED CONCRETE BEAMS DEPTH 12″ TO LESS THAN 14″—continued

5.

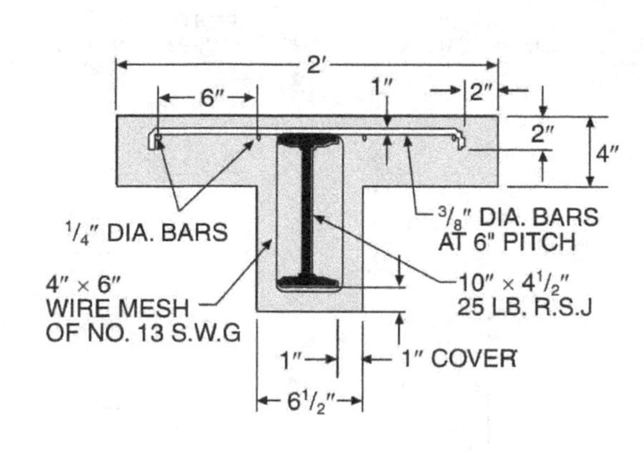

6.

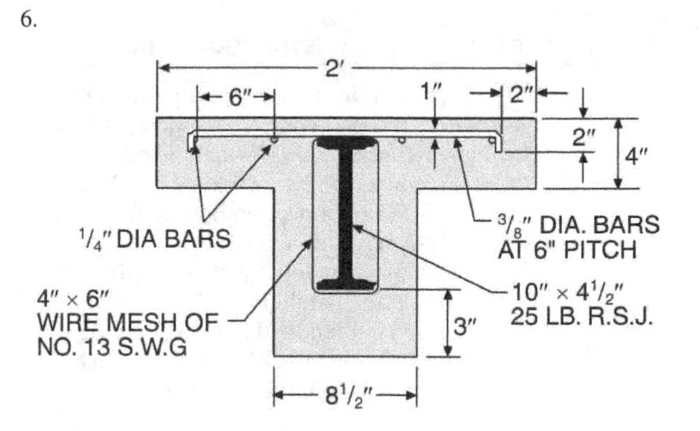

7.

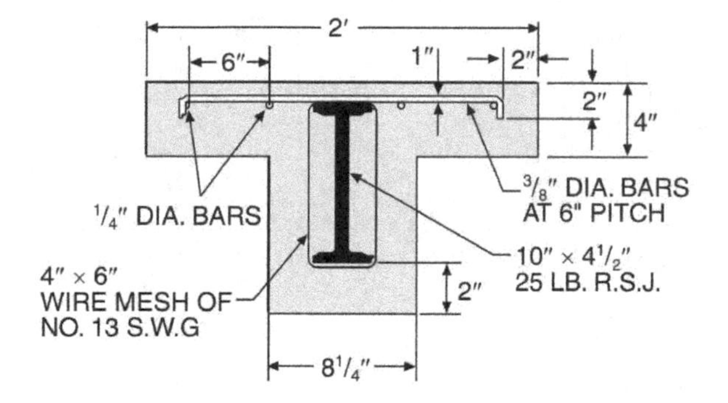

8. See Table 4.1.3, Note 5.

9. Hourly rating based upon B-12-RC-2 above.

TABLE 4.1.3—REINFORCED CONCRETE BEAMS DEPTH 14″ TO LESS THAN 16″

ITEM CODE	DEPTH	CONSTRUCTION DETAILS	PERFORMANCE		REFERENCE NUMBER			NOTES	REC. HOURS
			LOAD	TIME	PRE-BMS-92	BMS-92	POST-BMS-92		
B-15-RC-1	15″	Concrete flange: 4″ deep × 2′ wide (3290 psi) concrete; Concrete beam: 10″ deep × $8^1/_2$″ wide; "I" beam reinforcement: 10″ × $4^1/_2$″ × 25 lbs. R.S.J.; 4″ cover on bottom flange; 1″ cover on top flange; Flange reinforcement: $^3/_8$″ diameter bars at 6″ pitch parallel to "T"; $^1/_4$″ diameter bars perpendicular to "T"; Beam reinforcement: 4″ × 6″ wire mesh No. 13 SWG; Span: 11′ restrained.	10 tons	6 hrs.			7	1, 2, 3 5, 6	4
B-15-RC-2	15″	Concrete flange: 4″ deep × 2′ wide (4820 psi) concrete; Concrete beam: 10″ deep × $8^1/_2$″ wide; "I" beam reinforcement: 10″ × $4^1/_2$″ × 25 lbs. R.S.J.; 1″ cover over wire mesh on bottom flange; 1″ cover on top flange; Flange reinforcement: $^3/_8$″ diameter bars at 6″ pitch parallel to "T"; $^1/_4$″ diameter bars perpendicular to "T"; Beam reinforcement: 4″ × 6″ wire mesh No. 13 SWG; Span: 11′ restrained.	10 tons	6 hrs.			7	1, 2, 4, 5, 6	4

For SI: 1 inch = 25.4 mm, 1 foot = 305 mm, 1 pound = 0.004448kN, 1 pound per square inch = 0.00689 MPa, 1 ton = 8.896 kN.

Notes:

1. Load concentrated at mid span.
2. Achieved 6 hour fire rating (Grade "A," British).
3. 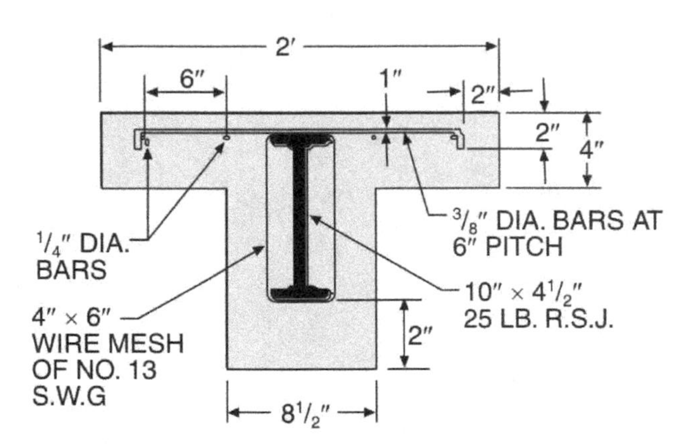

4. 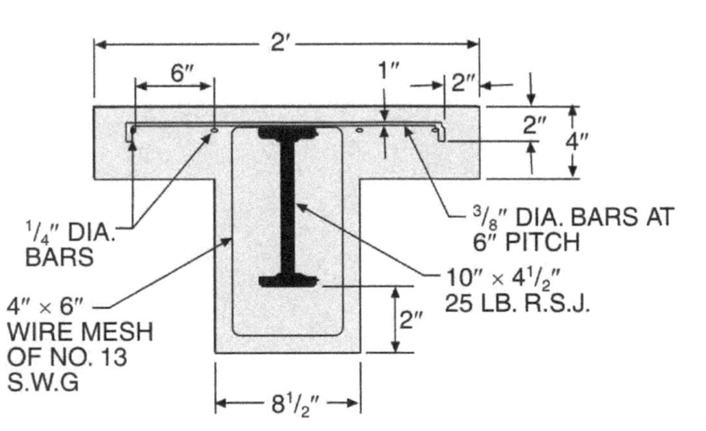

5. Section 43.147 of the 1979 edition of the *Uniform Building Code Standards* provides:

   "A restrained condition in fire tests, as used in this standard, is one in which expansion at the supports of a load-carrying element resulting from the effects of the fire is resisted by forces external to the element. An unrestrained condition is one in which the load-carrying element is free toexpand and rotate at its support.

   "(R)estraint in buildings is defined as follows: Floor and roof assemblies and individual beams in buildings shall be considered restrained when the surrounding or supporting structure is capable of resisting the thermal expansion throughout the range of anticipated elevated temperatures. Construction not complying . . . is assumed to be free to rotate and expand and shall be considered as unrestrained.

   "Restraint may be provided by the lateral stiffness of supports for floor and roof assemblies and intermediate beams forming part of the assembly. In order to develop restraint, connections must adequately transfer thermal thrusts to such supports. The rigidity of adjoining panels or structures shall be considered in assessing the capability of a structure to resist therm expansion."

   Because it is difficult to determine whether an existing building's structural system is capable of providing the required restraint, the lower hourly ratings of a similar but unrestrained assembly have been recommended.
6. Hourly rating based upon Table 4.2.1, Item B-12-RC-2.

TABLE 4.2.1—REINFORCED CONCRETE BEAMS—UNPROTECTED DEPTH 10″ TO LESS THAN 12″

ITEM CODE	DEPTH	CONSTRUCTION DETAILS	PERFORMANCE		REFERENCE NUMBER			NOTES	REC. HOURS
			LOAD	TIME	PRE-BMS-92	BMS-92	POST-BMS-92		
B-SU-1	10″	10″ × 4$^{1}/_{2}$″ × 25 lbs. "I" beam.	10 tons	39 min.			7	1	$^{1}/_{3}$

For SI: 1 inch = 25.4 mm, 1 pound = 0.004448 kN, 1 ton = 8.896 kN.

Notes:

1. Concentrated at mid span.

TABLE 4.2.2—STEEL BEAMS—CONCRETE PROTECTION DEPTH 10″ TO LESS THAN 12″

ITEM CODE	DEPTH	CONSTRUCTION DETAILS	PERFORMANCE		REFERENCE NUMBER			NOTES	REC. HOURS
			LOAD	TIME	PRE-BMS-92	BMS-92	POST-BMS-92		
B-SC-1	10″	10″ × 8″ rectangle; aggregate concrete (4170 psi) with 1″ top cover and 2″ bottom cover; No. 13 SWG iron wire loosely wrapped at approximately 6″ pitch about 7″ × 4″ × 16 lbs. "I" beam.	3.9 tons	3 hrs. 46 min.			7	1, 2, 3	$3^{3}/_{4}$
B-SC-1	10″	10″ × 8″ rectangle; aggregate concrete (3630 psi) with 1″ top cover and 2″ bottom cover; No. 13 SWG iron wire loosely wrapped at approximately 6″ pitch about 7″ × 4″ × 16 lbs. "I" beam.	5.5 tons	5 hrs. 26 min.			7	1, 4, 5, 6, 7	$3^{3}/_{4}$

For SI: 1 inch = 25.4 mm, 1 pound = 0.004448 kN, 1 pound per square inch = 0.00689 MPa, 1 ton = 8.896 kN.

Notes:

1. Load concentrated at mid span.
2. Specimen 10-foot 3-inch clear span simply supported.
3. Passed Grade "C" fire resistance (British) including hose stream and reload.
4. Specimen 11-foot clear span - restrained.
5. Passed Grade "B" fire resistance (British) including hose stream and reload.
6. See Table 4.1.3, Note 5.
7. Hourly rating based upon B-SC-1 above.

SECTION V—DOORS

FIGURE 5.1—RESISTANCE OF DOORS TO FIRE EXPOSURE

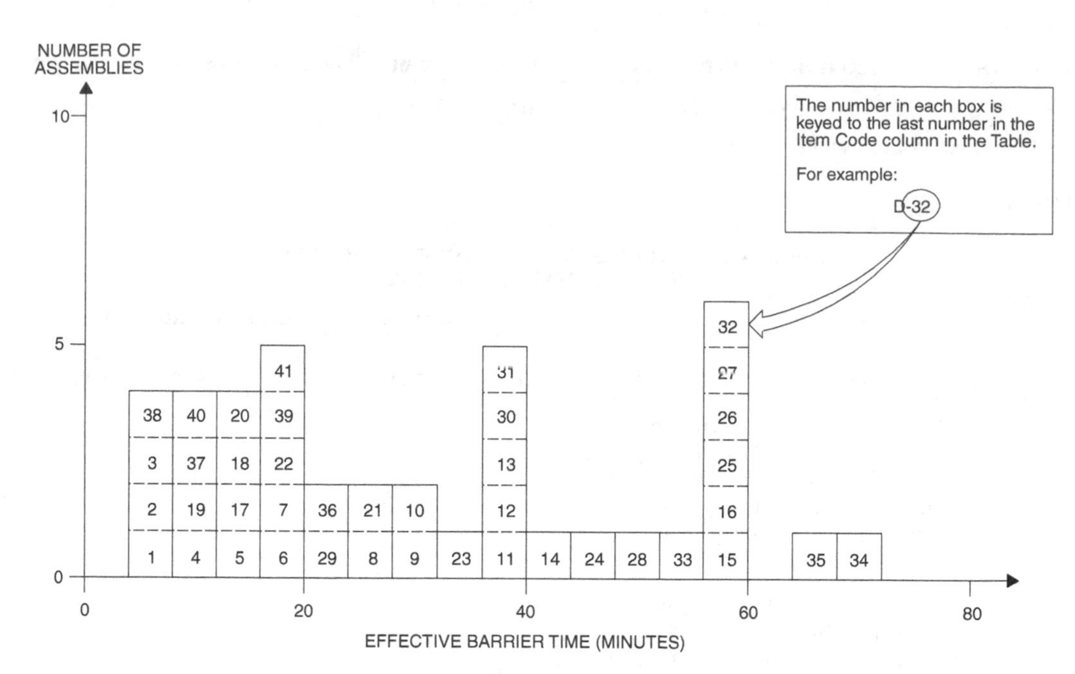

TABLE 5.1—RESISTANCE OF DOORS TO FIRE EXPOSURE

ITEM CODE	DOOR MINIMUM THICKNESS	CONSTRUCTION DETAILS	PERFORMANCE		REFERENCE NUMBER			NOTES	REC. (MIN.)
			EFFECTIVE BARRIER	EDGE FLAMING	PRE-BMS-92	BMS-92	POST-BMS-92		
D-1	$^3/_8''$	Panel door; pine perimeter ($1^3/_8''$); painted (enamel).	5 min. 10 sec.	N/A			90	1, 2	5
D-2	$^3/_8''$	As above, with two coats U.L. listed intumescent coating.	5 min. 30 sec.	5 min.			90	1, 2, 7	5
D-3	$^3/_8''$	As D-1, with standard primer and flat interior paint.	5 min. 55 sec.	N/A			90	1, 3, 4	5
D-4	$2^5/_8''$	As D-1, with panels covered each side with $^1/_2''$ plywood; edge grouted with sawdust filled plaster; door faced with $^1/_8''$ hardboard each side; paint see (5).	11 min. 15 sec.	3 min. 45 sec.			90	1, 2, 5, 7	10
D-5	$^3/_8''$	As D-1, except surface protected with glass fiber reinforced intumescent fire retardant coating.	16 min.	N/A			90	1, 3, 4, 7	15
D-6	$1^5/_8''$	Door detail: As D-4, except with $^1/_8''$ cement asbestos board facings with aluminum foil; door edges protected by sheet metal.	17 min.	10 min. 15 sec.			90	1, 3, 4	15
D-7	$1^5/_8''$	Door detail with $^1/_8''$ hardboard cover each side as facings; glass fiber reinforced intumescent coating applied.	20 min.	N/A			90	1, 3, 4, 7	20
D-8	$1^5/_8''$	Door detail same as D-4; paint was glass reinforced epoxy intumescent.	26 min.	24 min. 45 sec.			90	1, 3, 4, 6, 7	25

(Continued)

TABLE 5.1—RESISTANCE OF DOORS TO FIRE EXPOSURE—continued

ITEM CODE	DOOR MINIMUM THICKNESS	CONSTRUCTION DETAILS	PERFORMANCE		REFERENCE NUMBER			NOTES	REC. (MIN.)
			EFFECTIVE BARRIER	EDGE FLAMING	PRE-BMS-92	BMS-92	POST-BMS-92		
D-9	1 5/8″	Door detail same as D-4 with facings of 1/8″ cement asbestos board.	29 min.	3 min. 15 sec.			90	1, 2	5
D-10	1 5/8″	As per D-9.	31 min. 30 sec.	7 min. 20 sec.			90	1, 3, 4	6
D-11	1 5/8″	As per D-7; painted with epoxy intumescent coating including glass fiber roving.	36 min. 25 sec.	N/A			90	1, 3, 4	35
D-12	1 5/8″	As per D-4 with intumescent fire retardant paint.	37 min. 30 sec.	24 min. 40 sec.			90	1, 3, 4	30
D-13	1 1/2″ (nom.)	As per D-4, except with 24 ga. galvanized sheet metal facings.	39 min.	39 min.			90	1, 3, 4	39
D-14	1 5/8″	As per D-9.	41 min. 30 sec.	17 min. 20 sec.			90	1, 3, 4, 6	20
D-15	—	Class C steel fire door.	60 min.	58 min.			90	7, 8	60
D-16	—	Class B steel fire door.	60 min.	57 min.			90	7, 8	60
D-17	1 3/4″	Solid core flush door; core staves laminated to facings but not each other; Birch plywood facings 1/2″ rebate in door frame for door; 3/32″ clearance between door and wood frame.	15 min.	13 min.			37	11	13
D-18	1 3/4″	As per D-17.	14 min.	13 min.			37	11	13
D-19	1 3/4″	Door same as D-17, except with 16 ga. steel; 3/32″ door frame clearance.	12 min.	—			37	9, 11	10
D-20	1 3/4″	As per D-19.	16 min.	—			37	10, 11	10
D-21	1 3/4″	Doors as per D-17; intumescent paint applied to top and side edges.	26 min.	—			37	11	25
D-22	1 3/4″	Door as per D-17, except with 1/2″ × 1/8″ steel strip set into edges of door at top and side facing stops; matching strip on stop.	18 min.	6 min.			37	11	18
D-23	1 3/4″	Solid oak door.	36 min.	22 min.			15	13	25
D-24	1 7/8″	Solid oak door.	45 min.	35 min.			15	13	35
D-25	1 7/8″	Solid teak door.	58 min.	34 min.			15	13	35
D-26	1 7/8″	Solid (pitch) pine door.	57 min.	36 min.			15	13	35
D-27	1 7/8″	Solid deal (pine) door.	57 min.	30 min.			15	13	30
D-28	1 7/8″	Solid mahogany door.	49 min.	40 min.			15	13	45
D-29	1 7/8″	Solid poplar door.	24 min.	3 min.			15	13, 14	5
D-30	1 7/8″	Solid oak door.	40 min.	33 min.			15	13	35
D-31	1 7/8″	Solid walnut door.	40 min.	15 min.			15	13	20
D-32	2 5/8″	Solid Quebec pine.	60 min.	60 min.			15	13	60
D-33	2 5/8″	Solid pine door.	55 min.	39 min.			15	13	40
D-34	2 5/8″	Solid oak door.	69 min.	60 min.			15	13	60
D-35	2 5/8″	Solid teak door.	65 min.	17 min.			15	13	60
D-36	1 1/2″	Solid softwood door.	23 min.	8.5 min.			15	13	10
D-37	3/4″	Panel door.	8 min.	7.5 min.			15	13	5
D-38	5/16″	Panel door.	5 min.	5 min.			15	13	5

(Continued)

TABLE 5.1—RESISTANCE OF DOORS TO FIRE EXPOSURE—continued

ITEM CODE	DOOR MINIMUM THICKNESS	CONSTRUCTION DETAILS	PERFORMANCE		REFERENCE NUMBER			NOTES	REC. (MIN.)
			EFFECTIVE BARRIER	EDGE FLAMING	PRE-BMS-92	BMS-92	POST-BMS-92		
D-39	$^3/_4''$	Panel door, fire retardant treated.	$17^1/_2$ min.	3 min.			15	13	8
D-40	$^3/_4''$	Panel door, fire retardant treated.	$8^1/_2$ min.	$8^1/_2$ min.			15	13	8
D-41	$^3/_4''$	Panel door, fire retardant treated.	$16^3/_4$ min.	$11^1/_2$ min.			15	13	8

For SI: 1 inch = 25.4 mm, 1 foot = 305 mm.

Notes:

1. All door frames were of standard lumber construction.
2. Wood door stop protected by asbestos millboard.
3. Wood door stop protected by sheet metal.
4. Door frame protected with sheet metal and weather strip.
5. Surface painted with intumescent coating.
6. Door edge sheet metal protected.
7. Door edge intumescent paint protected.
8. Formal steel frame and door stop.
9. Door opened into furnace at 12 feet.
10. Similar door opened into furnace at 12 feet.
11. The doors reported in these tests represent the type contemporaries used as 20-minute solid-core wood doors. The test results demonstrate the necessity of having wall anchored metal frames, minimum cleaners possible between door, frame and stops. They also indicate the utility of long throw latches and the possible use of intumescent paints to seal doors to frames in event of a fire.
12. Minimum working clearance and good latch closure are absolute necessities for effective containment for all such working door assemblies.
13. Based on British tests.
14. Failure at door - frame interface.

Bibliography

1. Central Housing Committee on Research, Design, and Construction; Subcommittee on Fire Resistance Classifications, "Fire-Resistance Classifications of Building Constructions," *Building Materials and Structures,* Report BMS 92, National Bureau of Standards, Washington, Oct. 1942. (Available from NTIS No. COM-73-10974)
2. Foster, H. D., Pinkston, E. R., and Ingberg, S. H., "Fire Resistance of Structural Clay Tile Partitions," *Building Materials and Structures, Report* BMS 113, National Bureau of Standards, Washington, Oct. 1948.
3. Ryan, J. V., and Bender, E.W., "Fire Endurance of Open-Web Steel-Joist Floors with Concrete Slabs and Gypsum Ceilings," *Building Materials and Structures,* Report BMS 141, National Bureau of Standards, Washington, Aug. 1954.
4. Mitchell, N. D., "Fire Tests of Wood-Framed Walls and Partitions with Asbestos-Cement Facings," *Building Materials and Structures,* Report BMS 123, National Bureau of Standards, Washington, May 1951.
5. Robinson, H. E., Cosgrove, L. A., and Powell, F. J., "Thermal Resistance of Airspace and Fibrous Insulations Bounded by Reflective Surfaces," *Building Materials and Structures,* Report BMS 151, National Bureau of Standards,Washington, Nov. 1957.
6. Shoub, H., and Ingberg, S. H., "Fire Resistance of Steel Deck Floor Assemblies," *Building Science Series,* 11, National Bureau of Standards, Washington, Dec. 1967.
7. Davey, N., and Ashton, L. A., "Investigations on Building Fires, Part V: Fire Tests of Structural Elements," *National Building Studies,* Research Paper, No. 12, Dept. of Scientific and Industrial Research (Building Research Station), London, 1953.
8. National Board of Fire Underwriters, *Fire Resistance Ratings of Beam, Girder, and Truss Protections, Ceiling Constructions, Column Protections, Floor and Ceiling Constructions, Roof Constructions, Walls and Partitions,* New York, April 1959.
9. Mitchell, N.D., Bender, E.D., and Ryan, J.V., "Fire Resistance of Shutters for Moving-Stairway Openings," *Building Materials and Structures,* Report BMS 129, National Bureau of Standards, Washington, March 1952.
10. National Board of Fire Underwriters, *National Building Code; an Ordinance Providing for Fire Limits, and Regulations Governing the Construction, Alteration, Equipment, or Removal of Buildings or Structures,* New York, 1949.
11. Department of Scientific and Industrial Research and of the Fire Offices' Committee, Joint Committee of the Building Research Board, "Fire Gradings of Buildings, Part I: General Principles and Structural Precautions," *Post-War Building Studies,* No. 20, Ministry of Works, London, 1946.
12. Lawson, D. I.,Webster, C. T., and Ashton, L. A., "Fire Endurance of Timber Beams and Floors," *National Building Studies,* Bulletin, No. 13, Dept. of Scientific and Industrial Research and Fire Offices' Committee (Joint Fire Research Organization), London, 1951.
13. Parker, T. W., Nurse, R. W., and Bessey, G. E., "Investigations on Building Fires. Part I: The Estimation of the Maximum Temperature Attained in Building Fires from Examination of the Debris, and Part II: The Visible Change in Concrete or Mortar Exposed to High Temperatures," *National Building Studies,* Technical Paper, No. 4, Dept. of Scientific and Industrial Research (Building Research Station), London, 1950.
14. Bevan, R. C., and Webster, C. T., "Investigations on Building Fires, Part III: Radiation from Building Fires," *National Building Studies,* Technical Paper, No. 5, Dept. of Scientific and Industrial Research (Building Research Station), London, 1950.
15. Webster, D. J., and Ashton, L. A., "Investigations on Building Fires, Part IV: Fire Resistance of Timber Doors," *National Building Studies,* Technical Paper, No. 6, Dept. of Scientific and Industrial Research (Building Research Station), London, 1951.
16. Kidder, F. E., *Architects' and Builders' Handbook: Data for Architects, Structural Engineers, Contractors, and Draughtsmen,* comp. by a Staff of Specialists and H. Parker, editor-in-chief, 18th ed., enl., J. Wiley, New York, 1936.
17. Parker, H., Gay, C. M., and MacGuire, J. W., *Materials and Methods of Architectural Construction,* 3rd ed., J. Wiley, New York, 1958.
18. Diets, A. G. H., *Dwelling House Construction,* The MIT Press, Cambridge, 1971.
19. Crosby, E. U., and Fiske, H. A., *Handbook of Fire Protection,* 5th ed., The Insurance Field Company, Louisville, Ky., 1914.
20. Crosby, E. U., Fiske, H. A., and Forster, H.W., *Handbook of Fire Protection,* 8th ed., R. S. Moulton, general editor, National Fire Protection Association, Boston, 1936.
21. Kidder, F. E., *Building Construction and Superintendence,* rev. and enl., by T. Nolan, W. T. Comstock, New York, 1909-1913, 2 vols.
22. National Fire Protection Association, Committee on Fire-Resistive Construction, *The Baltimore Conflagration,* 2nd ed., Chicago, 1904.
23. Przetak, L., *Standard Details for Fire-Resistive Building Construction,* McGraw-Hill Book Co., New York, 1977.

24. Hird, D., and Fischl, C. F., "Fire Hazard of Internal Linings," *National Building Studies,* Special Report, No. 22, Dept. of Scientific and Industrial Research and Fire Offices' Committee (Joint Fire Research Organization), London, 1954.

25. Menzel, C. A., *Tests of the Fire-Resistance and Strength of Walls Concrete Masonry Units,* Portland Cement Association, Chicago, 1934.

26. Hamilton, S. B., "A Short History of the Structural Fire Protection of Buildings Particularly in England," *National Building Studies,* Special Report, No. 27, Dept. of Scientific and Industrial Research (Building Research Station), London, 1958.

27. Sachs, E. O., and Marsland, E., "The Fire Resistance of Doors and Shutters being Tabulated Results of Fire Tests Conducted by the Committee," *Journal of the British Fire Prevention Committee,* No. VII, London, 1912.

28. Egan, M. D., *Concepts in Building Firesafety,* J. Wiley, New York, 1978.

29. Sachs, E. O., and Marsland, E., "The Fire Resistance of Floors being Tabulated Results of Fire Tests Conducted by the Committee," *Journal of the British Fire Prevention Committee,* No. VI, London, 1911.

30. Sachs, E. O., and Marsland, E., "The Fire Resistance of Partitions being Tabulated Results of Fire Tests Conducted by the Committee," *Journal of the British Fire Prevention Committee,* No. IX, London, 1914.

31. Ryan, J. V., and Bender, E. W., "Fire Tests of Precast Cellular Concrete Floors and Roofs," *National Bureau of Standards Monograph,* 45, Washington, April 1962.

32. Kingberg, S. H., and Foster, H. D., "Fire Resistance of Hollow Load-BearingWall Tile," *National Bureau of Standards* Research Paper, No. 37, (Reprint from *NBS Journal of Research*, Vol. 2) Washington, 1929.

33. Hull, W. A., and Ingberg, S. H., "Fire Resistance of Concrete Columns," *Technologic Papers of the Bureau of Standards,* No. 272, Vol. 18, Washington, 1925, pp. 635-708.

34. National Board of Fire Underwriters, *Fire Resistance Ratings of Less than One Hour,* New York, Aug. 1956.

35. Harmathy, T. Z., "Ten Rules of Fire Endurance Rating," *Fire Technology,* Vol. 1, May 1965, pp. 93-102.

36. Son, B. C., "Fire Endurance Test on a Steel Tubular Column Protected with Gypsum Board," *National Bureau of Standards,* NBSIR, 73-165, Washington, 1973.

37. Galbreath, M., "Fire Tests of Wood Door Assemblies," *Fire Study,* No. 36, Div. of Building Research, National Research Council Canada, Ottawa, May 1975.

38. Morris, W. A., "An Investigation into the Fire Resistance of Timber Doors," *Fire Research Note,* No. 855, Fire Research Station, Boreham Wood, Jan. 1971.

39. Hall, G. S., "Fire Resistance Tests of Laminated Timber Beams," *Timber Association Research Report,* WR/RR/1, High Sycombe, July 1968.

40. Goalwin, D. S., "Fire Resistance of Concrete Floors," *Building Materials and Structures,* Report BMS 134, National Bureau of Standards, Washington, Dec. 1952.

41. Mitchell, N. D., and Ryan, J. V., "Fire Tests of Steel Columns Encased with Gypsum Lath and Plaster," *Building Materials and Structures,* Report BMS 135, National Bureau of Standards, Washington, April 1953.

42. Ingberg, S. H., "Fire Tests of BrickWalls," *Building Materials and Structures,* Report BMS 143, National Bureau of Standards, Washington, Nov. 1954.

43. National Bureau of Standards, "Fire Resistance and Sound-Insulation Ratings for Walls, Partitions, and Floors," *Technical Report on Building Materials,* 44, Washington, 1944.

44. Malhotra, H. L., "Fire Resistance of Brick and Block Walls," *Fire Note,* No. 6, Ministry of Technology and Fire Offices' Committee Joint Fire Research Organization, London, HMSO, 1966.

45. Mitchell, N. D., "Fire Tests of Steel Columns Protected with Siliceous Aggregate Concrete," *Building Materials and Structures,* Report BMS 124, National Bureau of Standards, Washington, May 1951.

46. Freitag, J. K., *Fire Prevention and Fire Protection as Applied to Building Construction; a Handbook of Theory and Practice,* 2nd ed., J. Wiley, New York, 1921.

47. Ingberg, S. H., and Mitchell, N. D., "Fire Tests of Wood and Metal-Framed Partition," *Building Materials and Structures,* Report BMS 71, National Bureau of Standards, Washington, 1941.

48. Central Housing Committee on Research, Design, and Construction, Subcommittee on Definitions, "A Glossary of Housing Terms," *Building Materials and Structures,* Report BMS 91, National Bureau of Standards, Washington, Sept. 1942.

49. Crosby, E. U., Fiske, H. A., and Forster, H.W., *Handbook of Fire Protection,* 7th ed., D. Van Nostrand Co., New York 1924.

50. Bird, E. L., and Docking, S. J., *Fire in Buildings,* A. & C. Black, London, 1949.

51. American Institute of Steel Construction, Fire Resistant Construction in Modern Steel-Framed Buildings, New York, 1959.

52. Central Dockyard Laboratory, "Fire Retardant Paint Tests - a Critical Review," CDL Technical Memorandum, No. P87/73, H. M. Naval Base, Portsmouth, Dec. 1973.

53. Malhotra, H. L., "Fire Resistance of Structural Concrete Beams," *Fire Research Note,* No. 741, Fire Research Station, Borehamwood, May 1969.

54. Abrams, M. S., and Gustaferro, A. H., "Fire Tests of Poke-Thru Assemblies," *Research and Development Bulletin,* 1481-1, Portland Cement Association, Skokie, 1971.

55. Bullen, M. L., "A Note on the Relationship between Scale Fire Experiments and Standard Test Results," *Building Research Establishment Note,* N51/75, Borehamwood, May 1975.

56. The America Fore Group of Insurance Companies, Research Department, *Some Characteristic Fires in Fire Resistive Buildings,* Selected from twenty years record in the files of the N.F.P.A. "Quarterly," New York, c. 1933.

57. Spiegelhalter, F., "Guide to Design of Cavity Barriers and Fire Stops," *Current Paper,* CP 7/77, Building Research Establishment, Borehamwood, Feb. 1977.

58. Wardle, T. M. "Notes on the Fire Resistance of Heavy Timber Construction," *Information Series,* No. 53, New Zealand Forest Service, Wellington, 1966.

59. Fisher, R. W., and Smart, P. M. T., "Results of Fire Resistance Tests on Elements of Building Construction," *Building Research Establishment Report,* G R6, London, HMSO, 1975.

60. Serex, E. R., "Fire Resistance of Alta Bates Gypsum Block Non-Load Bearing Wall," Report to Alta Bates Community Hospital, *Structural Research Laboratory Report,* ES-7000, University of Calif., Berkeley, 1969.

61. Thomas, F. G., and Webster, C. T., "Investigations on Building Fires, Part VI: The Fire Resistance of Reinforced Concrete Columns," *National Building Studies,* Research Paper, No. 18, Dept. of Scientific and Industrial Research (Building Research Station), London, HMSO, 1953.

62. Building Research Establishment, "Timber Fire Doors," *Digest,* 220, Borehamwood, Nov. 1978.

63. *Massachusetts State Building Code; Recommended Provisions,* Article 22: Repairs, Alterations, Additions, and Change of Use of Existing Buildings, Boston, Oct. 23, 1978.

64. Freitag, J. K., *Architectural Engineering; with Especial Reference to High Building Construction, Including Many Examples of Prominent Office Buildings,* 2nd ed., rewritten, J. Wiley, New York, 1906.

65. Architectural Record, *Sweet's Indexed Catalogue of Building Construction for the Year 1906,* New York, 1906.

66. Dept. of Commerce, Building Code Committee, "Recommended Minimum Requirements for Fire Resistance in Buildings," *Building and Housing,* No. 14, National Bureau of Standards, Washington, 1931.

67. British Standards Institution, "Fire Tests on Building Materials and Structures," *British Standards,* 476, Pt. 1, London, 1953.

68. Lönberg-Holm, K., "Glass," *The Architectural Record,* Oct. 1930, pp. 345-357.

69. Structural Clay Products Institute, "Fire Resistance," *Technical Notes on Brick and Tile Construction,* 16 rev., Washington, 1964.

70. Ramsey, C. G., and Sleeper, H. R., *Architectural Graphic Standards for Architects, Engineers, Decorators, Builders, and Draftsmen,* 3rd ed., J. Wiley, New York, 1941.

71. Underwriters' Laboratories, *Fire Protection Equipment List,* Chicago, Jan. 1957.

72. Underwriters' Laboratories, *Fire Resistance Directory; with Hourly Ratings for Beams, Columns, Floors, Roofs, Walls, and Partitions,* Chicago, Jan. 1977.

73. Mitchell, N. D., "Fire Tests of Gunite Slabs and Partitions," *Building Materials and Structures,* Report BMS 131, National Bureau of Standards, Washington, May 1952.

74. Woolson, I. H., and Miller, R. P., "Fire Tests of Floors in the United States," *Proceedings International Association for Testing Materials,* VIth Congress, New York, 1912, Section C, pp. 36-41.

75. Underwriters' Laboratories, "An Investigation of the Effects of Fire Exposure upon Hollow Concrete Building Units, Conducted for American Concrete Institute, Concrete Products Association, Portland Cement Association, Joint Submittors," *Retardant Report,* No. 1555, Chicago, May 1924.

76. Dept. of Scientific & Industrial Research and of the Fire Offices' Committee, Joint Committee of the Building Research Board, "Fire Gradings of Buildings. Part IV: Chimneys and Flues," *Post-War Building Studies,* No. 29, London, HMSO, 1952.

77. National Research Council of Canada. Associate Committee on the National Building Code, *Fire Performance Ratings,* Suppl. No. 2 to the National Building Code of Canada, Ottawa, 1965.

78. Associated Factory Mutual Fire Insurance Companies, The National Board of Fire Underwriters, and the Bureau of Standards, *Fire Tests of Building Columns; an Experimental Investigation of the Resistance of Columns, Loaded and Exposed to Fire or to Fire and Water, with Record of Characteristic Effects,* Jointly Conducted at Underwriters. Laboratories, Chicago, 1917-19.

79. Malhotra, H. L., "Effect of Age on the Fire Resistance of Reinforced Concrete Columns," *Fire Research Memorandum,* No. 1, Fire Research Station, Borehamwood, April 1970.

80. Bond, H., ed., *Research on Fire; a Description of the Facilities, Personnel and Management of Agencies Engaged in Research on Fire,* a Staff Report, National Fire Protection Association, Boston, 1957.

81. *California State Historical Building Code,* Draft, 1978.

82. Fisher, F. L., et al., "A Study of Potential Flashover Fires in Wheeler Hall and the Results from a Full Scale Fire Test of a Modified Wheeler Hall Door Assembly," *Fire Research Laboratory Report,* UCX 77-3; UCX-2480, University of Calif., Dept. of Civil Eng., Berkeley, 1977.

83. Freitag, J. K., *The Fireproofing of Steel Buildings,* 1st ed., J. Wiley, New York, 1906.

84. Gross, D., "Field Burnout Tests of Apartment Dwellings Units," *Building Science Series,* 10, National Bureau of Standards, Washington, 1967.

85. Dunlap, M. E., and Cartwright, F. P., "Standard Fire Tests for Combustible BuildingMaterials," *Proceedings of the American Society for Testing Materials,* vol. 27, Philadelphia, 1927, pp. 534-546.

86. Menzel, C. A., "Tests of the Fire Resistance and Stability of Walls of Concrete Masonry Units," *Proceedings of the American Society for Testing Materials,* vol. 31, Philadelphia, 1931, pp. 607-660.

87. Steiner, A. J., "Method of Fire-Hazard Classification of Building Materials," *Bulletin of the American Society for Testing and Materials,* March 1943, Philadelphia, 1943, pp. 19-22.

88. Heselden, A. J. M., Smith, P. G., and Theobald, C. R., "Fires in a Large Compartment Containing Structural Steelwork; Detailed Measurements of Fire Behavior," *Fire Research Note,* No. 646, Fire Research Station, Borehamwood, Dec. 1966.

89. Ministry of Technology and Fire Offices' Committee Joint Fire Research Organization, "Fire and Structural Use of Timber in Buildings; Proceedings of the Symposium Held at the Fire Research Station, Borehamwood, Herts on 25th October, 1967," *Symposium,* No. 3, London, HMSO, 1970.

90. Shoub, H., and Gross, D., "Doors as Barriers to Fire and Smoke," *Building Science Series,* 3, National Bureau of Standards, Washington, 1966.

91. Ingberg, S. H., "The Fire Resistance of Gypsum Partitions," *Proceedings of the American Society for Testing and Materials,* vol. 25, Philadelphia, 1925, pp. 299-314.

92. Ingberg, S.H., "Influence ofMineral Composition of Aggregates on Fire Resistance of Concrete," *Proceedings of the American Society for Testing and Materials,* vol. 29, Philadelphia, 1929, pp. 824-829.

93. Ingberg, S. H., "The Fire Resistive Properties of Gypsum," *Proceedings of the American Society for Testing and Materials,* vol. 23, Philadelphia, 1923, pp. 254-256.

94. Gottschalk, F.W., "Some Factors in the Interpretation of Small-Scale Tests for Fire-Retardant Wood," *Bulletin of the American Society for Testing and Materials,* October 1945, pp. 40-43.

95. Ministry of Technology and Fire Offices' Committee Joint Fire Research Organization, "Behaviour of Structural Steel in Fire; Proceedings of the Symposium Held at the Fire Research Station Borehamwood, Herts on 24th January, 1967," *Symposium,* No. 2, London, HMSO, 1968.

96. Gustaferro, A. H., and Martin, L. D., *Design for Fire Resistance of Pre-cast Concrete,* prep. for the Prestressed Concrete Institute Fire Committee, 1st ed., Chicago, PCI, 1977.

97. "The Fire Endurance of Concrete; a Special Issue," *Concrete Construction,* vol. 18, no. 8, Aug. 1974, pp. 345-440.

98. The British Constructional Steelwork Association, "Modern Fire Protection for Structural Steelwork," *Publication,* No. FPl, London, 1961.

99. Underwriters' Laboratories, "Fire Hazard Classification of Building Materials," *Bulletin,* No. 32, Sept. 1944, Chicago, 1959.

100. Central Housing Committee on Research, Design, and Construction, Subcommittee on Building Codes, "Recommended Building Code Requirements for New Dwelling Construction with Special Reference to War Housing; Report," *Building Materials and Structures,* Report BMS 88, National Bureau of Standards, Washington, Sept. 1942.

101. De Coppet Bergh, D., *Safe Building Construction; a Treatise Giving in Simplest Forms Possible Practical and Theoretical Rules and Formulae Used in Construction of Buildings and General Instruction,* new ed., thoroughly rev. Macmillan Co., New York, 1908.

102. *Cyclopedia of Fire Prevention and Insurance; a General Reference Work on Fire and Fire Losses, Fireproof Construction, Building Inspection...,* prep. by architects, engineers, underwriters and practical insurance men. American School of Correspondence, Chicago, 1912.

103. Setchkin, N. P., and Ingberg, S. H., "Test Criterion for an Incombustible Material," *Proceedings of the American Society for Testing Materials,* vol. 45, Philadelphia, 1945, pp. 866-877.

104. Underwriters' Laboratories, "Report on Fire Hazard Classification of Various Species of Lumber," *Retardant,* 3365, Chicago, 1952.

105. Steingiser, S., "A Philosophy of Fire Testing," Journal of Fire & Flammability, vol. 3, July 1972, pp. 238-253.

106. Yuill, C. H., Bauerschlag,W. H., and Smith, H. M., "An Evaluation of the Comparative Performance of 2.4.1 Plywood and Two-Inch Lumber Roof Decking under Equivalent Fire Exposure," *Fire Protection Section, Final Report,* Project No. 717A-3-211, Southwest Research Institute, Dept. of Structural Research, San Antonio, Dec. 1962.

107. Ashton, L. A., and Smart, P.M. T., *Sponsored Fire-Resistance Tests on Structural Elements,* London, Dept. of Scientific and Industrial Research and Fire Offices. Committee, London, 1960.

108. Butcher, E. G., Chitty, T. B., and Ashton, L. A., "The Temperature Attained by Steel in Building Fires," *Fire Research Technical Paper,* No. 15, Ministry of Technology and Fire Offices. Committee, Joint Fire Research Organization, London, HMSO, 1966.

109. Dept. of the Environment and Fire Offices' Committee, Joint Fire Research Organization, "Fire-Resistance Requirements for Buildings - a New Approach; Proceedings of the Symposium Held at the Connaught Rooms, London, 28 September 1971," *Symposium,* No. 5, London, HMSO, 1973.

110. Langdon Thomas, G. J., "Roofs and Fire," *Fire Note,* No. 3, Dept. of Scientific and Industrial Research and Fire Offices' Committee, Joint Fire Research Organization, London, HMSO, 1963.

111. National Fire Protection Association and the National Board of Fire Underwriters, *Report on Fire the Edison Phonograph Works,* Thomas A. Edison, Inc., West Orange, N.J., December 9, 1914, Boston, 1915.

112. Thompson, J. P., *Fire Resistance of Reinforced Concrete Floors,* Portland Cement Association, Chicago, 1963.

113. Forest Products Laboratory, "Fire Resistance Tests of Plywood Covered Wall Panels," Information reviewed and reaffirmed, *Forest Service Report,* No. 1257, Madison, April 1961.

114. Forest Products Laboratory, "Charring Rate of Selected Woods - Transverse to Grain," Forest Service *Research Paper,* FLP 69, Madison, April 1967.

115. Bird, G. I., "Protection of Structural Steel Against Fire," *Fire Note,* No. 2, Dept. of Scientific and Industrial Research and Fire Offices' Committee, Joint Fire Research Organization, London, HMSO, 1961.

116. Robinson, W. C., *The Parker Building Fire,* Underwriters' Laboratories, Chicago, c. 1908.

117. Ferris, J. E., "Fire Hazards of Combustible Wallboards," *Commonwealth Experimental Building Station Special Report,* No. 18, Sydney, Oct. 1955.

118. Markwardt, L. J., Bruce, H. D., and Freas, A. D., "Brief Description of Some Fire-Test Methods Used for Wood and Wood Base Materials," *Forest Service Report,* No. 1976, Forest Products Laboratory, Madison, 1976.

119. Foster, H. D., Pinkston, E. R., and Ingberg, S. H., "Fire Resistance of Walls of Gravel-Aggregate Concrete Masonry Units," *Building Materials and Structures,* Report, BMS 120, National Bureau of Standards, Washington, March 1951.

120. Foster, H. D., Pinkston, E.R., and Ingberg, S. H., "Fire Resistance of Walls of Lightweight-Aggregate Concrete Masonry Units," *Building Materials and Structures,* Report BMS 117, National Bureau of Standards, Washington, May 1950.

121. Structural Clay Products Institute, "Structural Clay Tile Fireproofing," *Technical Notes on Brick & Tile Construction,* vol. 1, no. 11, San Francisco, Nov. 1950.

122. Structural Clay Products Institute, "Fire Resistance Ratings of Clay Masonry Walls - I," *Technical Notes on Brick & Tile Construction,* vol. 3, no. 12, San Francisco, Dec. 1952.

123. Structural Clay Products Institute, "Estimating the Fire Resistance of Clay Masonry Walls - II," *Technical Notes on Brick & Tile Construction*, vol. 4, no. 1, San Francisco, Jan. 1953.

124. Building Research Station, "Fire: Materials and Structures," *Digest,* No. 106, London, HMSO, 1958.

125. Mitchell, N. D., "Fire Hazard Tests with Masonry Chimneys," *NFPA Publication,* No. Q-43-7, Boston, Oct. 1949.

126. Clinton Wire Cloth Company, *Some Test Data on Fireproof Floor Construction Relating to Cinder Concrete, Terra Cotta and Gypsum,* Clinton, 1913.

127. Structural Engineers Association of Southern California, Fire Ratings Subcommittee, "Fire Ratings, a Report," part of *Annual Report,* Los Angeles, 1962, pp. 30-38.

128. Lawson, D. I., Fox, L. L., and Webster, C. T., "The Heating of Panels by Flue Pipes," *Fire Research, Special Report,* No. 1, Dept. of Scientific and Industrial Research and Fire Offices' Committee, London, HMSO, 1952.

129. Forest Products Laboratory, "Fire Resistance of Wood Construction," Excerpt from 'Wood Handbook - Basic Information on Wood as a Material of Construction with Data for its Use in Design and Specification,' *Dept. of Agriculture Handbook,* No. 72, Washington, 1955, pp. 337-350.

130. Goalwin, D. S., "Properties of Cavity Walls," *Building Materials and Structures,* Report BMS 136, National Bureau of Standards, Washington, May 1953.

131. Humphrey, R. L., "The Fire-Resistive Properties of Various Building Materials," *Geological Survey Bulletin,* 370, Washington, 1909.

132. National Lumber Manufacturers Association, "Comparative Fire Test on Wood and Steel Joists," *Technical Report,* No. 1, Washington, 1961.

133. National Lumber Manufacturers Association, "Comparative Fire Test of Timber and Steel Beams," *Technical Report,* No. 3, Washington, 1963.

134. Malhotra, H. L., and Morris, W. A., "Tests on Roof Construction Subjected to External Fire," *Fire Note,* No. 4, Dept. of Scientific and Industrial Research and Fire Offices' Committee, Joint Fire Research Organization, London, HMSO, 1963.

135. Brown, C. R., "Fire Tests of Treated and Untreated Wood Partitions," *Research Paper,* RP 1076, part of *Journal of Research of the National Bureau of Standards,* vol. 20, Washington, Feb. 1938, pp. 217-237.

136. Underwriters' Laboratories, "Report on Investigation of Fire Resistance of Wood Lath and Lime Plaster Interior Finish," *Publication,* SP. 1. 230, Chicago, Nov. 1922.

137. Underwriters' Laboratories, "Report on Interior Building Construction Consisting of Metal Lath and Gypsum Plaster on Wood Supports," *Retardant,* No. 1355, Chicago, 1922.

138. Underwriters' Laboratories, "An Investigation of the Effects of Fire Exposure upon Hollow Concrete Building Units," *Retardant,* No. 1555, Chicago, May 1924.

139. Moran, T. H., "Comparative Fire Resistance Ratings of Douglas Fir Plywood," *Douglas Fir Plywood Association Laboratory Bulletin,* 57-A, Tacoma, 1957.

140. Gage Babcock & Association, "The Performance of Fire-Protective Materials under Varying Conditions of Fire Severity," Report 6924, Chicago, 1969.

141. International Conference of Building Officials, *Uniform Building Code* (1979 ed.), Whittier, CA, 1979.

142. Babrauskas,V., and Williamson, R. B., "The Historical Basis of Fire Resistance Testing, Part I and Part II," *Fire Technology,* vol. 14, no. 3 & 4, Aug. & Nov. 1978, pp. 184-194, 205, 304-316.

143. Underwriters' Laboratories, "Fire Tests of Building Construction and Materials," 8th ed., *Standard for Safety,* UL263, Chicago, 1971.

144. Hold, H. G., *Fire Protection in Buildings,* Crosby, Lockwood, London, 1913.

145. Kollbrunner, C. F., "Steel Buildings and Fire Protection in Europe," *Journal of the Structural Division,* ASCE, vol. 85, no. ST9, Proc. Paper 2264, Nov. 1959, pp. 125-149.

146. Smith, P., "Investigation and Repair of Damage to Concrete Caused by Formwork and Falsework Fire," *Journal of the American Concrete Institute,* vol. 60, Title no. 60-66, Nov. 1963, pp. 1535-1566.

147. "Repair of Fire Damage," 3 parts, *Concrete Construction,* March-May, 1972.

148. National Fire Protection Association, *National Fire Codes; a Compilation of NFPA Codes, Standards, Recommended Practices and Manuals,* 16 vols., Boston, 1978.

149. Ingberg, S. H. "Tests of Severity of Building Fires," *NFPA Quarterly,* vol. 22, no. 1, July 1928, pp. 43-61.

150. Underwriters' Laboratories, "Fire Exposure Tests of Ordinary Wood Doors," *Bulletin of Research,* no. 6, Dec. 1938, Chicago, 1942.

151. Parson, H., "The Tall Building under Test of Fire," *Red Book,* no. 17, British Fire Prevention Committee, London, 1899.

152. Sachs, E. O., "The British Fire Prevention Committee Testing Station," *Red Book,* no. 13, British Fire Prevention Committee, London, 1899.

153. Sachs, E. O., "Fire Tests with Unprotected Columns," *Red Book,* no. 11, British Fire Prevention Committee, London, 1899.

154. British Fire Prevention Committee, "Fire Tests with Floors a Floor by the Expended Metal Company," *Red Book,* no. 14, London, 1899.

155. *Engineering News,* vol. 56, Aug. 9, 1906, pp. 135-140.

156. *Engineering News,* vol. 36, Aug. 6, 1896, pp. 92-94.

157. Bauschinger, J., *Mittheilungen de Mech.-Tech. Lab. der K. Tech. Hochschule, München,* vol. 12, 1885.

158. *Engineering News,* vol. 46, Dec. 26, 1901, pp. 482-486, 489-490.

159. *The American Architect and Building News,* vol. 31, March 28, 1891, pp. 195-201.

160. British Fire Prevention Committee, First International Fire Prevention Congress, *Official Congress Report,* London, 1903.

161. American Society for Testing Materials, *Standard Specifications for Fire Tests of Materials and Construction (C19-18),* Philadelphia, 1918.

162. International Organization for Standardization, *Fire Resistance Tests on Elements of Building Construction (R834),* London, 1968.

163. *Engineering Record,* vol. 35, Jan. 2, 1897, pp. 93-94; May 29, 1897, pp. 558-560; vol. 36, Sept. 18, 1897, pp. 337-340; Sept. 25, 1897, pp. 359-363; Oct. 2, 1897, pp. 382-387; Oct. 9, 1897, pp. 402-405.

164. Babrauskas, Vytenis, "Fire Endurance in Buildings," PhD Thesis. *Fire Research Group,* Report, No. UCB FRG 76-16, University of California, Berkeley, Nov. 1976.

165. The Institution of Structural Engineers and The Concrete Society, *Fire Resistance of Concrete Structures,* London, Aug. 1975.

INDEX

E

F

G

H

I

L

M

O

P

R

S

T

U

V

W

EDITORIAL CHANGES – SECOND PRINTING

Page 1, Section 101.5, Exception, line 6 now reads . . . *alteration* as defined in Section 807.4.3. New structural

Page 1, Section 101.5.4.1, item 1 now reads . . . The *International Building Code* using 100 percent of the prescribed forces. The values of R, Ω_0 and C_d used for analysis in accordance with Chapter 16 of the *International Building Code* shall be those specified for structural systems classified as "Ordinary" in accordance with Table 12.2-1 of ASCE 7, unless it can be demonstrated that the structural system satisfies the proportioning and detailing requirements for systems classified as "Detailed," "Intermediate" or "Special."

Page 2, Table 101.5.4.1, column 2, row 4 now reads . . . Note a

Page 2, Table 101.5.4.1, Footnote now reads . . . a. Acceptable criteria for Occupancy Category III shall be taken as 80 percent of the acceptance criteria specified for Occupancy Category IV performance levels.

Page 2, Table 101.5.4.2, column 2, row 4 now reads . . . Notes a, b

Page 2, Table 101.5.4.2, Footnotes now read . . . a. Acceptable criteria for Occupancy Category III shall be taken as 80 percent of the acceptance criteria specified for Occupancy Category IV performance levels. b. For Occupancy Category III, the ASCE 31 screening phase checklists shall be based on the life safety performance level.

Page 29, Section 606.2.1, line 2 now reads . . . Where a permit is issued for reroofing more than 25 percent of the roof area of a building assigned to Seismic

Page 51, Section 1003.3.3, last line now reads . . . comply with Section 707.6.

Page 55, Section 1105.15, last line now reads . . . 1104.1.4 for those elements shall be permitted.

Page 63, Table 1301.6.10, Note a new reads . . . a. This value shall be 0 if compliance with Category d or e in Section 1301.6.8.1 has not been obtained.

Page 64, Section 1301.6.11.1, item 1, last line now reads . . . accordance with Section 305.

Page 65, Table 1301.6.15, column 1, row 1 now reads . . . NUMBER OF EXITS REQUIRED BY SECTION 1015 OF THE *INTERNATIONAL BUILDING CODE*

Page 81, Section A108.2, item 1, paragraph 2, line 2 now reads . . . in accordance with Section A106.3.3.5 and shall not

Page 120, Section A506.3, last line now reads . . . comply with the requirements of Section 101.5.4.1.

EDITORIAL CHANGES – THIRD PRINTING

Page 1, Section 101.5.4.1: item 1 now reads . . . One-hundred percent of the values in the *International Building Code*. Where the existing seismic force-resisting system is a type that can be designated as "Ordinary", values of R, Ω_0, and C_d used for analysis in accordance with Chapter 16 of the *International Building Code* shall be those specified for structural systems classified as "Ordinary" in accordance with Table 12.2-1 of ASCE 7, unless it is demonstrated that the structural system will provide performance equivalent to that of a "Detailed", "Intermediate" or "Special" system.

Page 2, TABLE 101.5.4.1: footnote a now reads . . . Acceptable criteria for Occupancy Category III shall be taken as 80 percent of the acceptance criteria specified for Occupancy Category II performance levels, but need not be less than the acceptance criteria specified for Occupancy Category IV performance levels.

Page 2, TABLE 101.5.4.2: footnote a now reads . . . Acceptable criteria for Occupancy Category III shall be taken as 80 percent of the acceptance criteria specified for Occupancy Category II performance levels, but need not be less than the acceptance criteria specified for Occupancy Category IV performance levels.

Page 2, Section 101.5.4.2: item 2.5 now reads . . . Seismic evaluation and design of concrete buildings in all occupancy categories are permitted to be based on the procedures specified in Chapter A5.

Page 4, new section added now reads . . . Whenever there is insufficient evidence of compliance with the provisions of this code or evidence that a material or method does not conform to the requirements of this code, or in order to substantiate claims for alternative materials or methods, the code official shall have the authority to require tests as evidence of compliance to be made at no expense to the jurisdiction. Test methods shall be as specified in this code or by other recognized test standards. In the absence of recognized and accepted test methods, the code official shall approve the testing procedures. Tests shall be performed by an approved agency. Reports of such tests shall be retained by the code official for the period required for retention.

Page 6, Section 106.1: line 4 now reads . . . shall be submitted in two or more sets with each application for

Page 15, Section 302.1: lines 5 now reads . . . structure together with the *addition* are no less conforming to the provisions of the *International Building Code*

Page 15, Section 302.2: line 5 now reads . . . 202, shall comply with the flood design requirements for new

Page 15, Section 302.2: second paragraph, line 5 now reads . . . tion 202, are not required to comply with the flood design

Page 15, Section 302.3: line 6 now reads . . . carry the increased load required by the *International Building Code*

Page 16, Section 302.4.1: line 1 now reads . . . Seismic requirements for additions shall

Page 16, Section 302.4.1: line 5 now reads . . . force-resisting system shall be those specified by the *International Building Code*

Page 16, Section 302.4.1: line 8 now reads . . . mance equivalent to that of a detailed, intermediate or spe-

Page 16, Section 303.1: line now reads . . . with the requirements of the *International Building Code*

Page 16, Section 303.1: line 5 now reads . . . building or structure is no less conforming to the provisions of the *International Building Code*

Page 16, Section 303.2: line 5 now reads . . . 202, shall comply with the flood design requirements for new

Page 16, Section 303.2: second paragraph, line 5 now reads . . . Section 202, are not required to comply with the flood design

Page 16, Section 303.3: line 6 now reads . . . gravity load required by the *International Building Code* for

Page 16, Section 303.3: line 10 now reads . . . resist the applicable design gravity loads required by the *International Building Code*

Page 16, Section 303.4.1: line 8 now reads . . . of a detailed, intermediate or special system.

Page 16, Section 303.5: line 2 now reads . . . ments are no less conforming to the provisions of the *International Building Code*

Page 17, Section 304.1: line 2 now reads . . . shall be repaired in conformance to this section and to Section

Page 17, Section 304.2.1: line 6 now reads . . . the *International Building Code* for wind and earthquake

Page 17, Section 304.2.3: line 5 now reads . . . cable provisions of the *International Building Code* for

Page 17, Section 304.2.3: line 11 now reads . . . required by the *International Building Code*, whichever are greater

Page 17, Section 304.2.3: line 19 now reads . . . detailing provisions of the *International Building Code* for

Page 17, Section 304.3: line 4 now reads . . . to comply with the applicable provisions of the *International Building Code*

Page 17, Section 304.3: line 15 now reads . . . shall comply with the detailing provisions of the *International Building Code*

Page 18, Section 304.4: line 6 now reads . . . comply with the detailing provisions of the *International Building Code*

Page 18, Section 304.5: now reads . . . **304.5 Flood hazard areas.** For buildings and structures in flood hazard areas established in Section 1612.3 of the *International Building Code*, any *repair* that constitutes *substantial improvement* of the existing structure, as defined in Section 202, shall comply with the flood design requirements for new construction, and all aspects of the existing structure shall be brought into compliance with the requirements for new construction for flood design.

For buildings and structures in flood hazard areas established in Section 1612.3 of the *International Building Code*, any repairs that do not constitute *substantial improvement* or *substantial damage* of the existing structure, as defined in Section 202, are not required to comply with the flood design requirements for new construction.

Page 18, Section 305: line 1 now reads . . . **[B] SECTION 305**

Page 18, Section 306: line 1 now reads . . . **[B] SECTION 306**

Page 18, [B] Section 307.4: line 1 now reads . . . **[B] 307.4 Seismic.** When a *change of occupancy* results in a

Page 18, [B] Section 307.4: line 7 now reads . . . force-resisting system shall be those specified by the *International Building Code*

Page 19, [B] Section 307.4: Exception 2, line 3 now reads . . . pancy Category III and the structure is located where the seismic coefficient, S_{DS}, is less than 0.33, compliance with the seismic requirements of Section 1613 of

Page 20, Section 310.8: line 2 now reads . . . 310.8.1 through 310.8.14 shall apply to alterations to existing

Page 29, Section 606.2: Exception 1, line 2 now reads . . . from the roofing or equipment does not increase the force in the element by more than 5 percent.

Page 48, Section 912.4.1: Exception 4, line1 now reads . . . Existing corridor walls constructed on both sides

Page 61, Section [B] 1301.6.3: line 2 now reads . . . ments created by fire barriers or horizontal assemblies

Page 61, Section [B] 1301.6.4.1: item 3, line 8 now reads . . . floor area.

Page 61, TABLE 1301.6.3: column 1, row 3 now reads . . . A-2

Page 63, Section [B] 1301.6.8.1: item 5, line 2 now reads . . . out the floor area.

Page 64, Section [B] 1301.6.10.1: item 5, line 7 now reads . . . other floor areas of the building under fire condi-

Page 64, TABLE 1301.6.11(1): footnote a, line 2 now reads . . . dance with Section 903.3.1.1 or 903.3.1.2 of the *International Building Code.*

Page 66, Section [B] 1301.6.19: line 8 now reads . . . 1301.6.19 for the building or floor area being evaluated and enter that value into Table 1301.7 under Safety Parameter 1301.6.19, Incidental Accessory Occupancy, for fire safety, means of egress and general safety.

Page 73, Section 1405: line 2 now reads . . . MEANS OF EGRESS

Page 73, Section 1405.1: now reads . . . **1405.1 Stairways required.** Where a building has been constructed to a *building height* of 50 feet (15 240 mm) or four *stories*, or where an *existing building* exceeding 50 feet (15 240 mm) in building height is altered, at least one temporary lighted *stairway* shall be provided unless one or more of the permanent stairways are erected as the construction progresses.

Page 73, Section 1405.2: now reads . . . **1405.2 Maintenance of means of egress.** Required *means of egress* shall be maintained at all times during construction, demolition, remodeling or *alterations* and *additions* to any building.

Exception: *Approved* temporary *means of egress* systems and facilities.

Page 73, Section 1406.1: now reads . . . **1406.1 Where required.** In buildings required to have standpipes by Section 905.3.1, not less than one standpipe shall be provided for use during construction. Such standpipes shall be installed where the progress of construction is not more than 40 feet (12 192 mm) in height above the lowest level of fire department vehicle access. Such standpipe shall be provided with fire department hose connections at accessible locations adjacent to usable stairs. Such standpipes shall be extended as construction progresses to within one floor of the highest point of construction having secured decking or flooring.

EDITORIAL CHANGES – FOURTH PRINTING

Page 18, Section [B]307.1: line 5 now reads . . . ply with the requirements of the *International Building Code*

Page 39, Section 708.2: line 4 now reads . . . Chapter 6.

Page 125, Section B102.2.3: line 4 now reads . . . with Section 605.2 from the point of connection to boarding

EDITORIAL CHANGES – FIFTH PRINTING

Page 29, Section 606.3.1: line 7 now reads . . . *tional Building Code* seismic forces level as specified in

Page 39, Section 708.1: line 4 now reads . . . ter 6.

Page 59, Section [B]1301.2.5: line 3 now reads . . . form to the accessibility provisions of Section 310.

Page 68, TABLE 1301.7: column 1, row 8, line 1 now reads . . . 1301.6.16 Mixed Occupancies

Page 68, TABLE 1301.7: column 1, row 8, line 2 now reads . . . 1301.6.17 Automatic Sprinklers

Page 68, TABLE 1301.7: column 1, row 8, line 3 now reads . . . 1301.6.18 Standpipes

Page 68, TABLE 1301.7: column 1, row 8, line 4 now reads . . . 1301.6.19 Incidental Accessory Occupancy

Page 73, Section 1405: now reads . . . **SECTION 1405 MEANS OF EGRESS**

[B] 1405.1 Stairways required. Where a building has been constructed to a building height of 50 feet (15 240 mm) or four stories, or where an *existing building* exceeding 50 feet (15 240 mm) in building height is altered, at least one temporary lighted stairway shall be provided unless one or more of the permanent stairways are erected as the construction progresses.

[B] 1405.2 Maintenance of means of egress. Required means of egress shall be maintained at all times during construction, demolition, remodeling or *alterations* and *additions* to any building.

Exception: Approved temporary means of egress systems and facilities.

Page 125, Section B101.3: line 12 now reads . . . tive requirements of Section 310.9 for that element are permit-

Page 125, Section B101.4: line 15 now reads . . . Section 310.9 for that element are permitted.

EDITORIAL CHANGES – SEVENTH PRINTING

Page 27, Section 604.1: line 1 now reads . . . **604.1 General.** Alterations shall be done in a manner that